NCA-Vol. 6

NUMERICAL TECHNIQUES IN ACOUSTIC RADIATION

presented at

THE WINTER ANNUAL MEETING OF
THE AMERICAN SOCIETY OF MECHANICAL ENGINEERS
SAN FRANCISCO, CALIFORNIA
DECEMBER 10–15, 1989

sponsored by

THE NOISE CONTROL AND ACOUSTICS DIVISION, ASME
ANALYSIS TECHNIQUES IN ACOUSTICS COMMITTEE

edited by

R. J. BERNHARD
PURDUE UNIVERSITY

R. F. KELTIE
NORTH CAROLINA STATE UNIVERSITY

THE AMERICAN SOCIETY OF MECHANICAL ENGINEERS
United Engineering Center 345 East 47th Street New York, N.Y. 10017

ISBN No. 0-7918-0430-5

Library of Congress
Catalog Number 89-46296

FOREWORD

The prediction of the sound radiating from an acoustical source is of significant interest in many machinery applications, both in fluid and gaseous media. In many applications, however, the geometry of the machine and/or the acoustical phenomena is so complex that analytical models are insufficient to describe important behavior. In such cases numerical techniques offer the only practical alternative for predicting acoustic radiation. Thus, the formulation and development of numerical techniques for prediction of acoustic radiation have been of considerable interest for the past several decades. At the present time a variety of circumstances, including continuing growth in the capability of computers, the attention these methods are receiving from funding agencies, and the length of time the methods have been studied, have resulted in an exciting variety of contributions being made by investigators from around the world.

This volume is a compilation of the papers presented at the 1989 ASME Winter Annual Meeting for the International Symposium on Numerical Techniques in Acoustic Radiation. The reader will find an excellent assortment of papers from an international group of contributors. One group of papers discusses new numerical formulations of harmonic and transient acoustic radiation problems. Several papers address the coupling of finite element structural dynamic methods with boundary element acoustic methods for structural/acoustic analysis. Other papers consider the improvement of the efficiency of the more established Helmholz Integral Equation methods and integration of these methods into engineering design analysis. Several papers discuss engineering applications of numerical radiation techniques. The range of the papers reflects the relative maturity of the discipline, but also highlights the significant issues yet to be resolved. It is hoped that this volume will serve for many years as a reference to numerical methods in acoustic radiation, and will also motivate new investigations.

The Symposium organizers express their appreciation to the reviewers who contributed this fine set of papers and to the reviewers who contributed time to the effort. Significant contributions have also been made by the Executive Committee of the Noise Control and Acoustics Division of the ASME, and by D. K. Hsu, Chairman of the Analysis Techniques in Acoustics Technical Committee.

Robert J. Bernhard
Richard F. Keltie

CONTENTS

CONTENTS

UNIQUENESS OF SOLUTIONS TO VARIATIONALLY FORMULATED ACOUSTIC RADIATION PROBLEMS

X.-F. Wu
Department of Mechanical Engineering and
Institute for Manufacturing Research
Wayne State University
Detroit, Michigan

A. D. Pierce
Pennsylvania State University
University Park, Pennsylvania

ABSTRACT

Determination of the surface acoustic pressure given the surface velocity of a vibrating body can be formulated in various ways. However, for some such formulations such as the surface Helmholtz integral equation, solutions are not unique at certain discrete frequencies. Such uniqueness problems can also be present for variational formulations of the problem, but the variational formulation based on the normal derivative of the Kirchhoff integral theorem has unique solutions for vibrating disks and plate-like bodies. For bodies of finite volume, but for which each surface point is vibrating in phase, the total radiated acoustic power is always unique, even though the pressure may not be. The latter conclusion is supported by numerical calculations based on the Rayleigh-Ritz technique for the case of a finite cylinder vibrating as a rigid body in the axial direction.

I. INTRODUCTION

A variational principle has been previously proposed by Pierce and Wu [1] for use in the prediction of acoustic radiation when the normal velocity is prescribed over the surface of a vibrating body. The detailed derivations of this principle, along with exemplary calculations for cases of a transversely vibrating disk and finite cylinders are given in Refs. [2, 3]. For the achievement of accuracy comparable to that previously obtained by Leitner [4], who made use of an expansion in terms of spheroidal wave functions, the Rayleigh-Ritz implementation of the variational principle is relatively simple. Also, in the Rayleigh-Ritz procedure, certain known features (such as behavior at edges) of the exact solution can be a priori built into the approximate solution with a careful construction of the basis functions, whereas in the expansion of spheroidal wave functions the slow convergence inhibits the satisfactory display of such features.

It is well known that there exist many ways to determine the surface acoustic pressure given the normal velocity on the vibrating surface, and for some such formulations derived from the standard Kirchhoff integral theorem, there are inherent difficulties in the numerical computations when the wavenumber k takes some particular values [5 – 7]. The variational formulation discussed in the present paper can be viewed as closely related to one of such formulations. It yields an integro-differential equation for the surface pressure as its Euler-Lagrange equation, and solutions of the latter are nonunique when the wavenumber corresponds to one of natural frequencies for the corresponding homogeneous interior Neumann problem (cavity with rigid walls). This shortcoming extends to the solution of the variational problem, but there is a wide class of cases where the solution of either formulation is unique. Such cases are those in which the vibrating body is a disk or a plate-like body whose two sides move together. Also, the variational principle can yield a unique answer for the total radiated acoustic power. The validity of these statements is demonstrated in the present paper with mathematical analysis and supporting calculations.

In what follows, we first give what we believe to be a new and improved proof of the circumstances for which surface integral formulations of acoustic radiation problems have unique or nonunique answers. Then we discuss the circumstances for which the variational principle under consideration can have nonuniqueness difficulties. An explicit example is a finite circular cylinder oscillating rigidly in the axial direction. The calculated surface acoustic pressures are shown to be nonunique at the interior Neumann eigenfrequencies. The latter part of the paper shows that the variational formulation always yields unique solutions for vibrating disks and plate-like bodies. Also, for the case when each surface point is vibrating either in phase or 180° out of phase, it is shown that the variationally formulated total radiated acoustic power is always unique even though the surface pressure is not.

II. KIRCHHOFF INTEGRAL THEOREMS

The integral theorems of interest to the present paper are briefly summarized here; the notation follows that previously used by Pierce [8].

With any given closed surface S one may associate *interior* and *exterior* acoustic problems. For the former, the complex amplitude of the acoustic pressure p_{int} satisfies the scalar Helmholtz equation throughout the volume V enclosed by S. While for the latter, the complex amplitude p_{ext} satisfies the Helmholtz equation throughout the region external to S and the radiation condition at large external distances. The present discussion assumes, for simplicity, that the external region is unbounded.

Following well-known procedures [9], one can derive the Kirchhoff integral corollaries to the wave equation for both *interior* and *exterior* problems. These can be written in such a form that they relate the appropriate acoustic pressure p on the surface to a quantity f which is the negative of the outward normal component of the gradient of p at the surface. Alternately, because of Euler's equation of motion and the implicit use of $e^{-i\omega t}$ time dependence, one has

$$-i\omega\rho\,\vec{v}(\vec{\xi}) = -\vec{\nabla}_\xi p(\vec{\xi}) \tag{1a}$$

$$f = -i\omega\rho\, v_n(\vec{\xi}) \tag{1b}$$

where $v_n(\vec{\xi})$ is the normal component of the fluid velocity $\vec{v}(\vec{\xi})$ at the surface. One can regard f as a convenient grouping of symbols, either as a constant times the normal velocity, or as a constant times the normal acceleration, or as the normal component of the apparent body force per unit volume exerted on the fluid at the surface. Because of the latter identification, the use of the symbol 'f' seems appropriate.

For the *interior* problem one has

$$\mathcal{M}\{\vec{x}, p_{\text{int}}, f_{\text{int}}\} = \begin{cases} -p_{\text{int}}(\vec{x}) & \text{if } \vec{x} \text{ inside } V; \\ 0 & \text{if } \vec{x} \text{ outside } V. \end{cases} \tag{2}$$

While for the *exterior* problem one has

$$\mathcal{M}\{\vec{x}, p_{\text{ext}}, f_{\text{ext}}\} = \begin{cases} p_{\text{ext}}(\vec{x}) & \text{if } \vec{x} \text{ outside } V; \\ 0 & \text{if } \vec{x} \text{ inside } V. \end{cases} \tag{3}$$

Both Eqs. (2) and (3) use the abbreviation

$$\mathcal{M}\{\vec{x}, p, f\} = \frac{1}{4\pi}\iint \left\{ f(\vec{\xi})G(\vec{x}|\vec{\xi}) + p(\vec{\xi})\left[\vec{n}(\vec{\xi})\bullet\vec{\nabla}_\xi G(\vec{x}|\vec{\xi})\right]\right\} dS_\xi \tag{4}$$

where

$$G(\vec{x}|\vec{\xi}) = \frac{e^{ikR}}{R} \tag{5}$$

is the so-called "free-space" Green's function with

$$R = |\vec{x} - \vec{\xi}| \tag{6}$$

denoting the distance between the "source" and the "receiver". In the integrand of Eq. (4), the point $\vec{\xi}$ (after evaluation of any requisite normal derivatives) is understood to range over the surface S, with the point $\vec{x}$ held fixed during the integration. The unit outward normal vector $\vec{n}(\vec{\xi})$ points out of the enclosed volume V at the surface point $\vec{\xi}$.

From either Eq. (2) or Eq. (3), one can derive two types of relations (distinguished by subscripts I and II) between the surface values of p and f. One such type of relation results [8] when the off-surface point $\vec{x}$ is allowed to approach an arbitrary but fixed surface point $\vec{\eta}$:

$$p_{\text{int}}(\vec{\eta}) + \mathcal{L}_{\text{I}}\{\vec{\eta}, p_{\text{int}}\} = -\mathcal{H}_{\text{I}}\{\vec{\eta}, f_{\text{int}}\} \tag{7}$$

$$p_{\text{ext}}(\vec{\eta}) - \mathcal{L}_{\text{I}}\{\vec{\eta}, p_{\text{ext}}\} = \mathcal{H}_{\text{I}}\{\vec{\eta}, f_{\text{ext}}\} \tag{8}$$

where

$$\mathcal{L}_{\text{I}}\{\vec{\eta}, p\} = \frac{1}{2\pi}\iint p(\vec{\xi})\left[\vec{n}(\vec{\xi})\bullet\vec{\nabla}_\xi G(\vec{\eta}|\vec{\xi})\right] dS_\xi \tag{9}$$

$$\mathcal{H}_{\text{I}}\{\vec{\eta}, f\} = \frac{1}{2\pi}\iint f(\vec{\xi})G(\vec{\eta}|\vec{\xi})\, dS_\xi \tag{10}$$

such that the symbols $\mathcal{L}_{\text{I}}$ and $\mathcal{H}_{\text{I}}$ can be regarded as linear operators which operate on the surface values of p and f, respectively, with the resultant in each case being a function of the position of the surface point $\vec{\eta}$.

The second type of surface relationship is obtained by taking the normal derivative of both sides of Eq. (2) or (3) at $\vec{\eta} + \epsilon\vec{n}(\vec{\eta})$, where $\vec{\eta}$ is an arbitrary point on the surface, using the vector identities and integration by parts to recast the second normal derivative of the free-space Green's function into a dot product of two tangential derivatives of some functions defined on the surface, and subsequently taking the limit as $\epsilon \to 0$. The details of derivations of the formulations are shown in Ref. [10] and the results are

$$\mathcal{L}_{\text{II}}\{\vec{\eta}, p_{\text{int}}\} = f_{\text{int}}(\vec{\eta}) - \mathcal{H}_{\text{II}}\{\vec{\eta}, f_{\text{int}}\} \tag{11}$$

$$-\mathcal{L}_{\text{II}}\{\vec{\eta}, p_{\text{ext}}\} = f_{\text{ext}}(\vec{\eta}) + \mathcal{H}_{\text{II}}\{\vec{\eta}, f_{\text{ext}}\} \tag{12}$$

where

$$\mathcal{L}_{\text{II}}\{\vec{\eta}, p\} = \left[\vec{n}(\vec{\eta})\times\vec{\nabla}_\eta\right]\bullet\frac{1}{2\pi}\iint\left[\vec{n}(\vec{\xi})\times\vec{\nabla}_\xi p(\vec{\xi})\right]G(\vec{\eta}|\vec{\xi})\, dS_\xi + \frac{k^2}{2\pi}\iint\left[\vec{n}(\vec{\eta})\bullet\vec{n}(\vec{\xi})\right]p(\vec{\xi})G(\vec{\eta}|\vec{\xi})\, dS_\xi \tag{13}$$

and

$$\mathcal{H}_{\text{II}}\{\vec{\eta}, f\} = \frac{1}{2\pi}\iint f(\vec{\xi})\left[\vec{n}(\vec{\eta})\bullet\vec{\nabla}_\eta G(\vec{\eta}|\vec{\xi})\right] dS_\xi. \tag{14}$$

It is noted that the operator $\vec{n}(\vec{\eta}) \times \vec{\nabla}_\eta$ in Eq. (13) involves only derivatives tangential to the surface, so that the integral on which it acts needs only be evaluated at the surface point $\vec{\eta}$.

The characteristic surface operators $\mathcal{L}_{\text{I}}$, $\mathcal{L}_{\text{II}}$, $\mathcal{H}_{\text{I}}$, and $\mathcal{H}_{\text{II}}$ have important properties in regard to their adjoints (denoted by the superscript †). Definitions of the adjoint operators are such that if $\Phi(\vec{\eta})$ and $\Psi(\vec{\eta})$ are functions defined over the surface S, then, for example,

$$\iint \Psi(\vec{\eta})\mathcal{L}_{\text{I}}\{\vec{\eta}, \Phi\}\, dS_\eta = \iint \Phi(\vec{\eta})\mathcal{L}_{\text{I}}^{\dagger}\{\vec{\eta}, \Psi\}\, dS_\eta. \tag{15}$$

Using such a definition of an adjoint operator, one derives [8]

$$\mathcal{L}_{\text{I}}^{\dagger}\{\vec{\eta}, p\} = \mathcal{H}_{\text{II}}\{\vec{\eta}, p\} \tag{16a}$$

$$\mathcal{H}_{\text{I}}^{\dagger}\{\vec{\eta}, f\} = \mathcal{H}_{\text{I}}\{\vec{\eta}, f\} \tag{16b}$$

$$\mathcal{L}_{\text{II}}^{\dagger}\{\vec{\eta}, p\} = \mathcal{L}_{\text{II}}\{\vec{\eta}, p\} \tag{16c}$$

$$\mathcal{H}_{\text{II}}^{\dagger}\{\vec{\eta}, f\} = \mathcal{L}_{\text{I}}\{\vec{\eta}, f\}. \tag{16d}$$

Thus the operators $\mathcal{H}_{\text{I}}$ and $\mathcal{L}_{\text{II}}$ are self-adjoint, while $\mathcal{L}_{\text{I}}$ and $\mathcal{H}_{\text{II}}$ are adjoints of each other (an adjoint pair).

III. UNIQUENESS THEOREMS

The oft-cited uniqueness theorems follow from the surface integral relations (7) and (8), the surface integro-differential relations (11) and (12), and the adjoint relations stated above.

The applicable mathematical theorem [11] is as follows. Let $\mathcal{L}$ be a linear operator, ψ be the sought function, and ϕ be the given function. Then the inhomogeneous linear equation

$$\mathcal{L}\{\psi\} = \phi \tag{17}$$

has a unique solution if and only if there are no nontrivial solutions of the homogeneous adjoint equation

$$\mathcal{L}^{\dagger}\{\psi\} = 0. \tag{18}$$

This theorem may be well-known to some readers, but a heuristic proof based on linear algebra considerations [12, 13] may be of interest to research engineers. If one takes the operator $\mathcal{L}$ to be an N-by-N matrix $[L]$, then its adjoint is the transposed matrix $[L]^T$. One knows that the solution of the matrix equation corresponding to Eq. (17) is unique if and only if $[L]$ has an inverse, which is true if and only if the determinant of $[L]$ is not zero. However, the determinant of $[L]$ is the same as that of its transpose, and the homogeneous adjoint equation has a nontrivial solution if and only if the determinant of $[L]^T$ vanishes. Thus if the homogeneous adjoint equation does not have a nontrivial solution, then the inhomogeneous equation corresponding to Eq. (17) must have a unique solution.

The *Neumann problem for the exterior region*, in which f on the surface S is given and p_{ext} is sought, can be posed by either Eq. (8) or Eq. (12). If Eq. (8), the usually cited Helmholtz surface integral equation, is used, then there is not a unique solution if there is a nontrivial solution of

$$\mathcal{L}_{\text{I}}\{\vec{\eta},\psi\} - \psi(\vec{\eta}) = 0 \tag{19}$$

which, according to the adjoint relation (16a), is the same as

$$\mathcal{H}_{\text{II}}\{\vec{\eta},\psi\} - \psi(\vec{\eta}) = 0. \tag{20}$$

The latter, however, corresponds to Eq. (11) with ψ identified as f_{int} and p_{int} set to zero, a condition known as the *Dirichlet boundary condition.* Consequently, Eq. (20) corresponds to the ***homogeneous Dirichlet problem for the interior region.*** It is well known that there exist certain characteristic values of the wavenumber k or *Dirichlet eigenfrequencies* for which this homogeneous equation (20) has nonzero solutions. Thus, Eq. (8) has a unique solution if and only if k does not correspond to one of the *Dirichlet eigenfrequencies.*

The alternative posing of the *Neumann problem for the exterior region*, represented by Eq. (12), has a unique solution for p_{ext} if and only if there are no nontrivial solutions of

$$\mathcal{L}_{\text{II}}\{\psi\} = 0. \tag{21}$$

The lack of such a solution, with reference to Eq. (11), is equivalent to the assertion that k is not a characteristic wavenumber or the *Neumann eigenfrequency* of the ***homogeneous Neumann problem for the interior region.***

The *Dirichlet problems for the exterior region*, in which p_{ext} is initially specified on the surface S, can also be posed by either Eq. (8) or Eq. (12). In such cases, the function f_{ext} is regarded as the unknown. The mathematical theorem cited above asserts that Eq. (8) has a unique solution for f_{ext} if and only if there are no nontrivial solutions of

$$\mathcal{H}_I^\dagger\{\vec{\eta},\psi\} = \mathcal{H}_I\{\vec{\eta},\psi\} = 0. \tag{22}$$

Such solutions exist, according to Eq. (7), if k corresponds to a characteristic wavenumber of the ***interior homogeneous Dirichlet problem.***

Similarly, the solution of Eq. (12) for f_{ext} is unique if and only if there are no nontrivial solutions of

$$\mathcal{H}_{\text{II}}^\dagger\{\vec{\eta},\psi\} + \psi(\vec{\eta}) = \mathcal{L}_{\text{I}}\{\vec{\eta},\psi\} + \psi(\vec{\eta}) = 0. \tag{23}$$

Such solutions, according to Eq. (7), will exist if k corresponds to a characteristic wavenumber of the ***interior homogeneous Neumann problem.***

IV. VARIATIONALLY FORMULATED SURFACE ACOUSTIC PRESSURE

The variational principle previously introduced [1 – 3] for the exterior Neumann problem can be written as

$$\delta J = 0 \tag{24}$$

where the stationary quantity J is given by

$$\begin{aligned} J = &\iint p(\vec{\eta})f(\vec{\eta})\,dS_\eta \\ &+\frac{1}{2\pi}\iiiint p(\vec{\eta})f(\vec{\xi})\left[\vec{n}(\vec{\eta})\bullet\vec{\nabla}_\eta G(\vec{\eta}|\vec{\xi})\right]dS_\xi dS_\eta \\ &-\frac{1}{4\pi}\iiiint\left[\vec{n}(\vec{\eta})\times\vec{\nabla}_\eta p(\vec{\eta})\right]\bullet\left[\vec{n}(\vec{\xi})\times\vec{\nabla}_\xi p(\vec{\xi})\right]G(\vec{\eta}|\vec{\xi})\,dS_\xi\,dS_\eta \\ &+\frac{k^2}{4\pi}\iiiint\left[\vec{n}(\vec{\eta})\bullet\vec{n}(\vec{\xi})\right]p(\vec{\eta})p(\vec{\xi})G(\vec{\eta}|\vec{\xi})\,dS_\xi\,dS_\eta \\ &+\frac{1}{4\pi}\iiiint f(\vec{\eta})f(\vec{\xi})G(\vec{\eta}|\vec{\xi})\,dS_\xi\,dS_\eta. \end{aligned} \tag{25}$$

Here, for brevity, the subscript "ext" has been deleted from f_{ext} (which is a priori specified) and p_{ext} (which is to be determined). The correct solution for p is that for which the variation in J is zero to the first order when p is replaced by $p + \delta p$. In terms of the operators introduced in the earlier sections of this paper, one can express the stationary quantity J as

$$\begin{aligned} J = &\iint p(\vec{\eta})f(\vec{\eta})\,dS_\eta + \iint p(\vec{\eta})\mathcal{H}_{\text{II}}\{\vec{\eta},f\}\,dS_\eta \\ &+\frac{1}{2}\iint p(\vec{\eta})\mathcal{L}_{\text{II}}\{\vec{\eta},p\}dS_\eta + \frac{1}{2}\iint f(\vec{\eta})\mathcal{H}_I\{\eta',f\}dS_\eta \end{aligned} \tag{26}$$

after an application of some vector identities. One can readily show that the Euler-Lagrange equation resulting from the variation of Eq. (26) is the integro-differential equation (12). Consequently, the variational principle must have the same inherent uniqueness limitations as does Eq. (12). Suppose, for example, that k is a characteristic wavenumber of the ***inhomogeneous interior Dirichlet problem*** and let

$$p(\vec{\eta}) \to p(\vec{\eta}) + \epsilon\psi(\vec{\eta}) \tag{27}$$

where ϵ is some small number and ψ is a nonzero solution (which must exist) of Eq. (21). It follows from Eq. (12) and the self-adjointness of $\mathcal{L}_{\text{II}}$ that such a ψ must satisfy

$$\iint \psi f\,dS_\eta + \iint \psi\mathcal{H}_{\text{II}}\{\vec{\eta},f\}\,dS_\eta = 0. \tag{28}$$

Moreover, since the operator $\mathcal{L}_{II}$ is linear, one has

$$\begin{aligned} \iint (p+\epsilon\psi)\mathcal{L}_{\text{II}}\{\vec{\eta},p+\epsilon\psi\}\,dS_\eta = &\iint p\mathcal{L}_{\text{II}}\{\vec{\eta},p\}\,dS_\eta \\ &+\epsilon\iint\psi\mathcal{L}_{\text{II}}\{\vec{\eta},p\}\,dS_\eta + \epsilon\iint p\mathcal{L}_{\text{II}}\{\vec{\eta},\psi\}\,dS_\eta \\ &+\epsilon^2\iint\psi\mathcal{L}_{\text{II}}\{\vec{\eta},\psi\}\,dS_\eta. \end{aligned} \tag{29}$$

The second term on the right side in the above is equal to the third term because $\mathcal{L}_{\text{II}}$ is self-adjoint, according to Eq. (16c). Equation (21) then guarantees that all but the first term on the right side of Eq. (29) are identically zero. Consequently, since p satisfies Eq. (12), one finds from Eqs. (28) and (29) that the change induced in J is

$$J\{p+\epsilon\psi\} - J\{p\} = 0 \tag{30}$$

to all orders in ϵ. The consequential interpretation is that, as regards variations in the "direction" of the eigenfunction ψ, the functional J is "flat." The variational principle is unable to distinguish between the correctness of p and $p + \epsilon\psi$.

The example shown below is a finite circular cylinder oscillating back and forth along the axial direction as a rigid body. The surface pressure is obtained by using the Rayleigh-Ritz technique in which the unknown surface pressure is approximated by an expansion of preselected basis functions. Here we want to show that solutions to such a formulation are nonunique when the driving frequencies coincide with one of the *interior Neumann eigenfrequencies* k_N.

The corresponding interior problem in this case resembles a one-dimensional wave propagating in the longitudinal direction of a duct with two rigid end caps

$$\frac{d^2p(x)}{dx^2} + k^2p(x) = 0 \qquad 0 \le x \le L \tag{31}$$

$$\frac{dp(x)}{dx} = 0 \qquad x = 0 \text{ and } L. \tag{32}$$

The eigenfrequencies k_N can be found by using the technique of separation of variables

$$k_N = \frac{(2n-1)\pi}{L} \qquad n = 1, 2, \cdots \tag{33}$$

where L is the length of the finite cylinder.

In the Rayleigh-Ritz procedure, the unknown surface pressure is approximated by the preselected trial functions,

$$p(s) = \sum_{n=0}^{N} C_n P_n(s) \tag{34}$$

where s is the position index along the generator of the finite cylinder, C_n are unknown coefficients to be determined by the variational principle, and $P_n(s)$ are the trial functions. As an example we use the sinusoidal function as the arbitrary admissible trial function

$$P_n(s) = \sin\left[\frac{(2n-1)\pi}{(2a+L)}s\right] \qquad n = 1, 2, \ldots, N \tag{35}$$

where a is the radius of the finite cylinder.

Figure 1 depicts the real and imaginary parts of the acoustic pressure at the center of the end cap of the finite cylinder at different wavenumbers k. The ill-conditioning in numerical computations in the vicinity of the first eigenfrequency, $k_1 = \pi/L$, is clearly exhibited.

This ill-conditioning, however, does not occur in cases of disks or plate-like bodies for which each surface point is vibrating either in phase or 180° out of phase. To verify this statement, let us consider a finite sized thin rectangular plate with dimensions $L_x \times L_y \times L_z$, where L_z is the thickness of the plate.

Using the technique of separation of variables, one readily finds the Helmholtz equations for the interior region in x, y, and z components,

$$\left.\begin{aligned} X'' + k_x^2 X &= 0 \\ Y'' + k_y^2 Y &= 0 \\ Z'' + k_z^2 Z &= 0 \end{aligned}\right\} \tag{36}$$

where the separation constants are related to each other through

$$k_x^2 + k_y^2 + k_z^2 = k^2. \tag{37}$$

Substituting the rigid-wall boundary conditions at $x = \pm L_x/2$, $y = \pm L_y/2$, and $z = \pm L_z/2$, respectively, one obtains

$$k = \pi\sqrt{\left(\frac{n_x}{L_x}\right)^2 + \left(\frac{n_y}{L_y}\right)^2 + \left(\frac{n_z}{L_z}\right)^2}. \tag{38}$$

For finite values of L_x and L_y, the lowest eigenfrequency k approaches infinity as the thickness of the plate L_z becomes infinitely small

$$k \to \infty \qquad \text{as} \qquad L_z \to 0. \tag{39}$$

Consequently, one will not encounter the ill-conditioning problem in numerical computations in the practical range of the wavenumber k and the external acoustic field is uniquely determined.

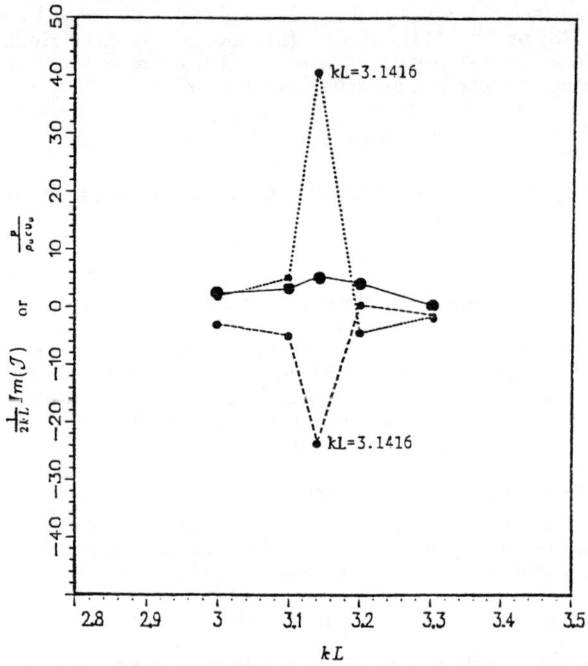

Figure 1. Dimensionless total radiated acoustic power $Im(J)/2kL$ and surface acoustic pressure $p/\rho_o c v_o$ evaluated at the center of the end cap of the finite cylinder as a function of kL. --•--: $Re(p/\rho_o c v_o)$; ··•··: $Im(p/\rho_o c v_o)$; and —●—: $Im(J)/2kL$.

V. VARIATIONALLY FORMULATED TOTAL RADIATED POWER

In many practical applications, our interests are often concerned with accurate estimations of some stationary physical quantities such as the total radiated acoustic power. A variational principle is thus advantageous in that it enables one to obtain an estimate of a required quantity with an error of second order by making a mediocre approximation of some other quantity with an error of the first order. Such a variational formulation can be derived by using the Gerjuoy-Rao-Spruch technique [14] as obtained by Pierce [8]. In terms of the integral operator notations used in this paper, one has

$$\begin{aligned} J = & \iint_S \tilde{p}(\vec{\eta}) f^*(\vec{\eta})\, dS_\eta \\ & + \iint_S \tilde{\Phi}_I \left[\tilde{p}(\vec{\eta}) - \mathcal{L}_I\{\vec{\eta}, \tilde{p}\} - \mathcal{M}_I\{\vec{\eta}, f\}\right] dS_\eta \\ & + \iint_S \tilde{\Phi}_{II} \left[\mathcal{L}_{II}\{\vec{\eta}, \tilde{p}\} + f(\vec{\eta}) + \mathcal{M}_{II}\{\vec{\eta}, f\}\right] dS_\eta \end{aligned} \tag{40}$$

where $\tilde{p}$ is the trial function of p, $\tilde{\Phi}_I$ and $\tilde{\Phi}_{II}$ are trial functions of Φ_I and Φ_{II} for the related problems, and f^* is the complex conjugate of f which is prescribed on the boundary. The Euler-Lagrange equations corresponding to Eq. (40) are obtained by independently varying $\tilde{p}$, $\tilde{\Phi}_I$, and $\tilde{\Phi}_{II}$.

$$\tilde{p}(\vec{\eta}) - \mathcal{L}_{\rm I}\{\vec{\eta}, \tilde{p}\} = \mathcal{M}_{\rm I}\{\vec{\eta}, f\} \tag{41}$$

$$-\mathcal{L}_{\rm II}\{\vec{\eta}, \tilde{p}\} = f(\vec{\eta}) + \mathcal{M}_{\rm II}\{\vec{\eta}, f\} \tag{42}$$

$$-\mathcal{L}_{II}^{\dagger}\{\vec{\eta}, \tilde{\Phi}_{II}\} = f^*(\vec{\eta}) + \tilde{\Phi}_I(\vec{\eta}) - \mathcal{L}_I^{\dagger}\{\vec{\eta}, \tilde{\Phi}_I\}. \tag{43}$$

Equations (41) and (42) are exactly the same as Eqs. (8) and (12). Derivations of the above equations show that they are mathematically consistent. Hence Eq. (43) implies that one can arbitrarily set either $\tilde{\Phi}_I$ or $\tilde{\Phi}_{II}$ to be zero, or choose independent values for $\tilde{\Phi}_I$ and $\tilde{\Phi}_{II}$ such that Eq. (43) is satisfied. As an example one can set

$$\tilde{\Phi}_I = -\frac{1}{2} f^* \tag{44a}$$

$$\tilde{\Phi}_{II} = \frac{1}{2} \tilde{p}_{aux} \tag{44b}$$

where $\tilde{p}_{aux}$ is the solution of an auxiliary exterior problem with the surface value f^* specified as

$$f^* = f_{aux}. \tag{45}$$

Substituting Eqs. (44) and (45) into Eq. (43) then gives

$$-\mathcal{L}_{II}\{\vec{\eta}, \tilde{p}_{aux}\} = f_{aux}(\vec{\eta}) + \mathcal{H}_{II}\{\vec{\eta}, f_{aux}\} \tag{46}$$

which is Eq. (42) for the auxiliary exterior problem.

It has been demonstrated earlier that Eqs. (41) and (42) suffer from nonuniqueness difficulties, the former at $k \in k_D$ (*Dirichlet eigenfrequencies*), while the latter at $k \in k_N$ (*Neumann eigenfrequencies*), respectively. Nevertheless, Eqs. (41) and (42) together always define unique solutions for all wavenumbers because Eqs. (41) and (42) have only one solution in common which is the required boundary value $\tilde{p}$ [15].

Substituting Eq. (44) into Eq. (40), one finds after some manipulations that the stationary quantity J reduces to

$$\begin{aligned} J = \frac{1}{2} \iint & \left[\tilde{p}(\vec{\eta}) f_{aux}(\vec{\eta}) + \tilde{p}_{aux}(\vec{\eta}) f(\vec{\eta})\right] dS_\eta \\ + \frac{1}{2} \iint & \left[f_{aux}(\vec{\eta}) \mathcal{L}_I\{\vec{\eta}, \tilde{p}\} + \tilde{p}_{aux}(\vec{\eta}) \mathcal{H}_{II}\{\vec{\eta}, f\}\right] dS_\eta \\ + \frac{1}{2} \iint & \tilde{p}_{aux}(\vec{\eta}) \mathcal{L}_{II}\{\vec{\eta}, \tilde{p}\} \, dS_\eta \\ + \frac{1}{2} \iint & f_{aux}(\vec{\eta}) \mathcal{H}_I\{\vec{\eta}, f\} \, dS_\eta . \end{aligned} \tag{47}$$

Note that the last term on the right side of Eq. (47) is merely a constant because both f_{aux} and f are prescribed on the boundary. The introduction of the acoustic pressure p_{aux} for the auxiliary exterior problem into the expression for J is to guarantee that the quantity related to the total radiated acoustic power be accurate to the second order. In particular, if the trial function $\tilde{p}$ happens to be the true function p, then the last two terms on the right side of Eq. (40) vanish identically and the total radiated power is given by

$$\mathbf{P} = \frac{1}{2\omega\rho_o} Im(J). \tag{48}$$

In Sec. IV, we have computed the surface acoustic pressure for the axially oscillating finite cylinder. So here we substitute these quantities into Eq. (47) for the stationary quantity J and further into Eq. (48) for $\mathbf{P}$. The results (See Fig. 1) show that the calculated total radiated power is stationary for all wavenumbers even though the surface pressure is not.

VI. CONCLUDING REMARKS

The interrelationship between the differentiated Kirchhoff integral equation and the Euler-Lagrange equation of the variational principle demonstrates that the variational formulation previously derived by the authors for the surface acoustic pressure suffers from nonuniqueness difficulties when the wavenumber k corresponds to some discrete values. This uniqueness problem is demonstrated by a rigid finite circular cylinder vibrating in the axial direction. The calculated surface acoustic pressures are not unique at the interior Neumann eigenfrequencies, although the radiated power is unique. The result that for disks and plate-like bodies solutions are unique for all frequencies should be of special interest. The general demonstration that the variationally formulated total radiated acoustic power is always unique regardless of frequencies, even though the surface pressure may not be, is of practical importance. This enables one to obtain an estimation of the total radiated acoustic power with an error of the second order from a rough estimation of surface pressure with an error of the first order.

ACKNOWLEDGEMENTS

This work was supported by the Office of Naval Research (ONR), code 1132-F and in part by the Institute for Manufacturing Research at Wayne State University.

REFERENCES

1. Pierce, A. D. and Wu, X.- F., "Variational Method for Prediction of Acoustic Radiation from Vibrating Bodies," *The Journal of The Acoustical Society of America, Suppl. 1*, **74**, Fall 1983, S107.
2. Wu, X.- F., Pierce, A. D., and Ginsberg, J. H., "Variational Method for Computing Surface Pressure on Vibrating Bodies, Applied to Transversely Oscillating Disks," *IEEE Journal of Oceanic Engineering*, **OE-12**, 1987, pp. 412 – 418.
3. Wu, X.- F., "Faster Calculations of Sound Radiation from Vibrating Cylinders Using Variational Formulations," *Transactions of The ASME Journal of Vibration, Acoustics, Stress, and Reliability in Design*, Vol. **111**, 1988, pp. 101 – 107.
4. Leitner, A., "Diffraction of Sound by a Circular Disk," *The Journal of The Acoustical Society of America*, Vol. **21**, 1949, pp. 331 – 334.
5. Lamb, H., *Hydrodynamics*, Cambridge University Press 1924, 5th ed., Sec. **290**.
6. Kupradze, W. D., "Fundamental Problems in the Mathematical Theory of Diffraction," (1935); translated by C. D. Benster, N. B. S. Rep. No. 2008 (1952); see also, "Randwertaufgaben der Schwingungstheorie und Integralgleichungen," Deutscher Verlag der Wissenschaften, Berlin (1956).
7. Schenck, H. A., "Improved Integral Formulation for Acoustic Radiation," *The Journal of The Acoustical Society of America*, Vol. **75**, 1968, pp. 363 – 375.
8. Pierce, A. D., "Stationary Variational Expressions for Radiated and Scattered Acoustic Power and Related Quantities," *IEEE Journal of Oceanic Engineering*, **OE-12**, 1987, pp.404 – 411.
9. Pierce, A. D., *Acoustics: An Introduction to Its Physical Principles and Applications*, New York: McGraw-Hill 1981, Chap. 4, pp. 180 – 182.
10. Wu, X.- F., *Variational Principles for Acoustic Radiation and Diffraction from Underwater Structures*, Ph. D. Thesis, Georgia Institute of Technology, 1987.
11. Smithies, F., *Integral Equations*, Cambridge University Press, Cambridge, England, 1985, pp. 51 – 52.
12. Birchoff, G. and Rota, G.- C., *Ordinary Differential Equations* (Blaisdell Publishing Company 1969), 2nd ed.
13. Garabedoam, P.R., *Partial Differential Equations*, New York, Wiley, 1964, pp. 353 – 358.
14. Gerjuoy, E., Rau, A. R. P., and Spruch, L., "A Unified Formulation of the Construction of Variational Principles," *Reviews of Modern Physics*, Vol. **55**, 1983, pp. 725 – 774.
15. Burton, A. J. and Miller, G. F., "The Application of Integral Equation Methods to the Numerical Solution of Some Exterior Boundary-Value Problems," *Proceedings of The Royal Society of London*, **A 323**, 1971, pp. 201 – 210.

A CONTRIBUTION TO DOUBLY ASYMPTOTIC APPROXIMATIONS: AN OPERATOR TOP-DOWN DERIVATION

B. Nicolas-Vullierme
Office National D'Etudes Et De Recherches Aerospatiales
Chatillon, France

ABSTRACT

As a contribution to Doubly Asymptotic Approximations (DAA's) a formal top-down derivation specialized to steady-state motions is proposed, which generalizes previously published ones. As opposed to the latters which relied on a modal approach [1] or on a scalar approach for a model problem [2], our derivation is based on an integral representation of the solution of the Helmholtz equation in an unbounded domain : first, two asymptotic expansions of the representation are obtained in the low- and the high-frequency ranges, then these expansions are matched.

This procedure allows to point out that some geometrical assumptions underlie the validity of high order continuous forms of the DAA's. Special attention is also devoted to their discretized counterparts. In both cases it suggests further investigations of some interesting open geometry and numerical analysis problems.

0. INTRODUCTION

In order to get some insight into Doubly Asymptotic Approximations, the response to prescribed harmonic motions of an homogeneous acoustic medium will be considered throughout this paper. All unknowns will be assumed to be harmonic in time with a time factor exp(+iωt) which will be omitted in the sequel. Expressed in terms of the fluid displacement potential field φ, the equilibrium equations, the boundary condition and the Sommerfeld radiation conditions read :

$$(P_k)\begin{cases}\Delta\varphi + k^2\varphi = 0 \text{ in } \Omega \\ \partial_n\varphi = g \text{ on } \Gamma \\ \left|\frac{\partial\varphi}{\partial r} + ik\varphi\right| = O(\frac{1}{r^2}) \text{ and } |\varphi| = O(\frac{1}{r}) \text{ as } r\to\infty.\end{cases}$$

where k is the wave number ω/c, Ω denotes (Figure 1) the exterior domain and Γ its boundary, n denotes the outer unit normal, ∂_n the normal derivative and the real function g is the prescribed normal displacement field on the surface Γ.

Figure 1

It is a well-known result that for all values of k this problem has a unique solution (cf. for example [3]) which restriction to Γ will be denoted by :

$\varphi_{|\Gamma}=T(g)$.

Our attention will be devoted to approximations of the "impedance" operator T and of the solution $\varphi_{|\Gamma}$ (in the following, the subscript $_{|\Gamma}$ will be omitted) by means of asymptotic expansions.

More precisely, the use of Green's second identity allows to get the so-called Helmholtz integral representation ot the solution. It reads :

$$\alpha\varphi(x)= \int_\Gamma [\ \varphi(y)\frac{\partial G}{\partial n_y}(x,y) - \frac{\partial\varphi}{\partial n_y}G(x,y)\ ]\ d\Gamma_y, \qquad (1)$$

where α=1/2 if x ∈ Γ, α=1 if x is outside Γ and α=0 if x is inside Γ. The Green function G reads :

$$G(x,y) = \frac{\exp(-ikd(x,y))}{4\pi d(x,y)} \qquad (2)$$

where d(x,y) denotes the distance between x and y and n_y is the outer normal to the surface Γ at point y.

For a point x lying on the surface Γ, the Helmholtz integral representation gives the exact impedance relation between φ and its normal derivative, i.e. the exact operator T.

It will be shown that in both low-k and high-k ranges, this representation provides successive approximations of the operator T. Both formal asymptotic expansions will be considered in the two next sections. Their matching will be studied in a third one in order to get continous forms of DAA's. In the last section their discretized counterparts will be considered.

1. LOW-FREQUENCY EXPANSION

When k is assumed to be small compared to 1, the Green function G(x,y) may be expanded as :

$$G = G_0 + kG_1 + k^2G_2 + ... + k^nG_n + ..., \qquad (3)$$

which reads, according to definition (2) :

$$4\pi G(x,y) = \frac{1}{d(x,y)} - ik - \frac{k^2\, d(x,y)}{2} + ... + \frac{(-ik)^n\, d(x,y)^{n-1}}{n!} + ...$$

$$4\pi \frac{\partial G}{\partial n_y}(x,y) = (x-y, n_y) \times \left(\frac{1}{d^3(x,y)} - \frac{k^2}{2d(x,y)} + ... - \frac{n(-ik)^{n+1}\, d(x,y)^{n-2}}{(n+1)!} + ...\right)$$

Classically the first term G_0 of expansion (3) is the Green function associated with problem (P_0) :

$$(P_0)\begin{cases} \Delta\varphi = 0 \text{ in } \Omega \\ \partial_n\varphi = g \text{ on } \Gamma \\ \left|\frac{\partial\varphi}{\partial r}\right| = O(\frac{1}{r^2}) \text{ and } |\varphi| = O(\frac{1}{r}) \text{ as } r\to\infty. \end{cases}$$

Similarly the restriction to Γ of the solution of (P_k) may be expanded as

$$\varphi = \varphi_0 + k\varphi_1 + k^2\varphi_2 + ... + k^n\varphi_n + ..., \qquad (4)$$

which may be written as

$$\varphi_i = \sum_i T_i(g). \qquad (5)$$

Substituting the expansions (3) and (4) into the Helmholtz integral representation (1) for x lying on Γ and cancelling the coefficients of k^i up to the second order yields :

$$\frac{1}{2}\varphi_0(x) = \int_\Gamma \varphi_0(y)\frac{\partial G_0}{\partial n_y}(x,y)d\Gamma_y - \int_\Gamma g(y)G_0(x,y)\, d\Gamma_y \qquad (6)$$

$$\frac{1}{2}\varphi_1(x) = \int_\Gamma \varphi_1(y)\frac{\partial G_0}{\partial n_y}(x,y)d\Gamma_y + \frac{i}{4\pi}\int_\Gamma g(y)\, d\Gamma_y \qquad (7)$$

$$\frac{1}{2}\varphi_2(x) = \int_\Gamma \varphi_2(y)\frac{\partial G_0}{\partial n_y}(x,y)d\Gamma_y - \frac{1}{8\pi}\int_\Gamma \varphi_0(y)\frac{\partial d(x,y)}{\partial n_y}d\Gamma_y + \frac{1}{8\pi}\int_\Gamma g(y)\, d(x,y)d\Gamma_y.$$

Not surprisingly, equation (6) for the zeroth order term is the Helmholtz integral representation of the solution of (P_0). Thus,

$$\varphi_0 = T_0(g) \qquad (8)$$

is the restriction to Γ of the solution of (P_0).

By use of a single layer representation of the solution of (P_k), it is in shown [4] that φ_1 does not depend on x. From (7), one gets :

$$\varphi_1(x) = T_1(g) = \frac{i}{4\pi}\int_\Gamma g(y)\, d\Gamma_y$$

so that the first order low-frequency approximation of φ reads :

$$\varphi = T_0(g) + kT_1(g) = T_0(g) + \frac{ik}{4\pi}\int_\Gamma g(y)\, d\Gamma_y, \qquad (9)$$

2. HIGH FREQUENCY EXPANSION

When k is assumed to be large compared to 1, the formal asymptotic expansion relies on the limiting absorption principle [5] and on the method of stationary phase [6] :

Recall first that the outgoing solution of problem (P_k) may be sought as the limit, when the real positive number ζ tends to 0_+, of the solution φ' of problem (P'_k) :

$$(P'_k)\begin{cases} \Delta\psi + (k-i\zeta)^2\psi = 0 \text{ in } \Omega \\ \partial_n\psi = g \text{ on } \Gamma \\ \int_\Omega |\psi|^2\, d\Omega < \infty \;;\; \int_\Omega |\nabla\psi|^2\, d\Omega < \infty \end{cases}$$

For such a solution, the Helmholtz integral representation (1) for a point x lying on Γ holds with the Green function G' :

$$G'(x,y) = \frac{\exp(-ik'd(x,y))}{4\pi d(x,y)},\quad k' = k - i\zeta.$$

Thus, the right-hand side of this representation is the sum of two integrals over a manifold of $\mathbf{R}^2$. They are both of the form :

$$I(x) = \int F(\xi)\exp(-ikf(\xi))\, d\xi,$$

and it is known that when k tends to infinity, the main contribution to such integrals results from stationary points of the phase function f(ξ).

As f(ξ) is here the distance d(x,y) between point x and another point y lying on Γ, it is stationary at point x itself and at points y_i (Figure 2) where, if n_i is the outer normal to Γ at point y_i,

$$(x-y_i).n_i = \pm 1.$$

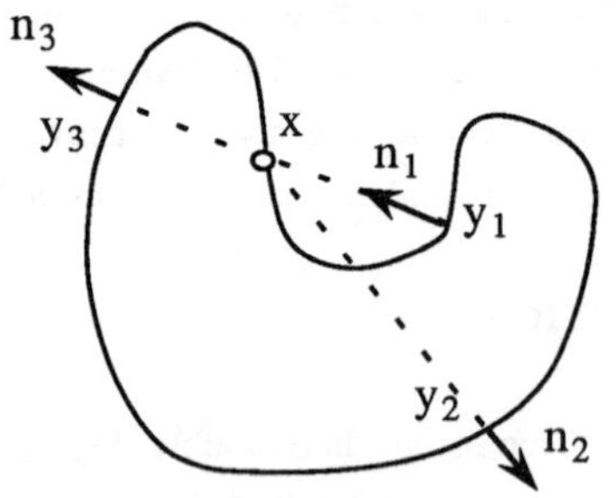

Figure 2

Thus the integral I(x) reads [6],

$$I(x) = \int_{N(x)\cup N(y_i)} F(\xi)\exp(-ikf(\xi))\, d\xi + 0\left(\frac{1}{k^n}\right), \forall n$$

where N(z) denotes a neighbourhood of point z.

Therefore, we are first left with the calculation of the single and double layer terms of

$$\frac{1}{2}\psi(x) \approx \int_{N(x)\cup N(y_i)} \left(\psi(y)\frac{\partial G'}{\partial n_y}(x,y) - g(y)G'(x,y) \right) d\Gamma_y, \quad (10)$$

and then with taking the limit of $\psi(x)$ when $\zeta \rightarrow 0_+$.

2.1 Contribution of Point x

The singularity of the Green function at point x makes necessary to approximate the surface Γ in the neighbourhood of point x.

2.1.1 Parametrization of the surface near x. Choosing the principal curvature directions as axes in the plane tangent to Γ at point x and the outer normal to Γ at point x as a third axis, any point y of the neighbourhood N(x) may be represented by its cylindrical coordinates as :

$$(\rho\cos\theta,\ \rho\sin\theta,\ -\rho^2 h(\rho,\theta)).$$

Thus, the distance d(x,y) between x and y and the elementary area at point y, $d\Gamma_y$ respectively read :

$$d^2(x,y) = \rho^2 + \rho^4 h^2(\rho,\theta),$$

$$d\Gamma_y = \rho \{1 + \rho^2 [h_\theta'^2 + (2h+\rho h_\rho')^2] \}^{1/2} d\rho d\theta,$$

where h_ρ and h_θ denote the derivatives of the function $h(\rho,\theta)$ with respect to ρ and θ, respectively.

In the neighbourhood N(x) of x, $h(\rho,\theta)$, d(x,y) and $d\Gamma_y$ may be expanded in powers of ρ. One gets :

$$h(\rho,\theta) = \gamma(\theta)(1 + 0(\rho)), \quad (11)$$

where if κ_1 and κ_2 denote the main curvatures of the surface Γ,

$$\gamma(\theta) = \frac{\kappa_1}{2}\cos^2\theta + \frac{\kappa_2}{2}\sin^2\theta;$$

$$d(x,y) = \rho(1+O(\rho^2)); \quad (12)$$

$$d\Gamma_y = \rho(1+O(\rho^2))d\rho d\theta. \quad (13)$$

Similarly, any function χ defined on Γ may be expanded as :

$$\chi(y) = \chi(x) + (y-x).\nabla_y\chi(x)+... = \chi(x)\ (1+ O(\rho)). \quad (14)$$

2.1.2 Single layer term. Using (10)-(14), the contribution of the point x to the single layer term in (10) :

$$S(x) = \int_{N(x)} g(y)G'(x,y)d\Gamma_y,$$

can be rewritten as :

$$S(x) = \frac{g(x)}{4\pi} \int_{0\le\theta\le 2\pi} d\theta \int_{0\le\rho\le\varepsilon} \exp(-ik'\rho)(1+O(\rho))d\rho$$

According to the theorem of stationary phase, the ρ-integral may be calculated as

$$\int_{0\le\rho<\infty} \exp(-ik'\rho)d\rho = \frac{-1}{ik'} = \frac{-1}{i(k-i\zeta)}.$$

Thus,

$$S(x) \approx -\frac{g(x)}{2i(k-i\zeta)} + O(\frac{1}{k'^2}). \quad (15)$$

2.1.3 Double layer term. The contribution of the point x to the double layer term in (10) reads :

$$D(x) = \int_{N(x)} \psi(y)\frac{\partial G'}{\partial n_y}(x,y)\, d\Gamma_y,$$

and involves the normal derivative of the Green function :

$$4\pi\frac{\partial G'}{\partial n_y}(x,y) = \frac{\exp(-ik'd(x,y))}{d(x,y)}[-ik' - \frac{1}{d(x,y)}]\frac{(y-x).n_y}{d(x,y)},$$

where, according to the parametrization (11) :

$$(y - x).n_y = \rho^2\gamma(\theta)(1+O(\rho^2)).$$

Expanding every function of y in the neighborhood N(x) of x, D(x) reads :

$$D(x) = \frac{\psi(x)}{4\pi} \int_{0\le\theta\le 2\pi} \gamma(\theta)d\theta \int_{0\le\rho\le\varepsilon} \exp(-ik'\rho)(-ik'\rho-1)(1+O(\rho))d\rho$$

The θ-integral is easily calculated as :

$$\int_{0\le\theta\le 2\pi} \gamma(\theta)d\theta = \int_{0\le\theta\le 2\pi} (\frac{\kappa_1}{2}\cos^2\theta + \frac{\kappa_2}{2}\sin^2\theta)\, d\theta = \pi\kappa_0,$$

where κ_0 is the mean curvature at $\rho = 0$, that is at point x :

$$\kappa_0 = \kappa(x) = \frac{\kappa_1 + \kappa_2}{2}.$$

Again, according to the theorem of stationary phase, the ρ-integral may be calculated as

$$\int_{0\le\rho<\infty} \exp(-ik'\rho)(-ik'\rho-1)d\rho = \frac{-2}{ik'} = \frac{-2}{i(k-i\zeta)}.$$

Therefore,

$$D(x) \approx -\frac{\psi(x)\kappa(x)}{2i(k-i\zeta)} + O(\frac{1}{k'^2}). \quad (16)$$

Remark : Note that the limiting absorption principle, i.e. the use of the real positive number ζ, is required in the two last sub-sections in order to get convergent integrals over $[0, \infty[$.

2.2 Contribution of Points y_i

For points such as y_i (Figure 2), standard 2-D stationnary phase results [5],[6] may be used for the single layer and the double layer terms involved in the Helmholtz integral representation. Both integrals are of the form $I(y_i)$:

$$I(y_i) = \int_{N(y_i)} F(\xi)\exp(-ikf(\xi))\, d\xi.$$

and, as k tends to ∞, they both yield :

$$I(y_i) = \frac{2\pi}{k}\frac{\exp[-ikf(y_i)]\, \exp[i\frac{\pi}{4}\mathrm{sign}[(\mathrm{Hess}(f))(y_i)]}{|\mathrm{Hess}(f)(y_i)|}F(y_i) + O(\frac{1}{k^2}) \quad (17)$$

where Hess(f) and sign[(Hess(f))(y_i)] respectively denote the Hessian of function f and its signature.

Their sum has also the form of the expression (17) where $F(y_i)$ may be written as the product of a factor : $Z(y_i)$, which is common to the single and the double layer integrals of the Helmholtz integral representation, by a factor $T(y_i)$ which is not :

$$Z(y_i) = \frac{\exp[-\zeta d(x,y_i)]}{d(x,y_i)} \,|\, J(y_i) \,|\,,$$

$$T(y_i) = -\, g(y_i) - (ik' + \frac{1}{d(x,y_i)}) \frac{(y_i - x).n_i}{d(x,y_i)} \psi(y_i), \tag{18}$$

where $J(y_i)$ denotes the Jacobian of the parametrization function on $N(y_i)$.

Notice that

$$\delta(y_i) = \frac{(y_i - x, n_i)}{d(x,y_i)} = \pm 1,$$

according to whether the outer normal to Γ at point y_i is directed towards point x (Points y_i, i=2,3, on Figure 2 : $\delta(y_i) = +1$) or not (Point y_1 : $\delta(y_1) = -1$). Both cases will be considered successively.

2.2.1 Contribution of points where $\delta(y_i)=+1$. For such points,

$$T(y_i) = -\, g(y_i) - (ik' + \frac{1}{d(x,y_i)}) \psi(y_i)$$

$$= -\, g(y_i) - (ik + \frac{1}{d(x,y_i)}) \psi(y_i) - \zeta\psi(y_i),$$

the first two terms of which appear in the Sommerfeld boundary condition as written by Bloom (1976) :

$$|\, g(y) + (ik + \frac{1}{r}) \psi(y) \,| = O(\frac{1}{r^2}) \text{ as } r \to \infty.$$

For our purpose, we shall admit that

$$|\, g(y_i) - (ik + \frac{1}{d(x,y_i)}) \psi(y_i) \,| = O(\frac{1}{k^2}) \text{ as } k \to \infty. \tag{19}$$

This heuristic assumption makes sense as the adimensionnal factor in the Sommerfeld condition is indeed the product kd. In both previous conditions kd tends to infinity : in the former for fixed k and r tending to infinity and in the latter for fixed y_i and k tending to infinity. Note also that the contributions of such points are classically not taken into account in geometrical acoustics [8] for physical reasons.

Therefore, from (19), the contribution of such points y_i may be written as :

$$I(y_i) = \zeta\psi(y_i)K(y_i) + O(\frac{1}{k^2}),$$

where $K(y_i)$ is bounded and involves terms from (17) and (18).

2.2.2 Contribution of points where $\delta(y_i)=-1$. As far as these points are concerned, from (18), results of previous subsection do not hold so that they may contribute to the asymptotic expansion of ψ with terms of the order k^{-1}.

In order to get a "local" relation between g(x) and $\psi(x)$, i.e. a relation which does not involve values of g and/or ψ at other points of the surface Γ, we shall assume that Ω is convex (cf. the treatment of multiple structures in [9]). This assumption is necessary for the following derivation to hold.

2.3 Expansion

With this geometrical assumption and for large k, the results of the previous subsections are summarized in :

$$\frac{\psi(x)}{2} = -\frac{g(x)}{2ik'} - \frac{\psi(x)\kappa(x)}{2ik'} + \sum_i \zeta\psi(y_i)K(y_i) + O(\frac{1}{k^2}).$$

Thus, as ζ tends to 0 we get the first and second high frequency approximations. From

$$\frac{\varphi(x)}{2} = -\frac{g(x)}{2ik} - \frac{\varphi(x)\kappa(x)}{2ik},$$

the first and the second order expansions respectively read :

$$g(x) = -ik\varphi(x), \tag{20}$$

$$g(x) = -ik\varphi(x) - \varphi(x)\kappa(x), \tag{21}$$

and the latter may be written as

$$g = -ikI(\varphi) - H(\varphi),$$

where I denotes the identity operator for functions defined on the surface Γ and where H associates to such a function φ the fonction $H(\varphi)$ defined by :

$$(H\varphi)(x)=\kappa(x)\varphi(x).$$

Not surprisingly, relations (19) and (20) are respectively the plane wave and the curved wave approximation as proposed by Mindlin and Bleich [10], Haywood [11] and Bedrosian and DiMaggio [12].

Remark : It is interesting to notice the similarities between this derivation and the derivation of high order dampers by Kando, Bettess and al. [13]. Their derivation pertains to a spherical surface Γ which simplicity makes possible to calculate any order of the expansion (cf. also [14]).

3 MATCHING

The expression "matched asymptotic expansion" usually applies to the search of uniformly valid approximate solutions of a given differential equation. On the contrary, DAAs are approximate equations whose solutions approach, in both low and high frequency ranges, the ones of the set of equations satisfied by the actual solution of (P_k).

The matching is obtained by assuming a relation of the form :

$$(I + kC_1 - ik^2C_2)(\varphi) = (T_0 + kC_2)(g). \tag{22}$$

where C_1 and C_2 are unknown operators. Moreover, the operator C_2 is supposed to be invertible (and this assumption will have to be checked later).

The unknowns operators C_1 and C_2 will be determined from the second order expansions for the low- and high-frequency range expansion, i.e respectively :

$$\varphi = T_0(g) + kT_1(g), \text{ as } k \to 0,$$

$$g = -ikI(\varphi) - H(\varphi), \text{ as } k \to \infty.$$

In the low frequency range (22) reads

$$\varphi = (I + kC_1 - ik^2C_2)^{-1}(T_0 + kC_2)(g),$$

so that, by inverting an operator of the form $(I+\varepsilon D)^{-1}$ for small ε,

$$\varphi = (T_0 + k(C_2 - C_1T_0) + O(k^2))(g). \tag{23}$$

The identification of (9) and (23) gives

$$C_1T_0 - C_2 = -iT_1. \tag{24}$$

In the high frequency range, (22) reads :

$$(-iC_2 + k^{-1}C_1 + k^{-2}I)(\varphi) = k^{-1}(C_2 + k^{-1}T_0)(g),$$

thus

$$g = k(C_2 + k^{-1}T_0)^{-1}(iC_2 + k^{-1}C_1 + k^{-2}I)(\varphi).$$

Since C_2 is assumed to be invertible, the previous equality becomes :

$$g = k(I + k^{-1}C_2^{-1}T_0)^{-1}(-iI + k^{-1}C_2^{-1}C_1 + k^{-2}C_2^{-1})(\varphi).$$

For small k^{-1}, the inversion gives :

$$g = -ik\varphi + (C_2^{-1}C_1 + i\, C_2^{-1}T_0)\varphi + O(k^2). \quad (25)$$

Identifying (21) and (25), one gets :

$$C_1 + C_2H = -iT_0. \quad (26)$$

Thus, from (24) and (26), C_1 and C_2 satisfy :

$$C_1T_0(T_0^{-1} + H) = -(iT_0 + T_1H), \quad (27)$$

$$C_2(T_0^{-1} + H)T_0 = -iT_0^2 + T_1. \quad (28)$$

At this point, recall that previous derivations required first Ω to be convex (in order to get the high-frequency expansion); moreover C_2 was assumed to be invertible (in order to identify (22) with high order expansions).

Since operator T_0 is related to the coercive form $\int|\nabla\varphi|^2 d\Omega$, it is invertible and so is T_0^{-1}. Thus, the invertibility of C_2 (28) requires that of $(T_0^{-1} + H)$ and $(-iT_0^2 + T_1)$. Giving necessary and/or sufficient conditions for these hypothesis to hold is, to author's knowledge, an open problem. Note also that in previous subsections these hypothesis have only been given as sufficient for our derivation of high order continuous forms of the DAA's to hold.

These forms will now be given according to the order of both expansions used for the matching.

a) A 1-low, 1-high DAA (the continuous form of the DAA1) is obtained by enforcing $C_2 = H = T_1 = 0$ in the derivation of the previous subsection :

$$[I - ikT_0](\varphi) = T_0(g). \quad (29)$$

b) A 1-low, 2-high DAA (the continuous DAA2c) is obtained by enforcing $T_1 = 0$. It requires the operator Θ,

$$\Theta = T_0^{-1} + H,$$

to be invertible. It reads :

$$[I - ikT_0\Theta^{-1}T_0^{-1} - k^2T_0\Theta^{-1}](\varphi) = [T_0 - ikT_0\Theta^{-1}](g). \quad (30)$$

c) A 2-low, 1-high DAA is obtained by enforcing $H = 0$ in the previous subsection. It requires $(-iT_0^2 + T_1)$ to be invertible.

With this assumption this approximation reads :

$$[I - ikT_0 - k^2(T_0^2 + iT_1)](\varphi) = [T_0 - ik(T_0^2 + iT_1)](g). \quad (31)$$

d) Finally, a 2-low, 2-high DAA is obtained which requires the operators $(-iT_0^2 + T_1)$ and $(T_0^{-1} + H)$ to be invertible. It reads :

$$[I-ik(T_0-iT_1H)\Theta^{-1}T_0^{-1}-k^2(T_0+iT_1T_0^{-1})\Theta^{-1}](\varphi)$$
$$= [T_0-ik(T_0+iT_1T_0^{-1})\Theta^{-1}](g). \quad (32)$$

Remark : The radiation damping r associated with these expansions is $r = -\pi \,\mathrm{Im} \int\bar{\varphi}.g\, d\Gamma$.

It is checked to be positive in both low and high frequency ranges where, respectively,

$$r = -\pi \,\mathrm{Im} \int\bar{\varphi}.g\, d\Gamma = \frac{k}{4}(\int g\, d\Gamma)^2 > 0,$$

$$r = -\pi \,\mathrm{Im} \int\bar{\varphi}.g\, d\Gamma = k\,\pi\int \bar{\varphi}\,\varphi\, d\Gamma > 0.$$

4. DISCRETIZED FORM OF DAAs

The discretized counterparts of the continuous forms of the DAAs derived above will now be addressed. To do so, consider a discretization of the surface Γ and the related discretized counterparts of the above defined operators (bold characters will stand for their representatives).

In order to recover the expressions used in the literature, introduce the two bilinear forms F and A involved in the variational formulation of (P_0) :

$$F(\varphi,\psi) = \int_\Omega \nabla\varphi\nabla\psi d\Omega,\ A_f(g,\psi) = \int_\Gamma g\psi d\Gamma,$$

$$F(\varphi,\psi) + A_f(g,\psi) = 0,\ \forall\psi, \quad (33)$$

The discretized counterpart of the variational formulation (33) reads

$$\mathbf{\Psi}^t\mathbf{F}\mathbf{\Phi} + \mathbf{\Psi}^t\mathbf{A_f}\mathbf{G} = 0,\ \forall\mathbf{\Psi}. \quad (34)$$

From (34) it is seen that the discretized counterpart of the operator T_0 which solves φ as a function of g, is :

$$\mathbf{T_0} = -\mathbf{F}^{-1}\mathbf{A_f}.$$

Remark : Note that the added mass matrix reads :

$$\mathbf{M_f} = -\rho\mathbf{A_f}^t\mathbf{T_0} = -\rho\mathbf{A_f}^t\mathbf{F}^{-1}\mathbf{A_f}. \quad (35)$$

This expression is similar to the one found in the literature if $\mathbf{A_f}$ is symmetric, that is for example if the same discretization is used for $\mathbf{\Psi}$ and $\mathbf{G}$. As the matrix $\mathbf{A_f}$ discretizes the operator A :

$$A(g,\psi) = \int_\Gamma g\psi d\Gamma \approx \mathbf{\Psi}^t\mathbf{A_f}\mathbf{G},$$

the condition "$\mathbf{A_f}$ is diagonal" as in Ref. [1] implies stronger conditions on the discretization chosen for g and ψ. This is the case for example for piecewise constant interpolations.

With these notations, the discretized counterpart of the 1-low, 2-high DAA (30) reads :

$$[\mathbf{I} - ik\mathbf{T_0}\mathbf{\Theta}^{-1}\mathbf{T_0}^{-1} - k^2\mathbf{T_0}\mathbf{\Theta}^{-1}]\mathbf{\Phi} = [\mathbf{T_0} - ik\mathbf{T_0}\mathbf{\Theta}^{-1}]\mathbf{G}.$$

or, multiplying it by $-\rho c^2\mathbf{A_f}\mathbf{\Theta}$,

$$[-\rho c^2\mathbf{A_f}\mathbf{\Theta} + i\omega\rho c\mathbf{A_f}\mathbf{\Theta}\mathbf{T_0}\mathbf{\Theta}^{-1}\mathbf{T_0}^{-1} + \rho\omega^2\mathbf{A_f}\mathbf{\Theta}\mathbf{T_0}\mathbf{\Theta}^{-1}]\mathbf{\Phi} =$$
$$[-\rho c^2\mathbf{A_f}\mathbf{\Theta}\mathbf{T_0} + i\omega\rho c\mathbf{A_f}\mathbf{\Theta}\mathbf{T_0}\mathbf{\Theta}^{-1}]\mathbf{G}. \quad (36)$$

This expression has to be compared with the currently used DAA2c which reads [1] :

$$(\rho c\Omega_f\mathbf{A_f} + i\omega\rho c\mathbf{A_f} - \omega^2\mathbf{M_f})\mathbf{P} = (i\omega\rho c\Omega_f\mathbf{M_f} - \rho c\omega^2\mathbf{M_f})\mathbf{V}. \quad (37)$$

In order to compare (36) and (37) it is necessary to assume that $\mathbf{T_0}$ and Θ commute in the former, or, as

$$\Theta = \mathbf{T_0}^{-1} + \mathbf{H},$$

that **H** and $\mathbf{T_0}$ commute.

Going back to the continuous operators H and T_0 it appears that it is not easy to prove this commutativity in a general case. However, it is seen that this assumption holds if H is a constant operator κI, that is if the mean curvature is constant over Γ. This is the case when Γ is a sphere or an infinite cylinder. The commutativity of their discretized counterparts will not be addressed here.

Thus, under the assumptions that Γ is convex and that $\mathbf{T_0}$ and **H** commute, the discretized form of the DAA2c is obtained :

$$[\rho c\mathbf{A_f}\Omega_f + i\omega\rho c\mathbf{A_f} - \omega^2\mathbf{M_f}]\Phi = -c[\mathbf{M_f}\Omega_f + i\omega\mathbf{M_f}]\mathbf{G}, \quad (38)$$

where

$$\Omega_f = -c\Theta = \rho c\mathbf{M_f}^{-1}\mathbf{A_f} - c\mathbf{H}.$$

This expression is similar to (36) up to some permutations in the matrix products :

$$\mathbf{M_f}^{-1}\mathbf{A_f} \text{ in } \mathbf{A_f}\mathbf{M_f}^{-1}, \ \mathbf{A_f}\Omega_f \text{ in } \Omega_f\mathbf{A_f} \text{ and } \mathbf{M_f}\Omega_f \text{ in } \Omega_f\mathbf{M_f}.$$

From the definition of all these matrices, these products do not seem to commute in general. A sufficient condition is that $\mathbf{A_f}$ commute with **F** and **H**. Note that the latter condition holds for piecewise constant interpolation as in this case both matrices are diagonal.

The others discretized low and high order expressions of the above mentioned DAAs are easily obtained :

The DAA1 (29) does not need any other geometric assumption to hold but the convexity of Γ. It reads :

$$[\rho c\mathbf{A_f} + i\omega\mathbf{M_f}]\Phi = -c\mathbf{M_f}\mathbf{G}.$$

The 2-low, 2-high DAA (31) reads :

$$[\rho c\mathbf{A_f}\Omega_f + i\omega\rho c\mathbf{A_f}(\mathbf{I}+c\mathbf{NH}) - \omega^2(\mathbf{M_f}+\rho c\mathbf{A_f}\mathbf{N})]\Phi = c[\mathbf{M_f}\Omega_f + i\omega(\mathbf{M_f}+\rho c\mathbf{A_f}\mathbf{N})]\mathbf{G},$$

where

$$\mathbf{T_0}^{-1}\mathbf{T_1} = c\mathbf{N},$$

which is equivalent to the one obtained by Felippa (1980), up to the above mentioned permutations in matrices products.

The 2-low, 1-high DAA (30) reads :

$$[\rho c\mathbf{A_f}\Lambda_f + i\omega\rho c\mathbf{A_f} - \omega^2(\mathbf{M_f} + \rho c\mathbf{A_f}\mathbf{N})]\Phi = -c[\rho c\mathbf{A_f} + i\omega(\mathbf{M_f} + \rho c\mathbf{A_f}\mathbf{N})]\mathbf{G},$$

where

$$\Lambda_f = \rho c\mathbf{M_f}^{-1}\mathbf{A_f}.$$

5. CONCLUSION

Continuous forms of DAA's have been formally obtained and some geometrical assumptions which underlie their validity have been pointed out. It is clear that in this formal derivation only sufficient conditions (and not necessary) have been given. A complete and rigorous derivation would involve some difficult numerical analysis and geometry problems.

However, it may be interesting to consider the analytically tractable case of the axisymmetric motions of an immersed spherical shell.

In this case,any solution of the problem (P_k) may be decomposed as a linear combination of Legendre polynomials on the surface Γ [15]. Moreover, all the above mentioned operators may be explicitely calculated.

For example, for a "breathing mode" normal displacement field, i.e. $g = P_0$:

$T_0(g)$ reduces to a similarity of ratio $-r_0$ (the minus sign is due to the orientation of the normal n) : $T_0(P_0) = -r_0P_0$ where r_0 is the radius of the sphere.

From (9), $T_1(P_0) = ir_0^2P_0$, so that $(T_0^2+iT_1)(P_0) = 0$ and $(T_0^2+iT_1)$ is not invertible.

From (21), H(g) reduces to a similarity of ratio r_0^{-1} : $H(P_0)=r_0^{-1}P_0$ so that $(T_0^{-1}+H)$ is not invertible as $(T_0^{-1}+H)(P_0) = 0$.

In this case, going back to the derivation of the previous subsections, it can be shown that the continuous DAA2c reduces to continuous DAA1. Moreover, the continuous DAA1 is exact. A similar feature is observed for the corresponding discretized forms.

For a $g = P_n$, n>1 normal displacement field, the operator T_1 equals 0 so that higher low-frequency expansion would be needed.

One may think that these peculiarities are strongly related to the spherical geometry, and that the invertibility of the operator C_2 holds in general.

As far as discretized DAA's are concerned, they were obtained through discretization of the continuous ones. Thus, they are more general than currentlty used forms and may be obtained whatever the interpolation of the unknowns is (for example for non diagonal $\mathbf{A_f}$). The differences pointed out between these derived forms and the usual ones should be studied numerically. Recall also that open numerical analysis and geometry problems have been mentioned.

ACKNOWLEDGEMENTS

A first and unpublished version of this work was part of a report written by the author at the end of a one-year stay as a visitor at the Applied Mechanics Laboratory of the Lockheed Missiles and Space Company in Palo Alto, California during the academic year 1984-1985. The author wishes to thank the Direction des Recherches et Etudes Techniques which sponsored this stay under contract Nº 84/1395. He also wishes to thank T.L. Geers, at the time head of the Applied Mechanics Laboratory, and Walter Craig of the Mathematics Department at Stanford University for suggestions, comments and valuable discussions on some topics mentioned in this paper.

REFERENCES

1. T.L. Geers, "Doubly Asymptotic Approximation for transient motions of submerged structures", J. Acoust. Soc. Am., Vol. 64, 1978, pp. 1500-1508.

2. C.A. Felippa, "Top-down derivation of doubly asymptotic approximations for structure-fluid interaction analysis", Innovative Numerical Analysis for the Engineering Sciences, edited by R.P. Shaw et al., U.P. of Virginia, Charlottesville, 1980, pp. 79-88.

3. R.E. Kleinman and G.F. Roach, "Boundary integral equations for the three-dimensional Helmholtz equations", S.I.A.M. Rev., Vol 16, 1974, pp. 214-235.

4. R. Ohayon and E. Sanchez-Palencia, "On the vibration problem for an elastic body surrounded by a slightly compressible fluid", R.A.I.R.O. Numerical Analysis, Vol 17, 1983, pp. 311-326.

5. C.H. Wilcox, "Scattering Theory for the d'Alembert Equation in Exterior Domains", Lecture Notes in Mathematics, Vol. 442, 1975, Springer Verlag.

6. C.M. Bender and S.A. Orszag, Advanced Mathematical Methods for Scientists and Engineers, McGraw-Hill Book Company, 1978.

7. C.O. Bloom and N.D. Kazarinoff, "Short Wave Radiation Problems in Inhomogeneous Media : Asymptotic Solutions", Lecture Notes in Mathematics, Vol. 522, 1976, Springer Verlag.

8. J. Lighthill, Waves in Fluids, Cambridge University Press, 1978.

9. T.L. Geers and G.C. Ruzicka, "Finite-Element / Boundary-Element analysis of multiple structures excited by transient acoustic waves", Numerical Methods for Transient and Coupled Problems, edited by R.W. Lewis, E. Hinton, P. Bettess and B.A. Schrefler, Pineridge Press, Swansea, UK, 1984, pp. 150-162.

10. R.D. Mindlin and H.H. Bleich, "Response of an elastic cylindrical shell to a transverse step shock wave", J. Appl. Mech. Vol. 20, 1953, pp. 189-195.

11. J.H. Haywood, "Response of an elastic cylindrical shell to a transverse shock wave", Quart. J. Mech. Appl. Math., Vol. 11, 1958, pp. 129-141.

12. B. Bedrosian and F.L. DiMaggio, "Acoustic approximations in fluid shell interaction", J. Eng. Mech. Div. ASCE, Vol. 98, 1972, pp. 731-742.

13. K. Bando, P. Bettess and C. Emson, "The effectiveness of dampers for the analysis of exterior scalar wave diffraction by cylinders and ellipsoids", Int. J. Num. Methods in Fluids, Vol. 4, 1984, pp. 599-617.

14. C.A. Felippa, "A family of early time approximations for fluid structure interaction", J. Appl. Mech., Vol. 47, 1980, pp. 703-708.

15. S. Hayek, "Vibration of a spherical shell in an acoustic medium", J. Acoust. Soc. Am., Vol 40, 1966, pp. 342-348 .

SOLUTION OF ELASTO-ACOUSTIC PROBLEMS USING A VARIATIONAL FINITE ELEMENT/BOUNDARY ELEMENT TECHNIQUE

J. P. Coyette and K. R. Fyfe
Dynamic Engineering N.V.
Heverlee, Belgium

ABSTRACT

The paper is concerned with the numerical modelling of exterior acoustic and elasto–acoustic problems in the frequency domain using an integral formulation. The selected approach lies in the variational solution of boundary integral equations. This unconventional method proposed by Hamdi is further extended in order to allow for consideration of symmetry and halfspace options. Also of interest is the development of a library incorporating triangular and quadrilateral surface elements (both linear and quadratic) and the use of adaptive regular and singular integration schemes. All these options allow to improve the efficiency of the variational approach. The main advantage of this formulation is the related symmetry of fluid matrices ensuring a natural coupling with finite element structural models for elasto–acoustic analysis.

INTRODUCTION

Numerical prediction procedures for acoustic and elasto–acoustic modelling are now widely used in various engineering areas such as automotive and aircraft industries, submarine applications, audio equipments, etc...

In the "low" frequency range, acoustic modelling usually relies on either finite element (FE) or boundary element (BE) methods. These methods have the powerful and flexible capabilities necessary to model complex geometries and boundary conditions.

It is well known that integral equations possess some intrinsic advantages over finite element methods for scattering and radiation problems : the dimensionality of the problem is reduced by one and the radiation condition is satisfied automatically.

Attention will be mainly restricted here to acoustic BE modelling. An indirect formulation in terms of simple and double layer potentials has been selected. A variational solution procedure proposed by Hamdi (1982) is used in order to ensure symmetry of resulting fluid matrices. This symmetry is exploited in the second part of the paper where coupling with a structural model allows to study elasto–acoustic problems.

Special attention is devoted to particular enhancements of this method dealing with implementation aspects. Extensive treatment of symmetry conditions and halfspace options are described for the first time. Adaptive regular and singular integration procedures are proposed to deal effectively with the calculation of various influence matrices involved in the computational process. Moreover an extended library of boundary elements is described including linear and quadratic triangles and quadrangles. All these facilities have been implemented in SYSNOISE (General purpose program for acoustics) and allow to speed up both model's generation and computation process.

BOUNDARY ELEMENT ACOUSTIC MODELLING

<u>Introduction</u>

Various boundary element formulations have been proposed for wave problems (Shaw, 1979). For acoustic problems governed by Helmholtz equation in the frequency domain, either direct or indirect boundary element methods are used.

The direct approach is based on Helmholtz integral equation and uses primary boundary variables (pressure and normal velocity) while the indirect approach relies on Huygen's principle and makes reference to secondary boundary variables (layer potentials) whose determination allows to compute field pressures and velocities. Many papers have been published on direct formulations (Chertock, 1964; Schenck, 1968; Bell, Meyer and Zinn, 1977; Meyer, Bell and Stallybrass, 1978; Terai, 1980; Amini and Wilton, 1986) and indirect ones (Kleinman and Roach, 1974; Kipp and Bernhard, 1987). Both formulations fail at some critical wavenumbers (non–uniqueness of solution for the direct formulation and non–existence for the indirect approach).

Various procedures have been proposed to remove these singularities. The so called "CHIEF" method (Schenck, 1968) consists to add zero pressure constraints at some internal locations and to solve the combined system by a least square method (Seybert, 1987; Latcha 1986) or via Lagrange multipliers (Seybert, 1987). More rational is the Burton–Miller (1971) approach based on some combination of Helmholtz formula and its normal derivative which has been proven to be a reliable method (Tzu–chu Lin, 1984). An excellent review of some recent results in integral equation solutions to scattering problems is summarized in the book of Colton and Kress (1983).

<u>Basic Equation and Boundary Conditions</u>

The equation governing the acoustic problem in the frequency domain is the Helmholtz equation

(1) $$\nabla^2 p + k^2 p = 0 \quad in \quad V$$

where p is the acoustic pressure and k the wavenumber given by

(2) $$k = \frac{\omega}{c}$$

with ω = circular frequency and c = sound speed. Hereafter, it is assumed that the time dependence is of the form $e^{+i\omega t}$.

The velocity vector v is related to the scalar pressure p through the relation

(3) $$v = \frac{\nabla p}{-i \varrho \omega}$$

Equation (1) has to be solved with the following boundary conditions (see figure 1).

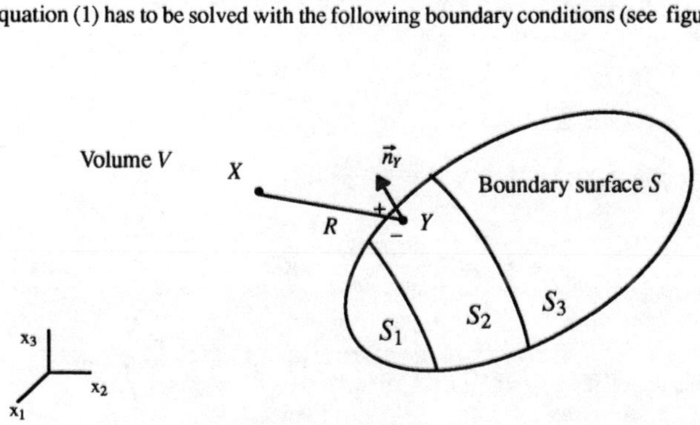

Figure 1 : Acoustic problem, boundary surfaces

(4) $$p = \bar{p} \quad on \quad S_1$$

(5) $$\frac{\partial p}{\partial n} = -i \varrho \omega \bar{v}_n \quad on \quad S_2$$

(6) $$\frac{\partial p}{\partial n} + ik \beta p = 0 \quad on \quad S_3$$

where $\bar{p}$, $\bar{v}_n$ and β denote fixed pressure, fixed normal velocity and fixed specific admittance respectively. These conditions are completed with the usual Sommerfeld radiation condition (for an unbounded domain) :

(7) $$\lim_{r \to \infty} r \left(\frac{\partial p}{\partial r} + ikp \right) = 0$$

Integral Representation

The selected integral representation is based on the use of single and double layer potentials defined on boundary surface S (union of S_1, S_2 and S_3) :

(8) $$p(X) = \int_S \left(\mu(Y) \frac{\partial G(X,Y)}{\partial n_Y} - \sigma(Y)\, G(X,Y) \right) dS(Y)$$

where the single layer potential σ is related to the jump of pressure normal derivative

(9) $$\sigma = \frac{\partial p^+}{\partial n} - \frac{\partial p^-}{\partial n}$$

and the double layer potential μ is defined as the jump of pressure across the surface S.

(10) $$\mu = p^+ - p^-$$

The kernel G is the fundamental solution of Helmholtz equation for a point source

(11) $$G(X,Y) = \frac{\exp(-ikR)}{4\pi R}$$

where $R = |X - Y|$ is the distance between points X and Y.

This integral representation is termed as indirect because acoustic field evaluation relies on preliminar computation of single and double layer potentials σ and μ defined on the surface.

Boundary Equations

Considering the above boundary conditions (4 – 6), it can be stated that

(12) $$\mu = 0 \quad on \quad S_1$$

(13) $$\sigma = 0 \quad on \quad S_2$$

(14) $$\sigma + ik \beta \mu = 0 \quad on \quad S_3$$

Starting from relation (8), boundary conditions can be written by letting field point X reach in turn each of the boundary surfaces.

Case 1 : $X \in S_1$

(15) $$\bar{p}(X) = -\int_{S_1} \sigma(Y)\, G(X,Y)\, dS_1(Y) + \int_{S_2} \mu(Y) \frac{\partial G(X,Y)}{\partial n_Y} dS_2(Y) + \int_{S_3} \mu(Y) \left(\frac{\partial G(X,Y)}{\partial n_Y} + ik\beta_Y\, G(X,Y) \right) dS_3(Y)$$

Case 2 : $X \in S_2$

(16) $$-i \varrho \omega \bar{v}_n(X) = -\int_{S_1} \sigma(Y) \frac{\partial G(X,Y)}{\partial n_X} dS_1(Y) + \int_{S_2} \mu(Y) \frac{\partial^2 G(X,Y)}{\partial n_X\, \partial n_Y} dS_2(Y) + \int_{S_3} \mu(Y) \left(\frac{\partial^2 G(X,Y)}{\partial n_X\, \partial n_Y} + ik\beta_Y \frac{\partial G(X,Y)}{\partial n_X} \right) dS_3(Y)$$

Case 3 : $X \in S_3$

(17) $$\begin{aligned} 0 = & -\int_{S_1} \sigma(Y) \frac{\partial G(X,Y)}{\partial n_X} dS_1(Y) + \int_{S_2} \mu(Y) \frac{\partial^2 G(X,Y)}{\partial n_X\, \partial n_Y} dS_2(Y) \\ & + \int_{S_3} \mu(Y) \left(\frac{\partial^2 G(X,Y)}{\partial n_X \partial n_Y} + ik\beta_Y \frac{\partial G(X,Y)}{\partial n_X} \right) dS_3(Y) \\ & + ik\, \beta_X \left[-\int_{S_1} \sigma(Y)\, G(X,Y)\, dS_1(Y) + \int_{S_2} \mu(Y) \frac{\partial G(X,Y)}{\partial n_Y} dS_2(Y) \right. \\ & \left. + \int_{S_3} \mu(Y) \left(\frac{\partial G(X,Y)}{\partial n_Y} + ik\, \beta_Y\, G(X,Y) \right) dS_3(Y) \right] \end{aligned}$$

In these relations, β_X and β_Y are the specific admittances at locations X and Y on S_3 surface. It must be emphasized that integrals involving second derivative of G need finite part integral computation.

The usual solution procedure relies on a collocation scheme and leads to unsymmetric matrices. An alternative solution procedure is described in the next section and involves minimization of suitable functional defined on the boundary surface.

Variational Formulation

The solution of above boundary equations can be obtained by minimization of the following functional, (Hamdi, 1982) :

(18) $$\begin{aligned} J = & \frac{1}{2} \int_{S_1}\int_{S_1} \sigma(X)\, \sigma(Y)\, G(X,Y)\, dS_1(X)\, dS_1(Y) \\ & - \int_{S_1}\int_{S_2} \sigma(X)\, \mu(Y) \frac{\partial G(X,Y)}{\partial n_Y} dS_2(Y)\, dS_1(X) \\ & - \int_{S_1}\int_{S_3} \sigma(X)\, \mu(Y) \left(\frac{\partial G(X,Y)}{\partial n_Y} + ik\, \beta_Y\, G(X,Y) \right) dS_3(Y)\, dS_1(X) \\ & + \frac{1}{2} \int_{S_2}\int_{S_2} \mu(X)\, \mu(Y) \frac{\partial^2 G(X,Y)}{\partial n_X\, \partial n_Y} dS_2(Y)\, dS_2(X) \\ & + \int_{S_2}\int_{S_3} \mu(X)\, \mu(Y) \left(\frac{\partial^2 G(X,Y)}{\partial n_X\, \partial n_Y} + ik\, \beta_Y \frac{\partial G(X,Y)}{\partial n_X} \right) dS_3(Y)\, dS_2(X) \\ & + \frac{1}{2} \int_{S_3}\int_{S_3} \mu(X)\, \mu(Y) \left(\frac{\partial^2 G(X,Y)}{\partial n_X\, \partial n_Y} + ik\, \beta_Y \frac{\partial G(X,Y)}{\partial n_X} + ik\, \beta_X \frac{\partial G(X,Y)}{\partial n_Y} \right. \\ & \left. - k^2\, \beta_X\, \beta_Y\, G(X,Y) \right) dS_3(X)\, dS_3(Y) \\ & + \int_{S_1} \bar{p}(X)\, \sigma(X)\, dS_1(X) \\ & - \int_{S_2} \varrho\, i\, \omega\, \bar{v}_n(X)\, \mu(X)\, dS_2(X) \end{aligned}$$

Regrouping the terms and defining :

(19) $$S_\sigma = S_1$$

(20) $$S_\mu = S_2 \cup S_3$$

where S_σ is the boundary surface where the single layer potential is defined while S_μ is the boundary surface where the double layer potential is defined, functional J can be written as :

(21) $$\begin{aligned} J = & \frac{1}{2} \int_{S_\sigma}\int_{S_\sigma} \sigma(X)\, \sigma(Y)\, G(X,Y)\, dS_\sigma(X)\, dS_\sigma(Y) \\ & + \frac{1}{2} \int_{S_\mu}\int_{S_\mu} \mu(X)\, \mu(Y) \left(\frac{\partial^2 G(X,Y)}{\partial n_X\, \partial n_Y} + ik\, \beta_Y \frac{\partial G(X,Y)}{\partial n_X} + ik\, \beta_X \frac{\partial G(X,Y)}{\partial n_Y} \right. \\ & \left. - k^2\, \beta_X\, \beta_Y\, G(X,Y) \right) dS_\mu(X)\, dS_\mu(Y) \end{aligned}$$

$$- \int_{S_\mu}\int_{S_\sigma} \sigma(X)\,\mu(Y)\left(\frac{\partial G(X,Y)}{\partial n_Y} + ik\,\beta_Y\,G(X,Y)\right) dS_\sigma(X)\,dS_\mu(Y)$$

$$+ \int_{S_1} \bar{p}(X)\,\sigma(X)\,dS_1(X)$$

$$- \int_{S_2} \varrho\, i\,\omega\,\bar{v}_n(X)\,\mu(X)\,dS_2(X)$$

where β_X (β_y) is assumed to be zero if X (or Y) belongs to S_2.

The second integral in (21) is highly singular, but using properties of Green's functions and continuity of μ, it can be shown (Hamdi, 1981) that this integral is also given by :

(22)
$$\int_{S_\mu}\int_{S_\mu} \mu(X)\,\mu(Y)\,\frac{\partial^2 G(X,Y)}{\partial n_X\,\partial n_Y}\,dS_\mu(X)\,dS_\mu(Y)$$
$$=$$
$$\int_{S_\mu}\int_{S_\mu} G(X,Y)\,[k^2\,\mu(X)\,\mu(Y)\,(n_X,n_Y) - (rot\,\mu(X), rot\,\mu(Y))]\,dS_\mu(X)\,dS_\mu(Y)$$

which contains only low level (integrable) singularities.

<u>Discretization by Surface Finite Elements</u>

The functional J, defined by (21) can be discretized by selecting a surface mesh and appropriate interpolation functions for both σ and μ potentials

(23) $$\sigma = N_\sigma\,.\,\bar{\sigma}$$

(24) $$\mu = N_\mu\,.\,\bar{\mu}$$

(25) $$\nabla\mu = B_\mu\,.\,\bar{\mu}$$

where $\bar{\sigma}$ and $\bar{\mu}$ denote the vectors of nodal values for single and double layer potentials and N_μ and N_σ the related sets of interpolation functions while B_μ contains cartesian derivatives of N_μ interpolation functions. Formerly, at least C^0 interpolation functions have to be selected for μ while only C^{-1} continuity is required for σ. Current implementation is based on iso–parametric triangular and quadrilateral elements (linear and quadratic) (figure 2).

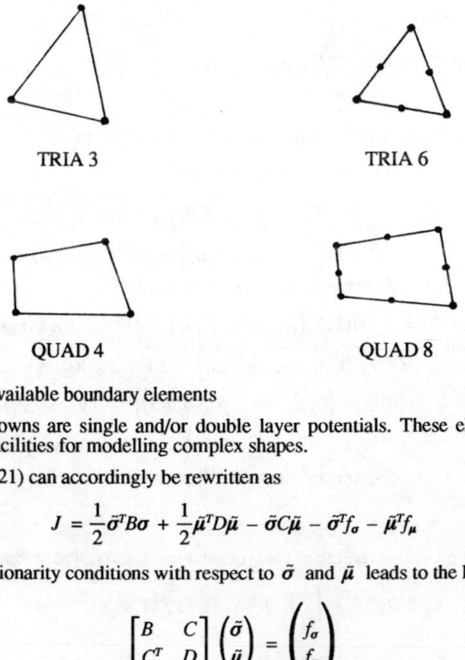

Figure 2 : Available boundary elements

Nodal unknowns are single and/or double layer potentials. These elements give additional facilities for modelling complex shapes.

Functional (21) can accordingly be rewritten as

(26) $$J = \frac{1}{2}\bar{\sigma}^T B\sigma + \frac{1}{2}\bar{\mu}^T D\bar{\mu} - \bar{\sigma}C\bar{\mu} - \bar{\sigma}^T f_\sigma - \bar{\mu}^T f_\mu$$

The stationarity conditions with respect to $\bar{\sigma}$ and $\bar{\mu}$ leads to the linear system

(27) $$\begin{bmatrix} B & C \\ C^T & D \end{bmatrix}\begin{pmatrix}\bar{\sigma} \\ \bar{\mu}\end{pmatrix} = \begin{pmatrix} f_\sigma \\ f_\mu \end{pmatrix}$$

where :

(28) $$B = \int_{S_\sigma}\int_{S_\sigma} N_\sigma^T(X)\,.\,N_\sigma(Y) G(X,Y)\,dS_\sigma(X)\,dS_\sigma(Y)$$

(29) $$C = \int_{S_\mu}\int_{S_\sigma} N_\sigma^T(X) N_\mu(X)\left(\frac{\partial G(X,Y)}{\partial n_Y} + ik\beta_Y\,G(X,Y)\right) dS_\sigma(X)\,dS_\mu(Y)$$

(30)
$$D = \int_{S_\mu}\int_{S_\mu}\Big\{N_\mu^T(X) N_\mu(Y)\Big[k^2(n_X,n_Y)G(X,Y) + ik\beta_Y\frac{\partial G(X,Y)}{\partial n_X}$$
$$+ ik\beta_X\frac{\partial G(X,Y)}{\partial n_Y} - k^2\beta_X\beta_Y G(X,Y)\Big] - \big(B_\mu(X)\wedge n_X,\; B_\mu(Y)\wedge n_Y\big)\Big\}\,dS_\mu(X)\,dS_\mu(Y)$$

(31) $$f_\sigma = -\int_{S_1}\bar{p}(X) N_\sigma(X)\,dS_1(X)$$

(32) $$f_\mu = \int_{S_2}\varrho i\omega\,\bar{v}_n(X)\,N_\mu(X)\,dS_2(X)$$

Incorporating additional conditions related to $\bar{\sigma}$ and/or $\bar{\mu}$ values, system (27) can be solved leading to the solution in terms of single and/or double layer potentials. From these potentials, the integral representation (8) allows evaluation of pressure at any field point. The derivative of equation (8) yields the integral representation for velocity at any field point.

<u>Treatment of Symmetry Conditions</u>

The existence of symmetry planes can be accounted for by modifications to the kernel involved in functional J.

As an example, let us consider how to take into account symmetry versus plane $x_1 = 0$ for the various terms of functional (21) starting with the first rewritten as :

(33) $$J_{\sigma\sigma} = \frac{1}{2}\int_{S_\sigma}\int_{S_\sigma}\sigma(X)\sigma(Y)\;G(X,Y)\;dS_\sigma(X)\;dS_\sigma(Y)$$

Owing to the symmetry, surface S_σ can be decomposed into two parts S_σ^* and S_σ^{**} on both sides of plane $x_1 = 0$:

(34) $$S_\sigma = S_\sigma^* \cup S_\sigma^{**}$$

With the notations $f(X,Y) = \sigma(X)\sigma(Y)\;G(X,Y)$, X and Y being points belonging to S_σ^* and ${}^1X$, ${}^1Y$ their symmetric counterparts on S_σ^{**}, $J_{\sigma\sigma}$ can be rewritten as :

(35)
$$J_{\sigma\sigma} = \frac{1}{2}\int_{S_\sigma^*}\int_{S_\sigma^*} f(X,Y)\;dS_\sigma^*(X)\;dS_\sigma^*(Y)$$
$$+\frac{1}{2}\int_{S_\sigma^*}\int_{S_\sigma^{**}} f(X,{}^1Y)\;dS_\sigma^*(X)\;dS_\sigma^{**}({}^1Y)$$
$$+\frac{1}{2}\int_{S_\sigma^{**}}\int_{S_\sigma^*} f({}^1X,Y)\;dS_\sigma^{**}({}^1X)\;dS_\sigma^*(Y)$$
$$+\frac{1}{2}\int_{S_\sigma^{**}}\int_{S_\sigma^{**}} f({}^1X,{}^1Y)\;dS_\sigma^{**}({}^1X)\;dS_\sigma^{**}({}^1Y)$$

Due to symmetry, the following relations hold :

(36)
$$\sigma({}^1X) = \sigma(X)$$
$$\sigma({}^1Y) = \sigma(Y)$$
$$G({}^1X,{}^1Y) = G(X,Y)$$
$$G(X,{}^1Y) = G({}^1X,Y)$$

so that :

(37)
$$f({}^1X,{}^1Y) = f(X,Y) = \sigma(X)\sigma(Y)\;G(X,Y)$$
$$f(X,{}^1Y) = f({}^1X,Y) = \sigma(X)\sigma(Y)\;G({}^1X,Y)$$

and (35) takes the form :

(38) $$J_{\sigma\sigma} = \frac{1}{2}\int_{S_\sigma^*}\int_{S_\sigma^*}\sigma(X)\sigma(Y)\;\big(2G(X,Y) + 2G({}^1X,Y)\big)\;dS_\sigma^*(X)\;dS_\sigma^*(Y)$$

This result can be extended to other symmetry conditions and gives the generalized form :

(39) $$J_{\sigma\sigma} = \frac{1}{2}\int_{S_\sigma^*}\int_{S_\sigma^*}\sigma(X)\sigma(Y)\;K(X,Y)\;\;dS_\sigma^*(X)\;dS_\sigma^*(Y)$$

where :

$$
\begin{aligned}
K(X,Y) &= G(X,Y) \quad \text{if } \textit{no symmetry} \\
&= 2(G(X,Y) + G({}^{l}X,Y)) \quad \text{if } \textit{symmetry about } x_l = 0 \ (l = 1,2,3) \\
&= 4(G(X,Y) + G({}^{l}X,Y) + G({}^{m}X,Y) + G({}^{lm}X,Y) \\
&\qquad \text{if } \textit{symmetry about } x_l = 0 \text{ and } x_m = 0 \ (l,m = 1,2,3) \\
&= 8(G(X,Y) + G({}^{1}X,Y) + G({}^{2}X,Y) + G({}^{3}X,Y) \\
&\quad + G({}^{12}X,Y) + G({}^{13}X,Y) + G({}^{23}X,Y) + G({}^{123}X,Y) \\
&\qquad \text{if } \textit{symmetry about } x_1 = 0,\ x_2 = 0 \text{ and } x_3 = 0
\end{aligned}
\tag{40}
$$

In all these expressions ${}^{\alpha\beta\gamma}X$ denotes the symmetric position of point X versus planes $\alpha = 0$, $\beta = 0$ and $\gamma = 0$ and S_σ^* is the remaining boundary surface owing to symmetry (original surface S_σ if no symmetry, one half if symmetry versus 1 plane, one fourth if symmetry versus 2 planes and one eighth if symmetry versus 3 planes).

Similarly taking into account (22), the second term of functional (21) can be written as:

$$
\begin{aligned}
J_{\mu\mu} = \frac{1}{2}\int_{S_\mu}\int_{S_\mu} \Big\{ \mu(X)\mu(Y) \Big[& k^2\, G(X,Y)\big((n_X,\ n_Y) - \beta_X\beta_Y\big) \\
& + ik\beta_Y \frac{\partial G(X,Y)}{\partial n_X} + ik\beta_X \frac{\partial G(X,Y)}{\partial n_Y} \Big] \\
& - \big(\mathrm{rot}\ \mu(X), \mathrm{rot}\ \mu(Y)\big)\ G(X,Y) \Big\}\ dS_\mu(X)\ dS_\mu(Y)
\end{aligned}
\tag{41}
$$

Let us consider separately the treatment of the various terms involved in (41) for symmetry versus plane $x_1 = 0$. The first term appears as :

$$
J_{\mu\mu}^{(1)} = \frac{1}{2}\int_{S_\mu}\int_{S_\mu} f^{(1)}(X,Y)\ dS_\mu(X)\ dS_\mu(Y) \tag{42}
$$

with :

$$
f^{(1)}(X,Y) = k^2 G(X,Y)\ \mu(X)\mu(Y)\big((n_X, n_Y) - \beta_X\beta_Y\big) \tag{43}
$$

Owing to the symmetry, surface S_μ can be decomposed into two parts S_μ^* and S_μ^{**} on both sides of plane $x_1 = 0$:

$$
S_\mu = S_\mu^* \cup S_\mu^{**} \tag{44}
$$

With the same notations as previously, (42) can be rewritten as :

$$
\begin{aligned}
J_{\mu\mu}^{(1)} &= \frac{1}{2}\int_{S_\mu^*}\int_{S_\mu^*} f^{(1)}(X,Y)\ dS_\mu^*(X)\ dS_\mu^*(Y) \\
&+ \frac{1}{2}\int_{S_\mu^*}\int_{S_\mu^{**}} f^{(1)}(X,{}^{1}Y)\ dS_\mu^*(X)\ dS_\mu^{**}({}^{1}Y) \\
&+ \frac{1}{2}\int_{S_\mu^{**}}\int_{S_\mu^*} f^{(1)}({}^{1}X,Y)\ dS_\mu^{**}({}^{1}X)\ dS_\mu^*(Y) \\
&+ \frac{1}{2}\int_{S_\mu^{**}}\int_{S_\mu^{**}} f^{(1)}({}^{1}X,{}^{1}Y)\ dS_\mu^{**}({}^{1}X)\ dS_\mu^{**}({}^{1}Y)
\end{aligned}
\tag{45}
$$

Again, due to symmetry (figure 3), the following relations hold :

$$
\mu({}^{1}X) = \mu(X) \tag{46.a}
$$

$$
\mu({}^{1}Y) = \mu(Y)
$$

$$
G({}^{1}X,{}^{1}Y) = G(X,Y) \tag{46.b}
$$

$$
G(X,{}^{1}Y) = G({}^{1}X,Y)
$$

$$
\beta_{1X} = \beta_X \tag{46.c}
$$

$$
\beta_{1Y} = \beta_Y
$$

$$
n_{1X,1} = -n_{X,1},\quad n_{1X,2} = +n_{X,2},\quad n_{1X,3} = +n_{X,3} \tag{46.d}
$$

$$
n_{1Y,1} = -n_{Y,1},\quad n_{1Y,2} = +n_{Y,2},\quad n_{1Y,3} = +n_{Y,3}
$$

$$
(n_{1X},\ n_{1Y}) = (n_X,\ n_Y) \tag{46.e}
$$

$$
(n_X,\ n_{1Y}) = (n_{1X},\ n_Y)
$$

so that :

$$
\begin{aligned}
f^{(1)}({}^{1}X,{}^{1}Y) &= f^{(1)}(X,Y) = k^2 G(X,Y)\ \mu(X)\mu(Y)\ \big((n_X,\ n_Y) - \beta_X\beta_Y\big) \\
f^{(1)}(X,{}^{1}Y) &= f^{(1)}({}^{1}X,Y) = k^2 G({}^{1}X,Y)\ \mu(X)\mu(Y)\ \big((n_{1X},\ n_Y) - \beta_X\beta_Y\big)
\end{aligned}
\tag{47}
$$

These results allow to write (45) under the form :

$$
\begin{aligned}
J_{\mu\mu}^{(1)} = \frac{1}{2}\int_{S_\mu^*}\int_{S_\mu^*} k^2\mu(X)\mu(Y) \Big[& 2G(X,Y)\ \big((n_X, n_Y) - \beta_X\beta_Y\big) \\
& + 2G({}^{1}X,Y)\big((n_{1X}, n_Y) - \beta_X\beta_Y\big) \Big]\ dS_\mu^*(X)\ dS_\mu^*(Y)
\end{aligned}
\tag{48}
$$

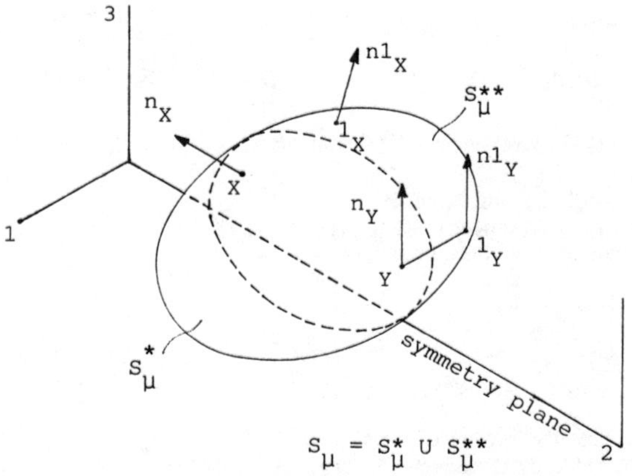

Figure 3. Symmetric boundary surface S_μ versus plane $x_1 = 0$

This result can be quite easily generalized and leads to the general expression :

$$
J_{\mu\mu}^{(1)} = \frac{1}{2}\int_{S_\mu^*}\int_{S_\mu^*} k^2 K(X,Y)\ \mu(X)\mu(Y)\ dS_\mu^*(X)\ dS_\mu^*(Y) \tag{49}
$$

where :

$$
\begin{aligned}
K(X,Y) &= G(X,Y)\big((n_X, n_Y) - \beta_X\beta_Y\big) \\
&\qquad \textit{if no symmetry} \\
K(X,Y) &= 2\ \Big[G(X,Y)\big((n_X, n_Y) - \beta_X\beta_Y\big) + G({}^{l}X,Y)\big((n_{lX}, n_Y) - \beta_X\beta_Y\big)\Big] \\
&\qquad \textit{if symmetry about } x_l = 0 \\
K(X,Y) &= 4\ \Big[G(X,Y)\big((n_X, n_Y) - \beta_X\beta_Y\big) + G({}^{l}X,Y)\big((n_{lX}, n_Y) - \beta_X\beta_Y\big) \\
&\quad + G({}^{m}X,Y)\big((n_{mX}, n_Y) - \beta_X\beta_Y\big) + G({}^{lm}X,Y)\big((n_{lmX}, n_Y) - \beta_X\beta_Y\big)\Big] \\
&\qquad \textit{if symmetry about } x_l = 0 \textit{ and } x_m = 0 \\
K(X,Y) &= 8\ \Big[G(X,Y)\big((n_X, n_Y) - \beta_X\beta_Y\big) + G({}^{1}X,Y)\big((n_{1X}, n_Y) - \beta_X\beta_Y\big) \\
&\quad + G({}^{2}X,Y)\big((n_{2X}, n_Y) - \beta_X\beta_Y\big) + G({}^{3}X,Y)\big((n_{3X}, n_Y) - \beta_X\beta_Y\big) \\
&\quad + G({}^{12}X,Y)\big((n_{12X}, n_Y) - \beta_X\beta_Y\big) + G({}^{23}X,Y)\big((n_{23X}, n_Y) - \beta_X\beta_Y\big) \\
&\quad + G({}^{13}X,Y)\big((n_{13X}, n_Y) - \beta_X\beta_Y\big) + G({}^{123}X,Y)\big((n_{123X}, n_Y) - \beta_X\beta_Y\big)\Big] \\
&\qquad \textit{if symmetry about } x_1 = 0,\ x_2 = 0 \textit{ and } x_3 = 0
\end{aligned}
\tag{50}
$$

The treatment of second term in (41) involves normal derivative of kernel G :

$$
J_{\mu\mu}^{(2)} = \frac{1}{2}\int_{S_\mu}\int_{S_\mu} f^{(2)}(X,Y)\ dS_\mu(X)\ dS_\mu(Y) \tag{51}
$$

with :

$$
f^{(2)}(X,Y) = ik\ \mu(X)\mu(Y)\left(\beta_Y \frac{\partial G(X,Y)}{\partial n_X} + \beta_Y \frac{\partial G(X,Y)}{\partial n_Y}\right) \tag{52}
$$

Again let us consider symmetry versus plane $x_1 = 0$. Boundary surface S_μ is broken into equivalent parts S_μ^* and S_μ^{**}, so that (51) can be written as :

(53)
$$J^{(2)}_{\mu\mu} = \frac{1}{2}\int_{S^*_\mu}\int_{S^*_\mu} f^{(2)}(X,Y)\ dS^*_\mu(X)\ dS^*_\mu(Y)$$
$$+ \frac{1}{2}\int_{S^*_\mu}\int_{S^{**}_\mu} f^{(2)}(X,{}^1Y)\ dS^*_\mu(X)\ dS^{**}_\mu({}^1Y)$$
$$+ \frac{1}{2}\int_{S^{**}_\mu}\int_{S^*_\mu} f^{(2)}({}^1X,Y)\ dS^{**}_\mu({}^1X)\ dS^*_\mu(Y)$$
$$+ \frac{1}{2}\int_{S^{**}_\mu}\int_{S^{**}_\mu} f^{(2)}({}^1X,{}^1Y)\ dS^{**}_\mu({}^1X)\ dS^{**}_\mu({}^1Y)$$

Symmetry condition implies (46a), (46c) and (46d) supplemented with :

(54)
$$\frac{\partial G({}^1X,{}^1Y)}{\partial n_{1X}} = \frac{\partial G(X,Y)}{\partial n_X},\quad \frac{\partial G({}^1X,{}^1Y)}{\partial n_{1Y}} = \frac{\partial G(X,Y)}{\partial n_Y}$$
$$\frac{\partial G(X,{}^1Y)}{\partial n_X} = \frac{\partial G({}^1X,Y)}{\partial n_{1X}},\quad \frac{\partial G(X,{}^1Y)}{\partial n_{1Y}} = \frac{\partial G({}^1X,Y)}{\partial n_Y}$$

so that the following relations are obtained :

(55)
$$f^{(2)}(X,{}^1Y) = f^{(2)}({}^1X,Y) = ik\ \mu(X)\mu(Y)\left(\beta_Y\frac{\partial G(X,{}^1Y)}{\partial n_X} + \beta_X\frac{\partial G({}^1X,Y)}{\partial n_Y}\right)$$
$$f^{(2)}({}^1X,{}^1Y) = f^{(2)}(X,Y) = ik\ \mu(X)\mu(Y)\left(\beta_Y\frac{\partial G(X,Y)}{\partial n_X} + \beta_X\frac{\partial G(X,Y)}{\partial n_Y}\right)$$

and $J^{(2)}_{\mu\mu}$ can be rewritten as :

(56)
$$J^{(2)}_{\mu\mu} = \frac{1}{2}\int_{S^*_\mu}\int_{S^*_\mu} ik\ \mu(X)\mu(Y)\left(2\beta_Y\frac{\partial G(X,Y)}{\partial n_X} + 2\beta_X\frac{\partial G(X,Y)}{\partial n_Y}\right.$$
$$\left. + 2\beta_Y\frac{\partial G(X,{}^1Y)}{\partial n_X} + 2\beta_X\frac{\partial G({}^1X,Y)}{\partial n_Y}\right) dS^*_\mu(X)\ dS^*_\mu(Y)$$

and the generalized form for this term is given by :

(57)
$$J^{(2)}_{\mu\mu} = \frac{1}{2}\int_{S^*_\mu}\int_{S^*_\mu} K(X,Y)\ \mu(X)\mu(Y)\ dS^*_\mu(X)\ dS^*_\mu(Y)$$

where the modified kernel appears as :

$$K(X,Y) = ik\left(\beta_Y\frac{\partial G(X,Y)}{\partial n_X} + \beta_X\frac{\partial G(X,Y)}{\partial n_Y}\right)$$

if no symmetry

$$K(X,Y) = 2\ ik\left(\beta_Y\frac{\partial G(X,Y)}{\partial n_X} + \beta_X\frac{\partial G(X,Y)}{\partial n_Y}\right.$$
$$\left. + \beta_Y\frac{\partial G(X,{}^lY)}{\partial n_X} + \beta_X\frac{\partial G({}^lX,Y)}{\partial n_Y}\right)$$

if symmetry about $x_l = 0$

$$K(X,Y) = 4\ ik\left(\beta_Y\frac{\partial G(X,Y)}{\partial n_X} + \beta_X\frac{\partial G(X,Y)}{\partial n_Y}\right.$$
$$+ \beta_Y\frac{\partial G(X,{}^lY)}{\partial n_X} + \beta_X\frac{\partial G({}^lX,Y)}{\partial n_Y}$$
$$+ \beta_Y\frac{\partial G(X,{}^mY)}{\partial n_X} + \beta_X\frac{\partial G({}^mX,Y)}{\partial n_Y}$$
$$\left. + \beta_Y\frac{\partial G(X,{}^{lm}Y)}{\partial n_X} + \beta_X\frac{\partial G({}^{lm}X,Y)}{\partial n_Y}\right)$$

if symmetry about $x_l = 0$ and $x_m = 0$

$$K(X,Y) = 8\ ik\left(\beta_Y\frac{\partial G(X,Y)}{\partial n_X} + \beta_X\frac{\partial G(X,Y)}{\partial n_Y}\right.$$
$$+ \beta_Y\frac{\partial G(X,{}^1Y)}{\partial n_X} + \beta_X\frac{\partial G({}^1X,Y)}{\partial n_Y}$$
$$+ \beta_Y\frac{\partial G(X,{}^2Y)}{\partial n_X} + \beta_X\frac{\partial G({}^2X,Y)}{\partial n_Y}$$
$$+ \beta_Y\frac{\partial G(X,{}^3Y)}{\partial n_X} + \beta_X\frac{\partial G({}^3X,Y)}{\partial n_Y}$$
$$+ \beta_Y\frac{\partial G(X,{}^{12}Y)}{\partial n_X} + \beta_X\frac{\partial G({}^{12}X,Y)}{\partial n_Y}$$
$$+ \beta_Y\frac{\partial G(X,{}^{13}Y)}{\partial n_X} + \beta_X\frac{\partial G({}^{13}X,Y)}{\partial n_Y}$$
$$+ \beta_Y\frac{\partial G(X,{}^{23}Y)}{\partial n_X} + \beta_X\frac{\partial G({}^{23}X,Y)}{\partial n_Y}$$
$$\left. + \beta_Y\frac{\partial G(X,{}^{123}Y)}{\partial n_X} + \beta_X\frac{\partial G({}^{123}X,Y)}{\partial n_Y}\right)$$

if symmetry about $x_1 = 0$, $x_2 = 0$ and $x_3 = 0$

(58)

Finally, the last term of (41) has also to be considered for symmetry treatment :

(59)
$$J^{(3)}_{\mu\mu} = \frac{1}{2}\int_{S_\mu}\int_{S_\mu} f^{(3)}(X,Y)\ dS_\mu(X)\ dS_\mu(Y)$$

where :

(60)
$$f^{(3)} = -\,G(X,Y)\left(rot\ \mu(X),\ rot\ \mu(Y)\right)$$

with :

(61)
$$rot\ \mu(X) = n_X \wedge \nabla\mu(X)$$

Symmetry versus plane $x_1 = 0$ allows to perform integration on complementary parts S^*_μ and S^{**}_μ :

(62)
$$J^{(3)}_{\mu\mu} = \frac{1}{2}\int_{S^*_\mu}\int_{S^*_\mu} f^{(3)}(X,Y)\ dS^*_\mu(X)\ dS^*_\mu(Y)$$
$$+ \frac{1}{2}\int_{S^*_\mu}\int_{S^{**}_\mu} f^{(3)}(X,{}^1Y)\ dS^*_\mu(X)\ dS^{**}_\mu({}^1Y)$$
$$+ \frac{1}{2}\int_{S^{**}_\mu}\int_{S^*_\mu} f^{(3)}({}^1X,Y)\ dS^{**}_\mu({}^1X)\ dS^*_\mu(Y)$$
$$+ \frac{1}{2}\int_{S^{**}_\mu}\int_{S^{**}_\mu} f^{(3)}({}^1X,{}^1Y)\ dS^{**}_\mu({}^1X)\ dS^{**}_\mu({}^1Y)$$

Symmetry condition implies (46a), (46c) and (46d) supplemented with :

(63)
$$\frac{\partial\mu}{\partial X_1}({}^1X) = -\frac{\partial\mu}{\partial X_1}(X),\quad \frac{\partial\mu}{\partial X_1}({}^1Y) = -\frac{\partial\mu}{\partial X_1}(Y)$$
$$\frac{\partial\mu}{\partial X_2}({}^1X) = +\frac{\partial\mu}{\partial X_2}(X),\quad \frac{\partial\mu}{\partial X_2}({}^1Y) = +\frac{\partial\mu}{\partial X_2}(Y)$$
$$\frac{\partial\mu}{\partial X_3}({}^1X) = +\frac{\partial\mu}{\partial X_3}(X),\quad \frac{\partial\mu}{\partial X_3}({}^1Y) = +\frac{\partial\mu}{\partial X_3}(Y)$$

so that :

(64)
$$f^{(3)}({}^1X,{}^1Y) = f^{(3)}(X,Y) = -\,G(X,Y)\left(rot\ \mu(X),\ rot\ \mu(Y)\right)$$
$$f^{(3)}(X,{}^1Y) = f^{(3)}({}^1X,Y) = -\,G({}^1X,Y)\left(rot\ \mu({}^1X),\ rot\ \mu(Y)\right)$$

and $J^{(3)}_{\mu\mu}$ can be rewritten as :

(65)
$$J^{(3)}_{\mu\mu} = -\frac{1}{2}\int_{S^*_\mu}\int_{S^*_\mu}\left[2\ G(X,Y)\left(rot\mu(X),\ rot\mu(Y)\right)\right.$$
$$\left. + 2G({}^1X,Y)\left(rot\mu({}^1X),\ rot\mu(Y)\right)\right] dS^*_\mu(X)\ dS^*_\mu(Y)$$

Generalization of this result leads immediately to the following expressions :

$$J^{(3)}_{\mu\mu} = -\frac{1}{2}\int_{S_\mu}\int_{S_\mu} G(X,Y)\left(rot\mu(X),\ rot\mu(Y)\right) dS_\mu(X)\ dS_\mu(Y)$$

if no symmetry

$$J^{(3)}_{\mu\mu} = -\frac{1}{2}\int_{S^*_\mu}\int_{S^*_\mu} 2\left[G(X,Y)\left(rot\mu(X),\ rot\mu(Y)\right)\right.$$
$$\left. + G({}^lX,Y)\left(rot\mu({}^lX),\ rot\mu(Y)\right)\right] dS^*_\mu(X)\ dS^*_\mu(Y)$$

if symmetry about $x_l = 0$

$$J^{(3)}_{\mu\mu} = -\frac{1}{2}\int_{S^*_\mu}\int_{S^*_\mu} 4\Big[\, G(X,Y)\,\big(rot\mu(X),\ rot\mu(Y)\big)$$
$$+\ G(^{l}X,Y)\,\big(rot\mu(^{l}X),\ rot\mu(Y)\big)$$
$$+\ G(^{m}X,Y)\,\big(rot\mu(^{m}X),\ rot\mu(Y)\big)$$
$$+\ G(^{lm}X,Y)\,\big(rot\mu(^{lm}X),\ rot\mu(Y)\big)\Big]\, dS^*_\mu(X)\ dS^*_\mu(Y)$$

if symmetry about $x_l = 0$ and $x_m = 0$

$$J^{(3)}_{\mu\mu} = -\frac{1}{2}\int_{S^*_\mu}\int_{S^*_\mu} 8\Big[\, G(X,Y)\,\big(rot\mu(X),\ rot\mu(Y)\big)$$
$$+\ G(^{1}X,Y)\,\big(rot\mu(^{2}X),\ rot\mu(Y)\big)$$
$$+\ G(^{2}X,Y)\,\big(rot\mu(^{2}X),\ rot\mu(Y)\big)$$
$$+\ G(^{3}X,Y)\,\big(rot\mu(^{3}X),\ rot\mu(Y)\big)$$
$$+\ G(^{12}X,Y)\,\big(rot\mu(^{12}X),\ rot\mu(Y)\big)$$
$$+\ G(^{23}X,Y)\,\big(rot\mu(^{23}X),\ rot\mu(Y)\big)$$
$$+\ G(^{13}X,Y)\,\big(rot\mu(^{13}X),\ rot\mu(Y)\big)$$
$$+\ G(^{123}X,Y)\,\big(rot\mu(^{123}X),\ rot\mu(Y)\big)\Big]\, dS^*_\mu(X)\ dS^*_\mu(Y)$$

if symmetry about $x_1 = 0$, $x_2 = 0$ and $x_3 = 0$

(66)

The third term in (21) can be treated following the same methodology. The following result can readily be obtained :

(67) $$J_{\sigma\mu} = -\int_{S^*_\mu}\int_{S^*_\sigma} \sigma(X)\mu(Y)\ K(X,Y)\ dS^*_\sigma(X)\ dS^*_\mu(Y)$$

where :

$$K(X,Y) = -\left(\frac{G(X,Y)}{\partial n_Y} + ik\ \beta_Y\ G(X,Y)\right)$$

if no symmetry

$$K(X,Y) = -2\left(\frac{\partial G(X,Y)}{\partial n_Y} + \frac{\partial G(^{l}X,Y)}{\partial n_Y} + ik\ \beta_Y\big(G(X,Y) + G(^{l}X,Y)\big)\right)$$

if symmetry about $x_l = 0$

$$K(X,Y) = -4\left(\frac{\partial G(X,Y)}{\partial n_Y} + \frac{\partial G(^{l}X,Y)}{\partial n_Y} + \frac{\partial G(^{m}X,Y)}{\partial n_Y} + \frac{\partial G(^{lm}X,Y)}{\partial n_Y}\right.$$
$$\left. +\ ik\ \beta_Y\big(G(X,Y) + G(^{l}X,Y) + G(^{m}X,Y) + G(^{lm}X,Y)\big)\right)$$

if symmetry about $x_l = 0$ and $x_m = 0$

$$K(X,Y) = -8\left(\frac{\partial G(X,Y)}{\partial n_Y} + \frac{\partial G(^{1}X,Y)}{\partial n_Y} + \frac{\partial G(^{2}X,Y)}{\partial n_Y} + \frac{\partial G(^{3}X,Y)}{\partial n_Y}\right.$$
$$+ \frac{\partial G(^{12}X,Y)}{\partial n_Y} + \frac{\partial G(^{13}X,Y)}{\partial n_Y} + \frac{\partial G(^{23}X,Y)}{\partial n_Y} + \frac{\partial G(^{123}X,Y)}{\partial n_Y}$$
$$+\ ik\ \beta_Y\ \big(G(X,Y) + G(^{1}X,Y) + G(^{2}X,Y) + G(^{3}X,Y)$$
$$\left. +\ G(^{12}X,Y) + G(^{13}X,Y) + G(^{23}X,Y) + G(^{123}X,Y)\big)\right)$$

if symmetry about $x_1 = 0$, $x_2 = 0$ and $x_3 = 0$

(68)

Also the two single surface integrals related to pressure and velocity constraints can be transformed in order to take into account symmetry.

(69) $$J_\sigma = +c\int_{S^*_1} \bar{p}(X)\ \sigma(X)\ dS^*_1(X)$$

(70) $$J_\mu = -c\int_{S^*_2} \varrho i\omega\ \bar{v}_n(X)\ \mu(X)\ dS^*_2(X)$$

where : $c = 1$ if no symmetry
$c = 2$ if symmetry versus one plane,
$c = 4$ if symmetry versus two planes,
$c = 8$ if symmetry versus three planes

with S^*_1 and S^*_2 corresponding to the remaining boundary surface.

<u>Half–space Options</u>

In many cases, the acoustic fluid can be considered as a semi–infinite medium bounded by a zero pressure or a zero velocity plane. The first case is usually encountered in submarine applications (free surface) while the second one is typical of environmental acoustics where acoustic propagation occurs above ground (assumed to be rigid).

When the region occupied by the fluid is bounded by a plane, the kernel G needs to be replaced by a modified kernel verifying prescribed conditions on that plane. If the boundary plane constitutes a free surface ($p = 0$, Dirichlet boundary condition), the modified kernel can be written as

(71) $$G_D\ (X,Y) = G\ (X,Y) - G\ (X,Y')$$

where Y' denotes the symmetric location of point Y versus boundary plane (figure 4). For rigid plane case ($v_n = 0$, Neumann boundary condition), the modified kernel can be written according to

(72) $$G_N\ (X,Y) = G\ (X,Y) + G\ (X,Y')$$

It can easily be shown that these kernels satisfy zero pressure and zero velocity conditions respectively on boundary half plane. The above defined integral (8) representation remains valid for half space cases provided kernels (71) and (72) are substituted into (8). It must be emphasized that treatment of halfspace with rigid plane conditions is strictly analogous to treatment of symmetry versus that plane.

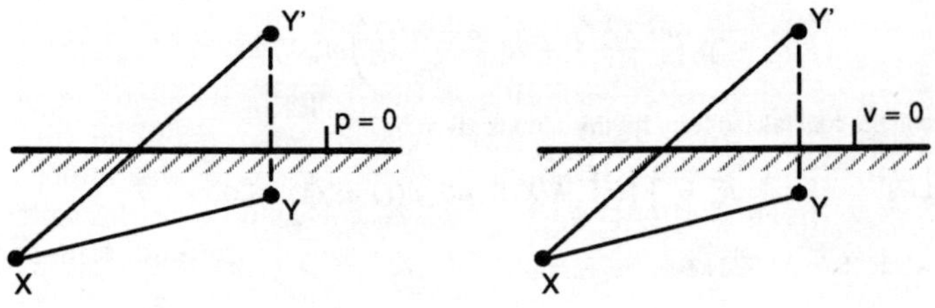

(a) free surface (b) rigid surface

Figure 4 : Halfspace options

<u>Integration Procedure</u>

The integration task is a crucial part of any boundary element program because generation of coefficient matrix is usually the most consuming part in the computational process.

To be effective, the integration procedure must recognize the special characters of integrands. This can be accomplished by selecting usual regular integration schemes when integrands are regular and resorting to singular integration procedure when singularity is encountered. This singularity occurs for auto–influence terms related to first and second term of functional (21) but is always integrable and can effectively be dealt with either by analytic or numerical integration procedures. The way selected here consists to perform singular integration using a non–product cubature formula of degree n (= 2 k – 1) using orthogonal polynomals of degree k (Haegemans and Wynendaele, 1988). These formulas have been set up for both triangle and quadrilateral shapes including corner or midside nodal singularity. The abscissae X_i, Y_i and weights W_i related to the unit triangular domain with both corner (TC) or midsided singularity (located at the origin) are given at table 1.

They allow to calculate the singular integral :

(73) $$\int_T \frac{f(X,Y)}{\sqrt{X^2 + Y^2}}\ dX\ dY = \sum_i W_i\ f(X_i, Y_i)$$

do-main	n	i	X_i	Y_i	W_i
TC	3	1	0.000000000000	0.400000000000	0.911386032310
		2	0.000000000000	2.000000000000	0.080201970843
		3	0.471506223759	1.500000000000	0.466629648543
		4	–0.471506223759	1.500000000000	0.466629648543
	5	1	0.000000000000	0.219641118964	0.519229470362
		2	0.000000000000	1.117166245626	0.373041126405
		3	0.000000000000	1.811605634129	0.205649674720
		4	0.350859639885	0.911696311978	0.254801111409
		5	–0.350859639885	0.911696311978	0.254801111409
		6	0.675387471206	1.754970354689	0.158662402967
		7	–0.675387471206	1.754970354689	0.158662402967
TM	3	1	0.000000000000	0.000000000000	0.754693885798
		2	0.000000000000	1.247042502161	0.609328587185
		3	0.468345117215	0.350509073400	1.039624527365
		4	–0.468345117215	0.350509073400	1.039624527365
	5	1	0.000000000000	0.078368399293	1.084233597733
		2	0.000000000000	0.570058268209	0.774413573238
		3	0.000000000000	1.533042936232	0.210735399059
		4	0.442960375887	0.859843508545	0.285417960310
		5	–0.442960375887	0.859843508545	0.285417960310
		6	0.654564097559	0.184704883757	0.401526518531
		7	–0.654564097559	0.184704883757	0.401526518531
	7	1	0.000000000000	0.054598584263	0.851454929337
		2	0.000000000000	0.369871354871	0.758201685219
		3	0.000000000000	0.968113513460	0.325863243483
		4	0.000000000000	1.702972841695	0.080342199050
		5	0.309703468766	1.211464847440	0.121361029053
		6	–0.309703468766	1.211464847440	0.121361029053
		7	0.541319426269	0.553620408865	0.317703017486
		8	–0.541319426269	0.553620408865	0.317703017486
		9	0.669781622992	0.115807892177	0.274630751298
		10	–0.669781622992	0.115807892177	0.274630751298
		11	3.193007129636	0.200646496729	0.000009937474
		12	–3.193007129636	0.200646496729	0.000009937474

Table 1. Non–product integration formulas for the domains TC, TM (degree n)

For regular terms (i.e. when typical field points X and Y do not belong to the same element), the integration procedure relies on regular quadrature (either Gauss–Legendre or Radan formulas can be selected) but the number of integration points is adapted such as to take into account integrand behavior. To deal effectively with this problem, a simple geometric criteria is introduced. This criteria consists to compare the distances between the gravity centers of both elements involved to the mean edge length of these elements and to define, according to this measure, three proximity status or attributes : "near", "intermediate" and "far". For each proximity status, a particular regular integration procedure is selected with quite more integration points for "near" elements and some less points for "far" elements.

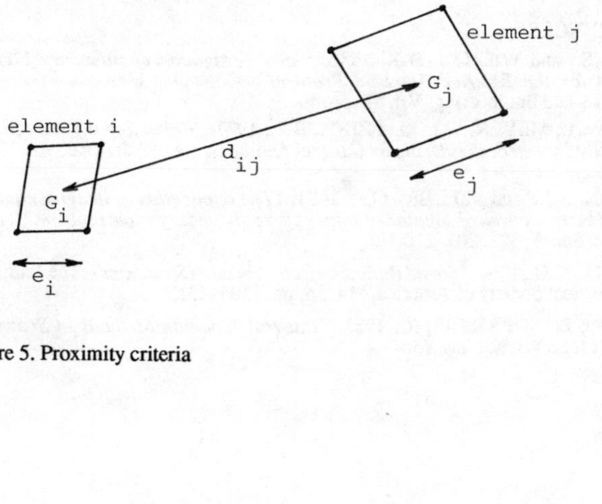

Figure 5. Proximity criteria

Quadrature schemes available are summarized for triangular and quadrilateral elements at table 2.

triangular elements		quadrilateral elements	
quadrature scheme	No. of Radan points	quadrature scheme	No. of Gauss–Legendre points
T_1	1	Q_1	1
T_2	4	Q_2	2 x 2 = 4
T_3	7	Q_3	3 x 3 = 9
		Q_4	4 x 4 = 16
		Q_5	5 x 5 = 25

Table 2. Available quadrature schemes for triangles

Table 3. Available quadrature schemes for quadrilaterals

Denoting by e_i and e_j the mean edges of elements i and j and by d_{ij} the distance between gravity centers of these elements (figure 5), one introduces the proximity factors r_d given by :

(74)
$$r_d = \frac{d_{ij}}{\left(\frac{e_i + e_j}{2}\right)}$$

and the integration scheme is selected according to the proximity status i_d given at table 4 :

proximity factor r_d	proximity status i_d
$r_d \leq r_n$	1 (near)
$r_n < r_d \leq r_f$	2 (intermediate)
$r_f < r_d$	3 (far)

Table 4. Proximity status versus proximity factor

Proximity factors r_n (near field) and r_f (far field) play an important roll in this procedure. Current default settings are summarized at table 5.

	proximity status ($r_n = 2, r_f = 5$)		
element type	1 (near)	2 (intermediate)	3 (far)
TRIA 3	T_2	T_2	T_1
TRIA 6	T_2	T_2	T_1
QUAD 4	Q_2	Q_2	Q_1
QUAD 8	Q_2	Q_2	Q_1

Table 5. Default quadrature schemes

Irregular frequencies

As outlined in the introduction, acoustic BE formulations fail at some critical frequencies (also termed "irregular" frequencies). The indirect variational method used here suffers from the same deficiencies but these can be overcome by adding some admittance conditions on an internal boundary. It can be demonstrated (Mebarek and Hamdi, 1986) that providing the real part of fixed admittance is different from zero, the only solution of the homogeneous problem is the trivial solution $p = 0$.

FINITE ELEMENT / BOUNDARY ELEMENT ELASTO–ACOUSTIC MODELLING

Introduction

Modelling coupled fluid–structure interaction in unbounded fluid domains usually relies on mixed formulations involving FE for the structure and BE for the fluid. The main difficulty is related to the lack of symmetry usually encountered with conventional BE formulations. Various symmetrization procedures have been proposed in the part to overcome this problem (Zienkiewicz, Kelly and Bettess, 1977). Usually this procedure consists to symmetrize the fluid matrix and to perform substitution into the structural set in order to get displacements. Such an approach has been described by Mathews (1984). Alternatively a fluid oriented procedure results from substitution of structure equation into the fluid set in order to get surface pressure (Wilton, 1978).

The numerical treatment of coupled elasto–acoustic problems is based on the selection of models describing the structure and the acoustic medium supplemented with appropriate interface conditions (kinematical and mechanical continuity conditions). Generally the fluid domain is assumed to be homogeneous and occupies either bounded (interior) or unbounded (exterior) volumes. In this context, the coupled BE/FE technique appears to be very attractive. Using this approach, the fluid is described using the above integral representation (BE formulation) while the structure is modeled using conventional finite elements (FE formulation).

Structural Functional

For the structure, the equilibrium equation results from stationarity of a functional J_S.

(75) $$J_S = U_S + V_S - T_S$$

where U_S, V_S and T_S are the strain energy, potential energy and kinetic energy, respectively. Using suitable finite elements to describe the structure, the structural functional may be expressed as (Craggs and Stead, 1976) :

(76) $$J_S = \frac{1}{2}\, \tilde{U}^T K \tilde{U} - \frac{\omega^2}{2}\, \tilde{U}^T M \tilde{U} - \tilde{U}^T F_F - \tilde{U}^T F_S$$

where K and M are the stiffness and mass matrices, $\tilde{U}$ the displacement vector and F the load vector.

Coupling Procedure

The coupling procedure relies on verification of mechanical and kinematic interface conditions. At the mechanical point of view, the load vector F can be partitioned into a structural load vector F_S and a fluid–originating load vector F_F. This last vector is due to the resulting pressure f_F coming from the fluid acting on both faces of the structure assumed to be thin. This resulting pressure is precisely the jump of pressure (or double layer potential) encountered in the integral description of acoustic field with velocity boundary condition.

(77) $$f_F = -(p^+ - p^-)\, n = -\mu n$$

where n is the normal to the structure (directed outwards) and p^+ and p^- are pressure outside and inside the structure respectively. Kinematic interface condition implies continuity of displacement at fluid–structure interface. Assuming only velocity conditions on the fluid boundary surface, ($S = S_\mu = S_2$) the above variational approach (in terms of double layer potentials) leads to the fluid functional

(78) $$J_F = \frac{1}{2\varrho\omega^2}\, \tilde{\mu}^T D \tilde{\mu} + \tilde{\mu}^T C U - \tilde{\mu}^T F_A$$

Taking into account (26) and the continuity of normal displacement, the stationarity conditions of both functionals lead to the coupled system.

(79.a)

(79.b)

$$\begin{bmatrix} K - \omega^2 M & C^T \\ C & \dfrac{D}{\varrho\omega^2} \end{bmatrix} \begin{bmatrix} \tilde{U} \\ \tilde{\mu} \end{bmatrix} = \begin{bmatrix} F_S \\ F_A \end{bmatrix}$$

Again this system is the starting point for various computational strategies. In the compressible case, only direct response analysis can be performed because the frequency is embedded in kernel G which appears in computation of D matrix. It leads to computation of nodal displacements and potentials from which other structural (strains, stresses) and acoustic results (pressures, velocities) can be evaluated. For incompressible cases ($c = \infty$ or $k = 0$), equation (79b) enables elimination of $\tilde{\mu}$ and evaluation of the added mass matrix

(80) $$M_a = \varrho\, C^T D\, C$$

to be used in equation (79a), rewritten as

(81) $$(K - \omega^2\, (M + M_a))\, .\, \tilde{U} = F_S$$

An eigenvalue extraction and forced response by modal superposition can then be performed using this relation.

APPLICATIONS

Immersed Cantilever Plate

The studied problem concerns vibrations of a cantilever plate (figure 6) in air and in water. The plate is rectangular (a = 0.4064 m, b = 0.2032 m, t = 0.00267 m) with elastic constitutive material (Young's modulus = 0.2068 10^{12} N/m^2, Poisson's ratio = 0.3 and volumetric mass = 7830 kg/m^3). Eigenfrequencies have to be computed for two cases (case 1 ; plate in vacuum, case 2 : plate immersed in water). Sound velocity in water is taken to be equal to 1500 m/s while volumetric mass is equal to 998 kg/m^3.

For the two cases solved, the clamping plane acts as a symmetry plane and jump of pressure is constrained to be zero on the free border of the plate. Results are summarized in table 6 together with theoretical and experimental results provided by Lindholm et al. (1965). The agreement between theoretical and experimental values is quite satisfactory.

	plate in vacuum		plate immersed in water	
mode	theory	numerical results	experimental results	numerical results
	(Hz)	(Hz)	(Hz)	(Hz)
1	13.8	13.8	5.1	6.1
2	59.3	59.4	29.8	34.8
3	83.9	89.1	34.4	40.8
4	194	198.5	99.1	120.2

Table 6. Natural frequencies of cantilever plate

Modeshapes related to case 2 are shown in figure 7.

Loudspeaker

This elasto–acoustic problem consists of a rigid piston (radius a = 0.08 m, mass m = 0.015 kg) elastically mounted on a rigid box (b = 0.50 m, h = 1 m) (figure 8).

The spring stiffness k is equal to 500 N/m. The fluid is the air (sound velocity = 340 m/s, density = 1.21 kg/m^3) The piston is loaded by a transversal force F (= 1N) at its center. Field acoustic pressure has to be computed at two typical frequencies (100 Hz and 1000 Hz).

This coupled problem has been modeled using rigid (but elastically mounted) elements for the piston and acoustic boundary elements for the air. Due to symmetry conditions, only one fourth has to be discretized. The mesh includes 155 nodes and 132 elements (figure 9).

At each frequency, the coupled system is described in terms of a structural variable (normal displacements of the piston) and acoustic variables (double layer potential at fluid boundary nodes). Determination of these variables allows to perform evaluation of field pressures at any location (both inside and outside the box). For representation purpose, a field point mesh has been selected in the horizontal plane located at mid–height. Pressure distributions (in dB, reference 2 10^{-5} N/m^2) in that plane at 100 and 1000 Hz are shown in figures 10 and 11. Directivity effects occurring at high frequencies can clearly be seen on figure 11.

Stretched Membrane Problem

A circular stretched membrane (radius a = 0.125 m, tension = 100 N/m, mass = 0.02 kg/m^3) is fixed on its edge (figure 12) and is subjected to a uniform transversal load (1 N/m^2). Response at point P located at 1 m from membrane's center must be determined in the frequency range 10 – 1000 Hz). Owing to symmetry, only one fourth of the structure needs to be discretized (figure 13). Spectral response at P is represented in figure 14.

CONCLUSIONS

An updated implementation of the variational boundary element formulation for exterior acoustics has been described. Based on earlier work of Hamdi, several extensions have been made in order to allow for consideration of symmetries and halfspace option. Details related to these features have been provided. The implementation has been shown to rely on an extended library of surface elements and uses adaptive integration rules to achieve better efficiency. These considerations and the symmetry of resulting fluid matrices allows to set up a powerful acoustic model which can be coupled quite easily with various FE structural models in order to perform elasto–acoustic analysis.

ACKNOWLEDGEMENTS

The authors wish to thank referees for their useful comments on the earlier version of this paper.

REFERENCES

AMINI, S., and WILTON, D.T., 1986, "*An Investigation of Boundary Element Methods for the Exterior Acoustic Problem*", Computer Methods in Applied Mechanics and Engineering, Vol. 54, pp. 49–56.

BELL, W.A., MEYER, W.L., and ZINN, B.T., 1977, "*Predicting the Acoustics of Arbitrarily Shaped Bodies Using an Integral Approach*", AIAA Journal, Vol. 15, pp. 813–820.

BURTON, A.J., and MILLER, G.F., 1971, "*The application of integral equation methods to the numerical solution of some exterior boundary value problems*", Proc. Roy. Soc. Ser. A, 323, 201–210.

CHERTOCK, G., 1964, "*Sound Radiation from Vibrating Structures*", The Journal of the Acoustical Society of America, Vol. 36, pp. 1305–1313.

COLTON, D., and KRESS, R., 1983, "*Integral Equation Methods in Scattering Theory*", John Wiley, New York

CRAGGS, A., and STEAD, G., 1976, *"Sound Transmission between Enclosures – A Study using Plate and Acoustic Finite Elements"*, Acoustica, Vol. 35, pp. 89–98.

HAEGEMANS, A., and WYNENDAELE, H., 1988, *"Cubature Formulas for Triangles and Squares with 1/r Singularity"*, submitted to Engineering Analysis

HAMDI, M.A., 1981 *"Une Formulation Variationnelle par Equations Intégrales pour la Résolution de l'Equation de Helmholtz avec des Conditions aux Limites Mixtes"*, C.R. Acad. Sc. Paris, Vol. 292, pp. 17–20.

HAMDI, M.A., 1982 *"Formulation Variationnelle par Equations Intégrales pour le Calcul de Champs Acoustiques Lineaires Proches et Lointaines"*, Thèse d'Etat, Université de Compiègne,

HUANG, H., 1984, *"Helmholtz Integral Equations for Fluid–Structure Interaction"*, Advances in Fluid–Structure Interaction, ASME, AMD, Vol. 64, pp. 39–48.

KIPP, C.R., and BERNHARD, R.J., 1987, *"Prediction of Acoustical Behavior in Cavities Using an Indirect Boundary Element Method"*, Journal of Vibration, Acoustics, Stress and Reliability in Design, Transactions of the ASME, Vol. 109, pp. 22–28.

KLEINMAN, R.E., and ROACH, G.F., 1974, *"Boundary integral equation for the three–dimensional Helmholtz equation"*, SIAM Review, 16, No. 2, pp. 214–236.

LATCHA, M.A., and AKAY, A., 1986, *"Applications of the Helmholtz Integral in Acoustics"*, Journal of Vibration, Stress and Reliability in Design, Vol. 108, pp. 447–453.

MATHEWS, I.C., 1984, *"A Symmetric Boundary Integral – Finite Element Approach for 3–D Fluid–Structure Interaction"*, Advances in Fluid–Structure Interaction, ASME, AMD, Vol. 64, pp. 19–38.

MEBAREK, L., and HAMDI, M.A., 1986, *"Résolution du Problème des Fréquences Irrégulières dans la Méthode des Equations Intégrales"*, Communication presented at 10eme Colloque d'Acoustique Aéronautique et Navale, Marseille.

MEYER, W.L., BELL, W.A., ZINN, B.T., and STALLYBRASS, M.P., 1978, *"Boundary Integral Solutions of Three Dimensional Acoustic Radiation Problems"*, Journal of Sound and Vibration, Vol. 59, pp. 245–262.

SCHENCK, H.A., 1968, *"Improved integral formulation for acoustic radiation problems"*, J. Acous. Soc. Am., 44, 41–58.

SEYBERT, A.F., and RENGARAJAN, T.K., 1987, *"The Use of CHIEF to obtain unique solutions for Acoustic Radiation using Boundary Integral Equations"*, J. Acoust. Soc. Am., Vol. 81, pp. 1299–1306.

SEYBERT, A.F., and RENGARAJAN, T.K., 1987, *"The High–Frequency Radiation of Sound from Bodies of Arbitrary Shape"*, Journal of Vibration, Acoustics, Stress and Reliability in Design, Vol. 109, pp. 381–387.

SHAW, R.P., 1979, *"Boundary Integral Equation Method Applied to Wave Problems"*, Development in Boundary Element Methods, Chapter 6, Banerjee, P.K. and Butterfield, R., eds., Applied Science Publishers Ltd, London.

STRENG, J.H., 1988, *"Sound Radiation from Circular Stretched Membrane in Free Space"*, Audio Engineering Society, 84th Convention, Paris (Preprint 2573 (C2)).

TERAI, T., 1980, *"On Calculation of Sound Fields around Three–dimensional Objects by Integral Equation Methods"*, Journal of Sound and Vibration, Vol. 69, pp. 71–100.

TZU–CHU LIN, 1984, *"A Proof for the Burton and Miller Integral Equation Approach for the Helmholtz Equation"*, Journal of Mathematical Analysis and Applications, Vol. 103, pp. 565–574.

WILTON, D.T., 1978, *"Acoustic Radiation and Scattering from Elastic Structures"*, International Journal for Numerical Methods in Engineering, Vol. 13, pp. 123–138.

ZIENKIEWICZ, O.C., KELLY, D., and BETTESS, P., 1977, *"The Coupling of Finite Element and Boundary Solution Procedures"*, International Journal for Numerical Methods in Engineering, Vol. 13, pp. 355–375.

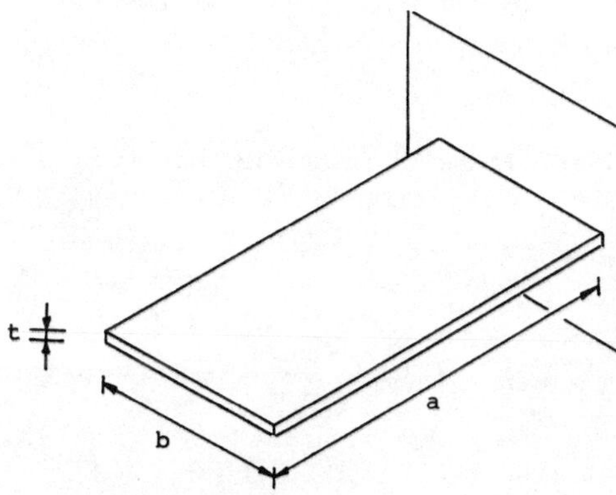

Figure 6. Cantilever plate geometry

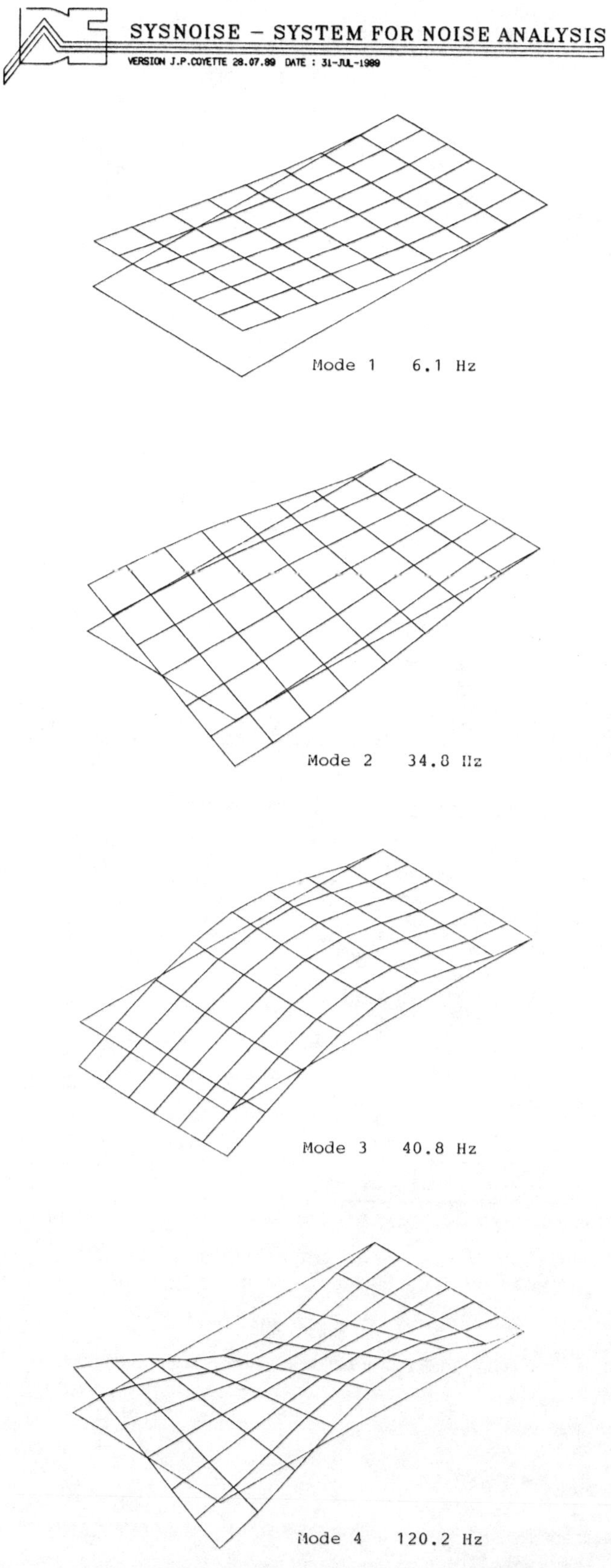

Figure 7. Immersed cantilever plate, modeshapes

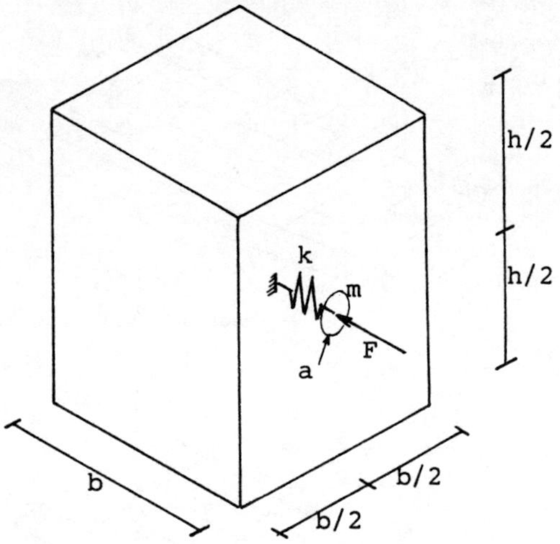

Figure 8. Loudspeaker's geometry

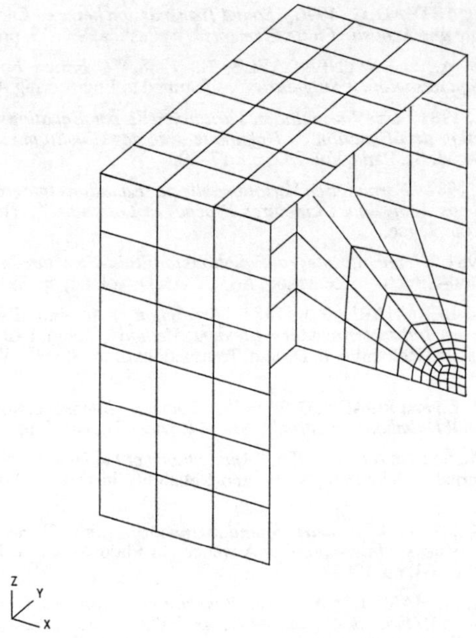

Figure 9. Loudspeaker's mesh

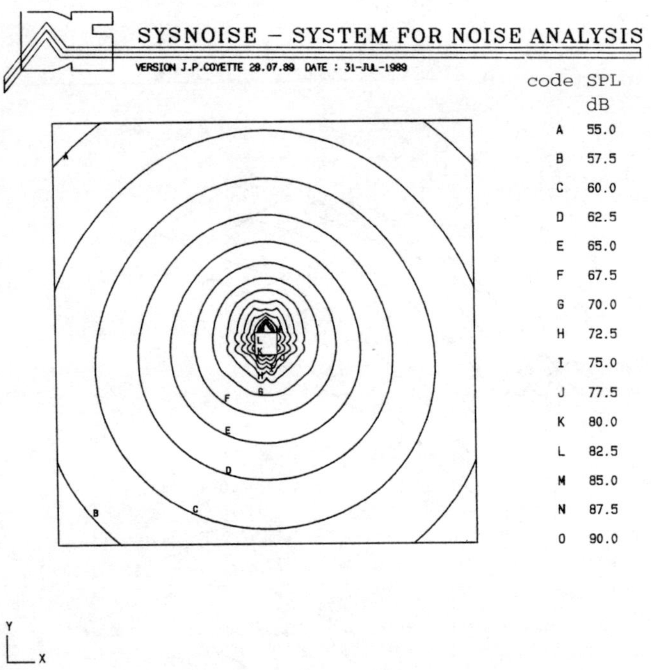

Figure 10. Pressure distribution at 100 Hz in horizontal plane at mid-height

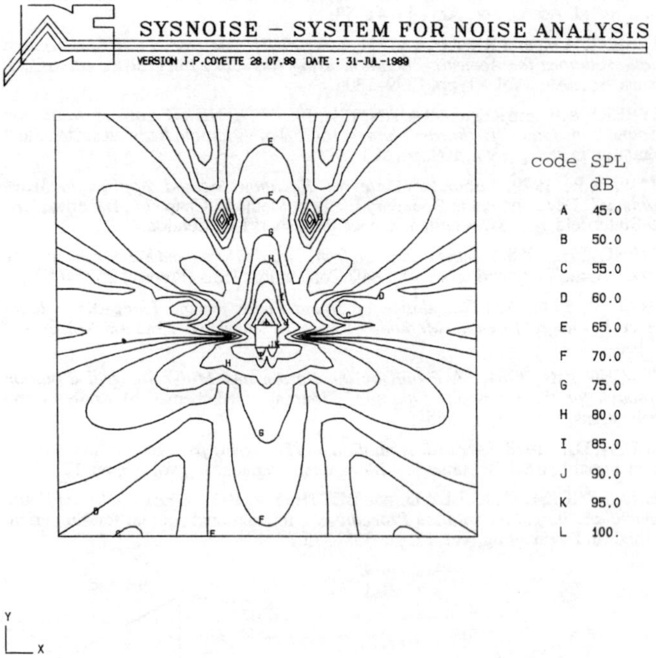

Figure 11. Pressure distribution at 1000 Hz in horizontal plane at mid-height

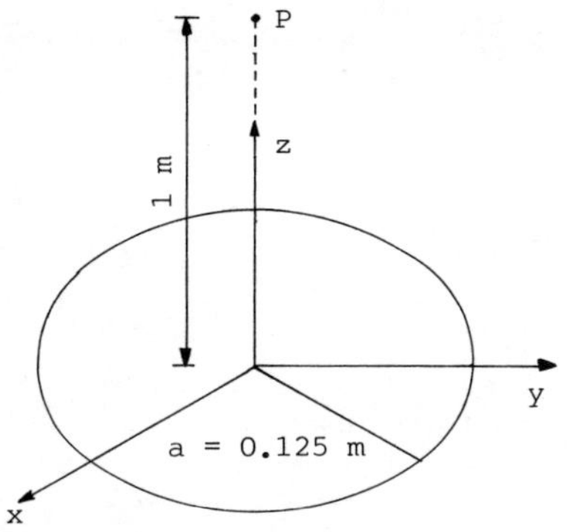

Figure 12. Membrane's geometry

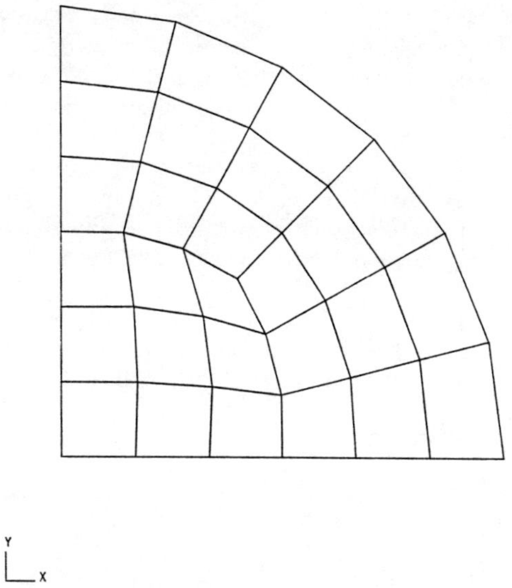

Figure 13. Membrane's mesh

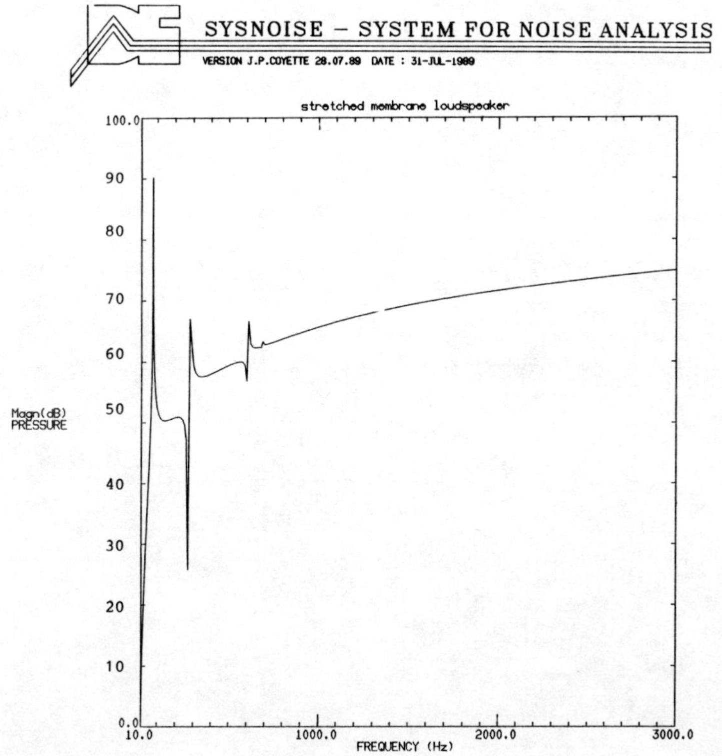

Figure 14. Pressure at point P

ACOUSTIC FIELD AND RADIATED POWER FROM VIBRATING SURFACES IN MOTION

X.-F. Wu and A. Akay
Department of Mechanical Engineering
Wayne State University
Detroit, Michigan

Abstract — This paper presents the acoustic field and radiated power from a noncompact vibrating surface in motion. Numerical results are obtained by using the extended Kirchhoff integral formulation recently proposed by Wu and Akay [1989]. One novelty of this formulation is that the volume integrals of higher-order poles are transformed into surface integrals, thus making numerical computations manageable. These higher-order poles result from the inclusion of the *local convection* in the equation of conservation of momentum. The effects of these higher-order poles are prominent at larger amplitudes of the surface oscillation and higher ka values. The integral formulation also shows that the source strength density is modified by the *source convection* by a factor of $(1-M_R)^{-1}$, where M_R is the component of the Mach number in the direction of wave propagation. Accordingly, the acoustic pressure is amplified in the forward direction while reduced in the reverse direction. The radiated acoustic power is found to increase with the Mach number and ka. Specifically, it increases as the square of the Mach number at low Mach numbers and small ka values, while increases in a complex manner at higher Mach numbers and ka values. Numerical examples are demonstrated by the transversely and radially oscillating spheres moving at a constant speed along a straight line.

I Introduction

There have been many formulations proposed for predicting sound radiation from a moving source, for example from a moving monopole and dipole [Morse and Ingard, 1968; Lowson, 1956; Morfey and Tanna, 1971; Tanna and Morfey, 1971], moving multipole [Gaunaurd and Eisler, 1972], and from a simple source in an unsteady rectilinear motion [Dowling, 1976; Leppington and Levine, 1987]. The analogous radiation problems are also well known in electromagnetism for stationary and moving electric charges [Stratton, 1941; Jones, 1964; Jackson, 1975]. Most of these formulations are concerned with a compact source, *i.e.* the wavelength of the sound emitted by the source is much larger than the source dimension. Under such circumstances, the sound field can be obtained by a direct solution of the inhomogeneous wave equation. However, when the source is of a finite size or when the wavelength is comparable to the source dimension, it is in general not a trivial task to solve the inhomogeneous wave equation to determine the sound field. One mathematical theorem which facilitates the calculation of the sound field is due to Kirchhoff [1883]. (For a modern derivation of this theorem, readers are referred to Pierce [1981].) Since this theorem was originally developed for a stationary source, it is not applicable when the source is in motion. Consequently, one needs to extend this theorem to take into account the effect of source convection.

The first extension of the Kirchhoff integral theorem was shown by Morgans [1930] for the interior problem. Although Morgans' formulation was questioned by Ffowcs-Williams and Hawkings [1969], Munro [1980] and Farassat and Myers [1988] showed that it was indeed correct. Another early version of the extended Kirchhoff integral formulation was given by Khromov [1963], but that formulation was in error as pointed by Ffowcs-Williams and Hawkings [1969]. In 1988, Farassat and Myers [1988] presented a modern derivation of the extended Kirchhoff integral formulation for the exterior region. Both Ffowcs-Williams and Hawkings [1969] and Farassat and Myers [1988] considered a field function defined for the entire space, but with discontinuities across the surface of the vibrating body. These discontinuities are taken as acoustic sources based on the generalized function theory and the acoustic field is obtained by expressing this inhomogeneous wave equation in an integral form. Attempts of applying the extended Kirchhoff integral formulation have been made to predict the noise generated by high speed propellers and helicopter rotors [Hawkings, 1979; Lyrintzis and George, 1987]. Because of the complexity involved in the formulation, applications of the extended Kirchhoff integral theorem are not as straightforward as in the stationary cases.

Recently, Wu and AKay [1989] proposed a different version of the extended Kirchhoff integral formulation. One novelty of that formulation is that the volume integrals associated with higher-order poles are transformed into surface integrals, thus making numerical computations manageable. The present paper uses such an integral formulation to calculate the acoustic field and radiated power from a noncompact vibrating surface in motion. In particular, contributions from higher-order poles are explicitly demonstrated and compared with those from monopole and dipole sources under different amplitudes of surface vibrations, ka values, and Mach numbers. The total radiated acoustic power is expressed in two parts, one corresponding to the stationary case while the other corresponding to that due to source convection. Numerical results confirm the previous results [Morse and Ingard, 1968; Gaunaurd and Eisler, 1972] that the total radiated acoustic power increases as the square of the Mach number at low Mach numbers and small ka values. At higher Mach numbers and larger ka values, however, the total radiated acoustic power increases in a complex manner. The examples considered are the transversely and radially oscillating spheres moving at a constant speed along a straight line. These simple sources are selected for the purposes of comparing the calculated results with the well-known results in the stationary cases and examining the effects of the magnitude of the surface oscillation, the noncompactness of the source, and the source convection speed on the resultant acoustic field and radiated power.

II Extended Kirchhoff Integral Formulation

It is emphasized here that the intent of the present paper is to demonstrate applications of the extended Kirchhoff integral formulation for predicting the acoustic radiation from a moving noncompact vibrating body. Derivations of such an integral formulation have been shown in detail in a separate paper [Wu and Akay, 1989]. For the sake of completeness, however, we shall briefly describe the derivations as follows.

Consider a closed, impermeable surface oscillating at a constant angular frequency ω and at the same time moving along a specified path in the medium. The medium is assumed to be unbounded, isentropic, and inviscid. There are no other sources in the medium except the vibrating body and there are no external nonconservative body forces exerted on the surface. To describe the wave motion thus generated in the medium, one considers the fundamental fluid-dynamic equations, *i.e.* conservation of mass, conservation of momentum, and equation of state for an isentropic fluid,

$$\frac{\partial \rho}{\partial t} + \vec{\nabla} \bullet (\rho \vec{v}) = 0 \tag{1}$$

$$\frac{\partial (\rho \vec{v})}{\partial t} + (\vec{v} \bullet \vec{\nabla})\rho \vec{v} = -\vec{\nabla} p \tag{2}$$

$$p = c^2 \rho \tag{3}$$

where ρ is the fluid density, c is the speed of sound in the medium, $\vec{v}$ is the fluid particle velocity, and p is the acoustic pressure. The convective derivative on the left side of Eq. (2) measures the rate of change of $\rho\vec{v}$ caused by fluid flow at a particular location and time. For the sake of clarity, we shall use in this paper the terms *local convection* to describe the effect of fluid flow around the boundary resulting from high-amplitude oscillations and *source convection* to describe the motion of the vibrating body as a whole in the medium.

The wave equation can be obtained by eliminating $\rho\vec{v}$ in Eqs. (1) and (2) and replacing ρ by p/c^2, the result being

$$\frac{1}{c^2}\frac{\partial^2 p}{\partial t^2} - \nabla^2 p = \vec{\nabla} \bullet (\vec{v} \bullet \vec{\nabla})\rho \vec{v}. \tag{4}$$

It should be pointed out here that the source term on the right side of Eq. (4) is normally neglected in the linear approximation, $|\,\vec{v}\,| << c$ [Pierce, 1981].

To derive the Kirchhoff integral formulation, one makes use of the Green's function [Pierce, 1981] which corresponds to a unit impulsive point source. Such a function satisfies the inhomogeneous wave equation

$$\frac{1}{c^2}\frac{\partial^2 G}{\partial t^2} - \nabla^2 G = 4\pi \delta(t_o - t)\delta(\vec{x}_o - \vec{x}) \tag{5}$$

where $\delta(\tau)$ is the Dirac delta function or generalized function [Jones, 1964].

Multiply Eq. (5) by p and Eq. (4) by G, use the chain-rule, and then subtract the two equations. The resulting equation involves both temporal and spatial generalized functions. To proceed with the derivation, one integrates the resulting equation over time and volume V between the vibrating surface S and the surface of some large sphere extended to infinity. In a more general case, one may replace S by some *control surface* S_c which completely encloses S, and even some fluid, at all times. If one assumes that this *control surface* S_c does not expand or contract as it moves in the medium, then V is constant. Consequently, one can freely interchange the order of temporal and spatial integrations. This allows one to transform one of the volume integrals to the surface integral by Gauss' theorem, thus reducing the order of integration by one.

$$4\pi p(\vec{x}_o, t_o) = \iiint_V \left[\int_{-t_1}^{t_2} G\vec{\nabla} \bullet (\vec{v} \bullet \vec{v})\rho\vec{v}\, dt\right] dV - \iint_S \left[\int_{-t_1}^{t_2} \left(\rho\frac{\partial G}{\partial n} - G\frac{\partial p}{\partial n}\right) dt\right] dS \tag{6}$$

where the symbol $\partial/\partial n$ represents the normal derivative taken at the surface. To facilitate the time integrals in Eq. (6), one introduces the Lagrange coordinate system [Morse and Ingard, 1986]

$$\vec{x} = \vec{\eta} + \int_0^t c\vec{M}(\vec{\eta}, \tau)\, d\tau \tag{7}$$

where $\vec{\eta}$ is the coordinate system moving with the surface, $\vec{M}$ is the Mach number in the direction of source convection motion, and τ is defined as

$$\tau = t_o - t - \frac{R}{c}. \tag{8}$$

where R is the distance between the source and the receiver.

With the change of the integration variable from t to τ and use of properties of the generalized functions, one can carry out the time integrals in Eq. (6) and obtain

$$4\pi p(\vec{x}_o, t_o) = -\iiint_V \left[\frac{1}{(1-M_R)R}\vec{\nabla} \bullet (\vec{v} \bullet \vec{\nabla})\rho\vec{v}\right]_{t^*} dV - \iint_S \left\{\frac{1}{(1-M_R)}\left[\frac{1}{R}\frac{\partial p}{\partial n} + \frac{p}{R^2}\frac{\partial R}{\partial n} - \frac{\partial}{\partial t}\left(\frac{p}{cR}\frac{\partial R}{\partial n}\right)\right]\right\}_{t^*} dS \tag{9}$$

where $t^* = t_o - R/c$ is the retarded time and M_R is the component of the Mach number in the direction of sound wave propagating from the source to the receiver, defined as

$$M_R = \frac{\vec{x}_o - \vec{x}}{|\,\vec{x}_o - \vec{x}\,|} \bullet \vec{M}. \tag{10}$$

Next, we use the vector identity [Selby, 1973] to rewrite the divergence of the convective derivative in the integrand of the volume integral on the right side of Eq. (9), with the fluid density ρ replaced by sum of the ambient density ρ_o and the density fluctuation ρ' and keeping only the first order terms,

$$\vec{\nabla} \bullet (\vec{v} \bullet \vec{\nabla})\rho\vec{v} \approx \rho_o \vec{\nabla} \bullet \vec{\nabla}\left(\frac{\vec{v} \bullet \vec{v}}{2}\right) - \rho_o \vec{\nabla} \bullet [\vec{v} \times (\vec{\nabla} \times \vec{v})]. \tag{11}$$

The second term on the right side of Eq. (11) represents the divergence of fluid vorticity. Since the fluid is assumed to be irrotational before the first disturbance arrives, it will remain irrotational afterwards [Currie, 1974]. Consequently, the second term vanishes identically. Such a result is consistent with Lighthill's acoustic analogy [Lighthill, 1952] for an inviscid and isentropic fluid.

Substituting Eq. (11) into the integrand of the volume integral in Eq. (9) and using the chain-rule, one obtains

$$\iiint_V \left[\frac{1}{(1-M_R)R}\vec{\nabla} \bullet (\vec{v} \bullet \vec{\nabla})\rho\vec{v}\right]_{t^*} dV = \iiint_V \vec{\nabla} \bullet \left[\frac{1}{(1-M_R)R}\vec{\nabla}\left(\frac{\rho_o v^2}{2}\right)\right]_{t^*} dV - \iiint_V \left[\frac{1}{(1-M_R)}\vec{\nabla}\left(\frac{1}{R}\right) \bullet \vec{\nabla}\left(\frac{\rho_o v^2}{2}\right)\right]_{t^*} dV. \tag{12}$$

Here it is assumed that the variation of the value of the Mach number across the finite-sized surface is small as in rectilinear motions. Hence the factor $(1 - M_R)^{-1}$ is considered constant in taking the derivative with respect to the coordinates moving with the surface.

Applying the chain-rule once again recasts the second integrand on the right side of Eq. (12) into two terms, one involving the quantity $\nabla^2(1/R)$ while the other being the divergence of a complex vector. The former vanishes identically because $1/R$ is a Laplacian within the volume V. The latter along with the first term on the right side of Eq. (12) can be transformed to surface integrals by Gauss' theorem. Doing so leads to an extended Kirchhoff integral formulation

$$p(\vec{x}_o, t_o) = -\frac{1}{4\pi}\iint_S \left[\frac{\nu}{(1-M_R)R}\right]_{t^*} dS \qquad (13)$$

where ν is defined as

$$\nu = \frac{\partial}{\partial n}\left(\frac{\rho_o v^2}{2}\right) + \frac{\rho_o v^2}{2R}\frac{\partial R}{\partial n} + \frac{\partial p}{\partial n} + \frac{p}{R}\frac{\partial R}{\partial n} - R\frac{\partial}{\partial t}\left(\frac{p}{cR}\frac{\partial R}{\partial n}\right). \qquad (14)$$

The first term on the right side of Eq. (14) is the rate of change of the kinetic energy density in the normal direction at the vibrating surface. While the second term is the kinetic energy density per unit distance between the source and receiver in the direction of wave propagating from the source to the receiver. These two terms reflect the effects of higher-order poles on the resulting acoustic field. Although the contribution of the second term is negligible in the far-field, it can be important in the near-field as will be shown in the numerical examples and hence is retained in the formulation. The third term represents the normal component of the body force per unit volume exerted on the fluid by the surface and the last two terms stand for the volume displacement (thickness) effect of the vibrating surface [Goldstein, 1976]. When the magnitude of the surface oscillation is much smaller than the speed of sound, $|\vec{v}| << c$, the first two terms are negligible. Under such circumstances, Eq. (13) reduces to Ffowcs-Williams and Hawking's integral formulation [1969] for an impermeable surface. As the Mach number tends to zero, Eq. (13) further reduces to the conventional Kirchhoff integral formulation [Pierce, 1981].

III Acoustic Field and Radiated Power

In this section the extended Kirchhoff integral formulation (13) obtained above is rewritten for the purpose of numerical computations. The acoustic field and the total radiated acoustic power are explicitly expressed in two parts, one corresponding to the stationary case while the other corresponding to source convection so as to show their changes with the source convection speed.

In the previous section it has been shown that the field acoustic pressure $p(\vec{x}_o, t_o)$ can be expressed in terms of integrals involving the surface pressure and velocity. For a specific radiation problem, the surface velocity with respect to the laboratory coordinates is given. Hence the first step in determining the field pressure is to solve for the surface pressure. The integral equation for solving the surface pressure can be obtained by taking the limit as the field point approaches the surface. Once this limit has been taken, both the receiver and the source are stationary with respect to the coordinate system moving with the surface. The resultant integral equation is different from that of the stationary case; however, it can be solved in the same manner as in the stationary case.

Once the surface pressure is obtained, one can determine the field pressure by using Eq. (13). The quantity R is now time-dependent, and from Eq. (7) one finds

$$R = \left|\vec{x}_o - \vec{\eta} - \int_o^t c\vec{M}(\vec{\eta}, \tau)d\tau\right|. \qquad (15)$$

The derivative of R with respect to time is then given by

$$\frac{\partial R}{\partial t} = -cM_R(t). \qquad (16)$$

Suppose that the vibrating body moves along a specified path in the medium, then the total velocity of a surface element is given by

$$\vec{v}(t) = \vec{v}_s(t) + c\vec{M}(t) \qquad (17)$$

where $\vec{v}_s(t)$ depicts the velocity at a surface point and $c\vec{M}(t)$ describes the source convection motion. The normal component of the acceleration for a surface element is accordingly written as

$$\frac{\partial v_n}{\partial t} = -c\left[1 + M_R(t)\right]\left[ik\vec{n} \bullet \vec{v}_s - \dot{M}_n(t)\right] \qquad (18)$$

where the dot over M_n indicates a time derivative of the component of the Mach number in the direction of unit normal on the surface

$$M_n(t) = \vec{n} \bullet \vec{M}(t). \qquad (19)$$

Substituting Eqs. (16) – (18) into Eq. (14) and replacing the normal derivative of the surface pressure by the negative of the left side of Eq. (2) in Eq. (14), one obtains,

$$\begin{aligned}\nu = &\frac{\partial}{\partial n}\left(\frac{\rho_o v^2}{2}\right) + \frac{\rho_o v^2}{2R}\frac{\partial R}{\partial n} \\ &+ \rho_o c(1 + M_R)(ik\vec{n} \bullet \vec{v}_s - \dot{M}_n) - \vec{n} \bullet (\rho_o \vec{v} \bullet \nabla)\vec{v} \\ &+ \left\{M_n + \left[1 - 2M_R + ikR(1 + M_R)\right]\frac{\partial R}{\partial n} - \frac{\vec{n} \bullet \vec{R}}{c}\right\}\frac{p}{R} \end{aligned} \qquad (20)$$

where a harmonic time dependence of $e^{-i\omega t}$ is assumed in the acoustic pressure p. It is understood here that all quantities are evaluated at the retarded time. With the substitution of the source strength density ν given by Eq. (20) into Eq. (13), the acoustic field is completely determined.

<u>Total Radiated Acoustic Power</u>

Now we calculate the total radiated acoustic power from the noncompact vibrating surface in motion. In particular we want to demonstrate the change in the total radiated acoustic power as a result of source convection. For this purpose we will write the radiated power in two parts, one corresponding to the stationary case while the other correspond to the source convection motion.

The expression for the total radiated acoustic power is well-known [Morse and Ingard, 1968]

$$\mathbf{P} = \iint_S \frac{|p|^2}{\rho_o c} dS \qquad (21)$$

where p is the acoustic pressure at the far-field and $\rho_o c$ is the *specific acoustic impedance* of the medium.

In order to explicitly demonstrate the effect of source convection on the radiated acoustic power, we rewrite the far-field pressure as

$$p = p_o + \Delta p \qquad (22)$$

where p_o corresponds to the acoustic pressure radiated when the vibrating body is stationary and Δp is the change of the acoustic pressure due to source convection.

$$p_o = -\frac{1}{4\pi}\iint_S \frac{\nu_o}{R} dS \qquad (23)$$

$$\Delta p = -\frac{1}{4\pi}\iint_S \left[M_R(1+M_R+M_R^2+\cdots)\frac{\nu_o}{R}\right]_{t^*} dS$$
$$-\frac{1}{4\pi}\iint_S \left[\frac{\Delta\nu}{(1-M_R)R}\right]_{t^*} dS. \tag{24}$$

Here we have approximated the factor $(1-M_R)^{-1}$ by a power series so as to separate the contribution from the stationary source from those induced by source convection. The quantities ν_o and $\Delta\nu$ correspond to the source strength densities of the stationary source and those due to source convection, respectively,

$$\nu_o = \frac{\partial}{\partial n}\left(\frac{\rho_o v_s^2}{2}\right) + \frac{\rho_o v_s^2}{2R}\frac{\partial R}{\partial n} + i\rho_o ck\vec{n}\bullet\vec{v}_s$$
$$-\rho_o\vec{n}\bullet(\vec{v}_s\bullet\nabla)\vec{v}_s + \left[(1+ikR)\frac{\partial R}{\partial n} - \frac{\vec{n}\bullet\vec{R}}{c}\right]\frac{p}{R} \tag{25}$$
$$\Delta\nu = \rho_o cM\frac{\partial v_s}{\partial n} + \rho_o cv_s\frac{\partial M}{\partial n} + \frac{\rho_o cMv_s}{R}\frac{\partial R}{\partial n} + \frac{\rho_o c^2M^2}{2R}\frac{\partial R}{\partial n}$$
$$+ i\rho_o cM_Rk\vec{n}\bullet\vec{v}_s - \rho_o c(1+M_R)\dot{M}_n - \rho_o c\vec{n}\bullet(\vec{M}\bullet\nabla)\vec{v}_s$$
$$-\rho_o c\vec{n}\bullet(\vec{v}\bullet\nabla)\vec{M} + [M_n - (2-ikR)M_R]\frac{\partial R}{\partial n}\frac{p}{R}. \tag{26}$$

where v_s is the magnitude of the surface velocity. It is noted that since the fluid is assumed to be inviscid, only the normal component of the surface velocity contributes to the sound generation and propagation.

The total radiated acoustic power can then be written as

$$\mathbf{P} = \mathbf{P}_o + \Delta\mathbf{P} \tag{27}$$

where $\mathbf{P}_o$ corresponds to the radiated acoustic power at $M = 0$ and the quantity $\Delta\mathbf{P}$ is proportional to

$$\Delta\mathbf{P} \propto p_o\Delta p^* + p_o^*\Delta p + |\Delta p|^2\,. \tag{28}$$

where the superscript $*$ represents a complex conjugate. Equation (27) shows that the total radiated acoustic power is increased by $\Delta\mathbf{P}$ when the vibrating body is in motion. The increment $\Delta\mathbf{P}$ vanishes only when M becomes zero.

IV Numerical Examples

In this section, we illustrate some numerical examples using the extended Kirchhoff integral formulation developed in this paper to calculate the acoustic field and total radiated power. For simplicity we consider the transversely and radially oscillating spheres moving at a constant speed along a straight line in air at room temperature. The reasons of selecting these simple cases are for comparison of the present results with the existing ones and examine effects of source convection and the noncompactness of the source. In all examples, the acoustic fields are evaluated at the same emission time $t_o = 1$ and the acoustic pressures are nondimensionalized with respect to $\rho_o cv_s$, where $v_s = cM_s$ is the magnitude of the surface velocity with M_s being the surface Mach number. The calculated acoustic fields extend from $1.2a$ to a distance of $20a$, where a is the radius of the sphere. In what follows, we will first demonstrate the effect of higher-order poles on the sound generation and propagation from a stationary source.

Figure 1 shows the field pressure distribution from a stationary sphere oscillating back and forth along the z-axis at $ka = 1$ without the inclusion of higher-order poles. Figure 2 depicts the acoustic field from the same source with the higher-order poles taken into account and with $M_s = 0.1$. The effects of higher-order poles are obvious. The higher-order poles are actually quadrupoles oscillating at twice the angular frequency ω. Consequently, for a fixed ω the amplitude of one of the two pressure lobes along the z-axis of the dipole source is always reinforced while the other is always suppressed. Figure 2 also shows that there is sound radiation in the perpendicular direction. This is caused by the pressure lobes of the quadrupoles along the y-axis. As the surface Mach number M_s increases, the effects of higher-order poles become more and more pronounced. Figure 3 shows the pressure distribution at $M_s = 1$. The resultant pressure distribution from the dipole source is almost completely changed.

Figures 4 and 5 show the comparison of pressure fields at $ka = 5$ for cases with and without the inclusion of higher-order poles. The effects of higher-order poles are evident (See Fig. 5). Similar to Fig. 2, one of pressure lobes along the z-axis is amplified while the other is reduced. Further, more ripples are generated in the pressure field. The above examples clearly demonstrate the importance of contributions from higher-order poles at a larger surface oscillation and a higher ka value.

Figure 6 shows the pressure distribution when the transversely oscillating sphere (with $M_s = 0.1$) is moving at a constant speed, $M = 0.2$, in the positive z-axis direction. Because of the source convection the front pressure lobe becomes much higher than the back lobe. The amplitudes of the field pressures on the whole are greatly amplified in the forward direction while reduced in the reverse direction. Particularly, the acoustic pressures are of finite values along the y-axis. This is caused by both the inclusion of higher-order poles and the source convection.

Figure 7 depicts the acoustic field from the radially oscillating sphere (with $M_s = 0.1$) at $M = 0$. The effects of higher-order poles are seen in two places, namely, (1) the amplitudes of the acoustic pressures become higher along the z-axis than those along the y-axis and (2) ripples are generated in the field. These changes become more prominent with the increase of the surface Mach number M_s and ka. Figure 8 shows the field pressure distribution from the same sphere moving at a constant Mach number, $M = 0.2$, in the positive z-axis direction. Again, the amplitudes of the field pressures are greatly amplified in the forward direction but reduced in the reverse direction.

Figures 9 and 10 show the acoustic pressures measured at the origin of the fixed coordinate system as the transversely and radially oscillating spheres move along the z-axis from $kz = -10$ to 10 at different Mach numbers. As a reference, we also plot the field pressure distribution of the stationary case ($M = 0$). The gap (zero pressure) in Figs. 9 and 10 coincide with the size of the sphere. As the Mach number increases, changes in the field pressure distributions become more and more obvious.

Next, we calculate the total radiated acoustic power from the transversely and radially oscillating spheres in motion. Figures 11 and 12 demonstrate the ratios of the radiated power with respect to stationary cases. Numerical results are compared with theoretical predictions [Morse and Ingard, 1986; Gaunaurd and Eisler, 1972] for a point monopole and point dipole, respectively. The agreements are good when the Mach number is low and ka value is small as expected. The effects of increasing ka on the total radiated acoustic power are seen in Figs. 11 and 12 as the present results deviate from those for point sources. The differences are primarily due to the noncompactness of the source. Figures 14 and 15 demonstrate the relationship between the total radiated acoustic power and the Mach number and the ka value. Specifically, for a given ka value the radiated power increases with the Mach number and for a given Mach number the radiated power increases with ka. In particular, the radiated power increases as the square of the Mach number at a low Mach number and a small ka value, whereas it increases in a complex manner at higher Mach numbers and larger ka values.

V Concluding Remarks

This paper shows the numerical results of acoustic fields and total radiated acoustic power from a moving noncompact vibrating body based on the extended Kirchhoff integral formulation recently developed by Wu and Akay [1989]. Since the higher-order poles are included in the derivations to account for the effect of local convection, the integral formulation are valid for large surface oscillations as well. It is shown that at larger surface oscillations and higher ka values, the effects of higher-order poles are prominent. Particularly, in the case of the transversely oscillating sphere, the higher-order poles are actually quadrupoles oscillating at twice the angular frequency ω. Hence for a constant angular frequency ω, the amplitude of

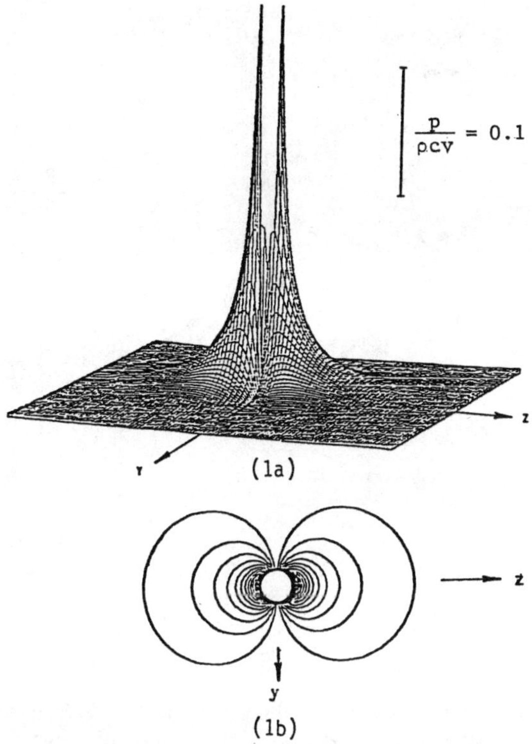

Figure 1. (a) Acoustic field pressure distributions from the transversely oscillating sphere (without the inclusion of higher-order poles) at $ka = 1$. (b) Topography of magnitudes of the dimensionless acoustic pressure.

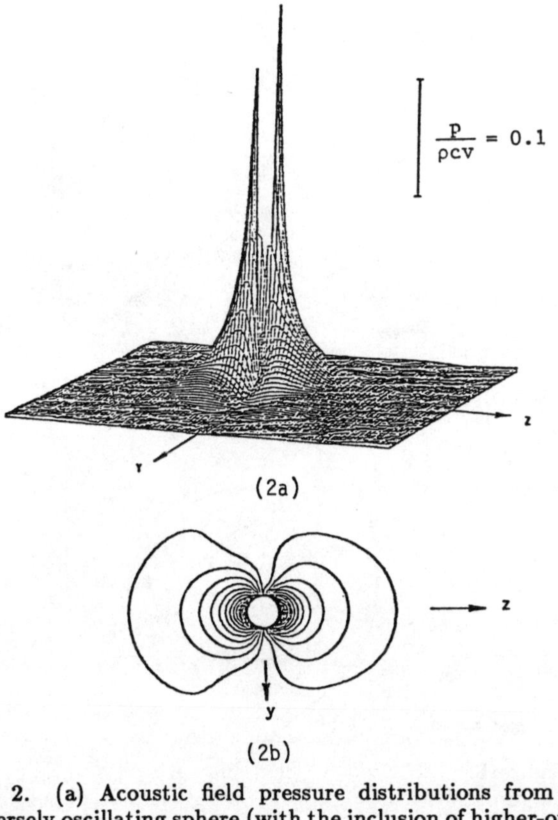

Figure 2. (a) Acoustic field pressure distributions from the transversely oscillating sphere (with the inclusion of higher-order poles and with $M_s = 0.1$) at $ka = 1$. (b) Topography of magnitudes of the dimensionless acoustic pressure.

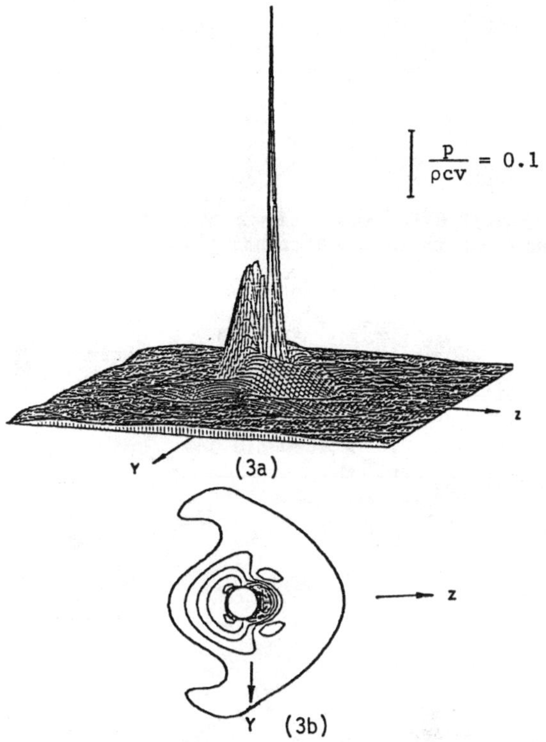

Figure 3. (a) Acoustic field pressure distributions from the transversely oscillating sphere (with the inclusion of higher-order poles and with $M_s = 1$) at $ka = 1$. (b) Topography of magnitudes of the dimensionless acoustic pressure.

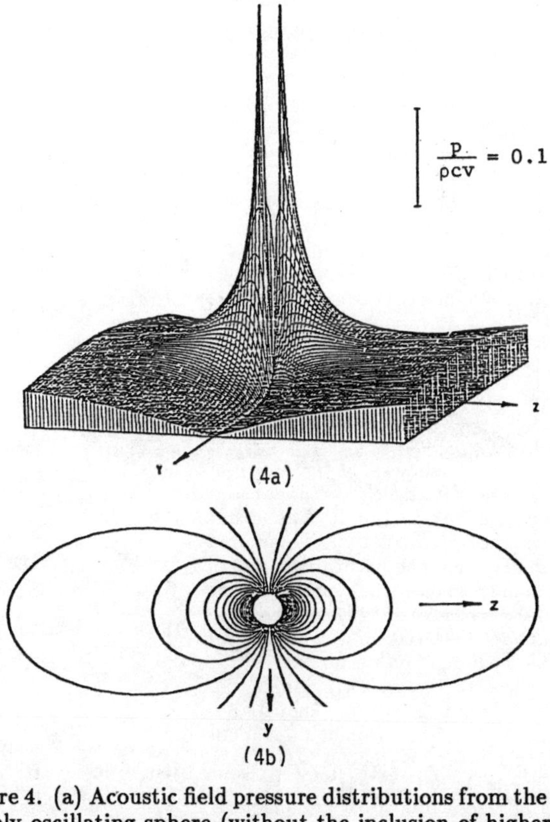

Figure 4. (a) Acoustic field pressure distributions from the transversely oscillating sphere (without the inclusion of higher-order poles) at $ka = 5$. (b) Topography of magnitudes of the dimensionless acoustic pressure.

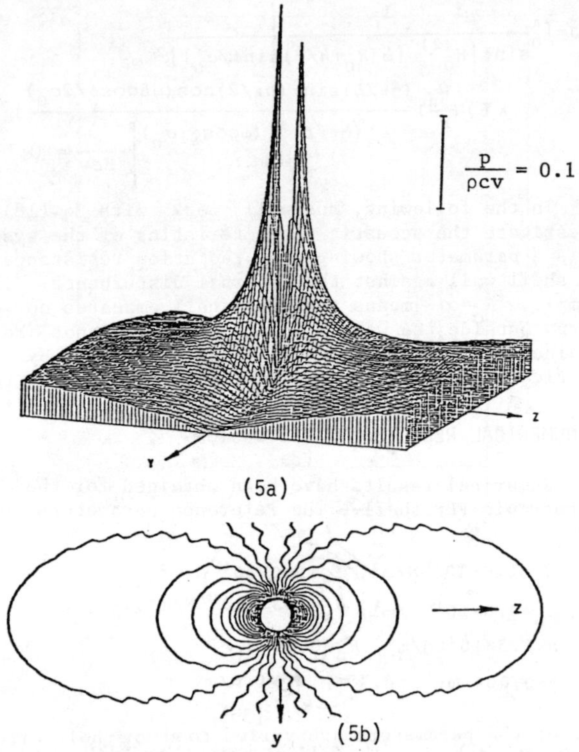

Figure 5. (a) Acoustic field pressure distributions from the transversely oscillating sphere (with the inclusion of higher-order poles and with $M_s = 0.1$) at $ka = 5$. (b) Topography of magnitudes of the dimensionless acoustic pressure.

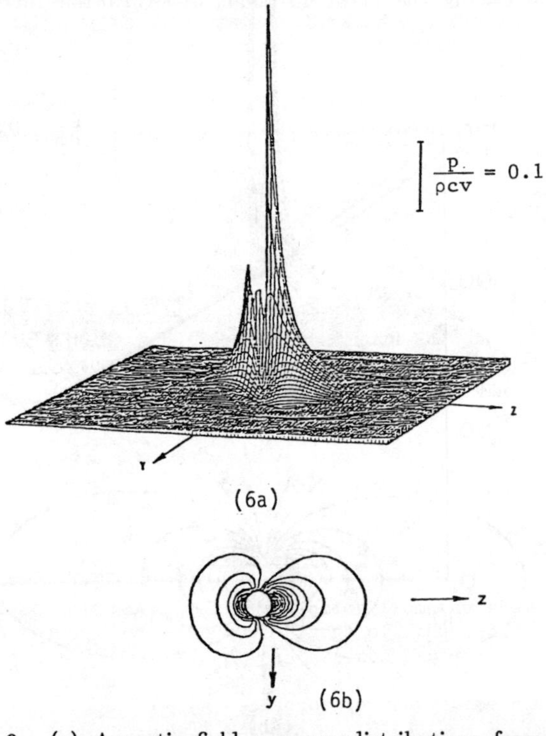

Figure 6. (a) Acoustic field pressure distributions from the transversely oscillating sphere (with the inclusion of higher-order poles and with $M_s = 1$) at $ka = 1$ moving in the positive z-axis direction at $M = 0.2$. (b) Topography of magnitudes of the dimensionless acoustic pressure.

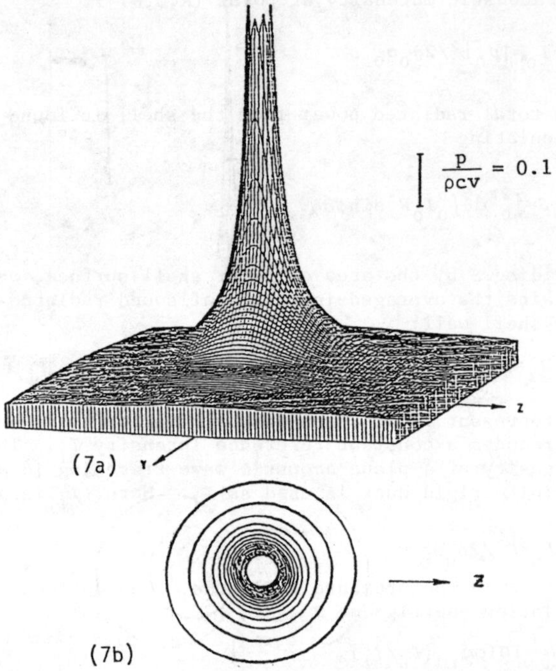

Figure 7. (a) Acoustic field pressure distributions from the radially oscillating sphere (with the inclusion of higher-order poles and with $M_s = 1$) at $ka = 1$. (b) Topography of magnitudes of the dimensionless acoustic pressure.

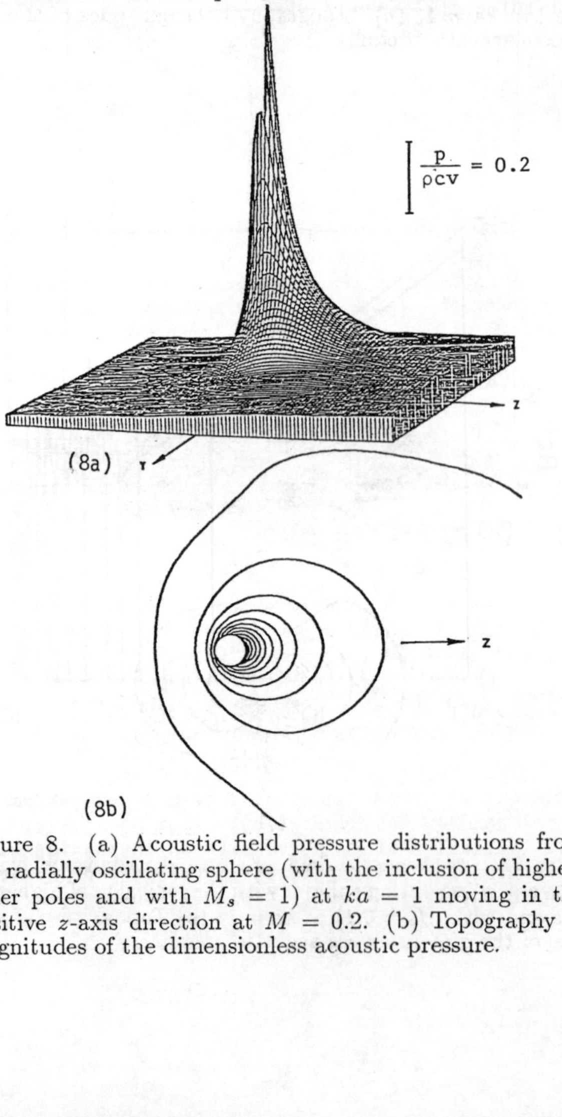

Figure 8. (a) Acoustic field pressure distributions from the radially oscillating sphere (with the inclusion of higher-order poles and with $M_s = 1$) at $ka = 1$ moving in the positive z-axis direction at $M = 0.2$. (b) Topography of magnitudes of the dimensionless acoustic pressure.

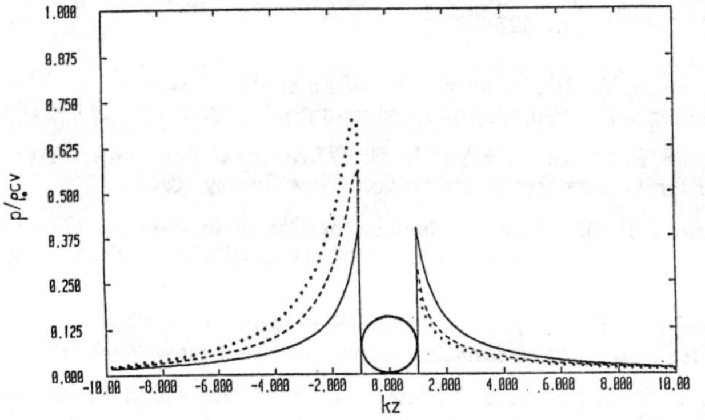

Figure 9. Acoustic field pressures measured at a fixed point as the transversely oscillating sphere moves along the z-axis from $kz = -10$ to 10. —— : $M = 0$; - - - - : $M = 0.2$; ······ : $M = 0.3$.

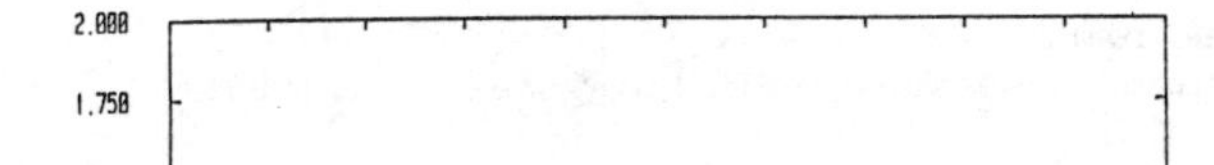

Figure 10. Acoustic field pressures measured at a fixed point as the radially oscillating sphere moves along the z-axis from $kz = -10$ to 10. —— : $M = 0$; - - - - : $M = 0.2$; ······ : $M = 0.3$.

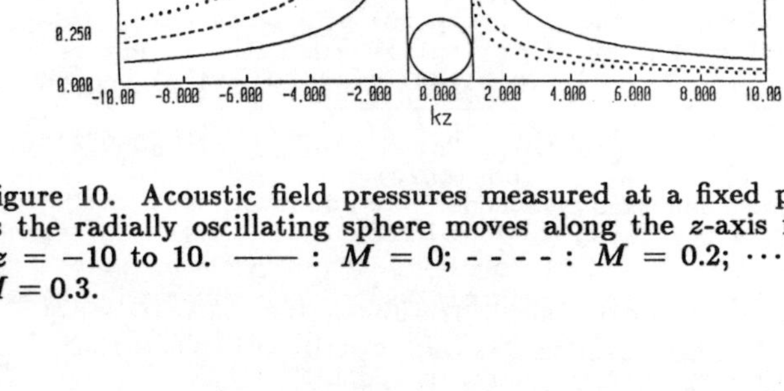

Figure 11. Ratios of total radiated acoustic power from the transversely oscillating sphere with respect to the stationary case. —— : Theoretical prediction; - - - - : Numerical results $ka = 0.3$; ······ : Numerical results $ka = 1.5$;

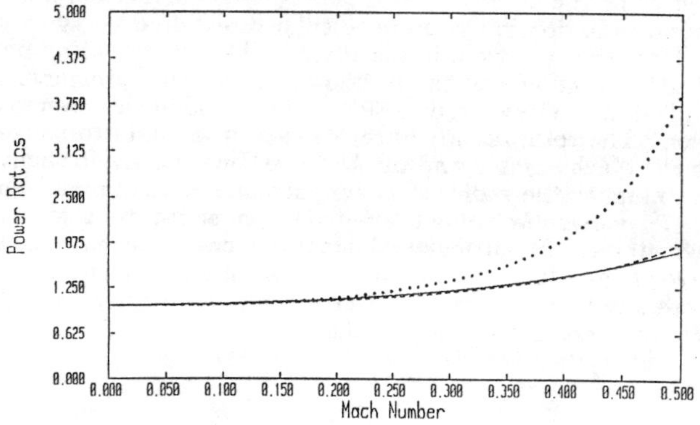

Figure 12. Ratios of total radiated acoustic power from the radially oscillating sphere with respect to the stationary case. —— : Theoretical prediction; - - - - : Numerical results $ka = 0.5$; ······ : Numerical results $ka = 2.5$;

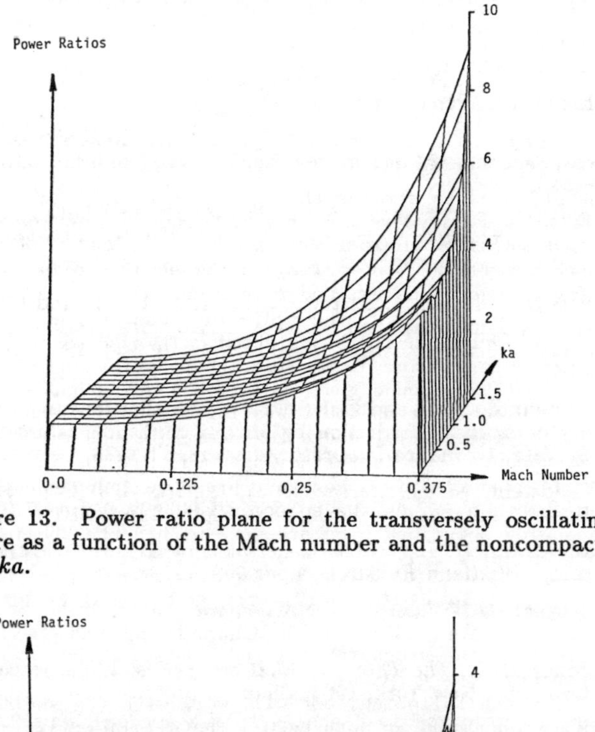

Figure 13. Power ratio plane for the transversely oscillating sphere as a function of the Mach number and the noncompactness ka.

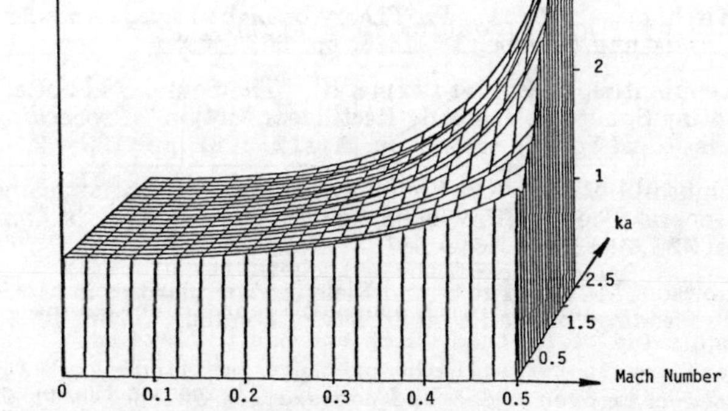

Figure 14. Power ratio plane for the radially oscillating sphere as a function of the Mach number and the noncompactness ka.

one of two pressure lobes of the dipole source is always reinforced while the other is always suppressed. Furthermore, there is sound radiation in the perpendicular direction even when the vibrating surface is fixed in the space. When the vibrating body is in motion, numerical results show that acoustic pressures are amplified in the forward direction and reduced in the reverse direction. The total radiated acoustic power is found to increase with the Mach number and the *ka* value. Specifically, for a given Mach number the radiated power increases with the *ka* value, while for a given *ka* value the radiated power increases with the Mach number. In particular, the radiated power increases as the square of the Mach number at a low Mach number and a small *ka* value, whereas it increases in a complex manner at higher Mach numbers and larger *ka* values.

Acknowledgements

This work was supported in part by the Dynamic Systems and Control Program of the National Science Foundation and the Institute for Manufacturing Research at Wayne State University.

References

Currie, I. G., *Fundamental Mechanics of Fluids*, McGraw-Hill, Inc. 1974, Chap. 3, pp. 50 – 52.

Dowling, A. P., "Convective Amplification of Real Simple Sources," *Journal of Fluid Mechanics*, Vol. **74**, No. 3, 1976, pp 529 – 546.

Farassat, F. and Myers, M. K., "Extension of Kirchhoff's Formula to Radiation from Moving Surfaces," *Journal of Sound and Vibration*, Vol. **123**, No. 3, 1988, pp. 451 – 460.

Ffowcs-Williams, J. E. and Hawkings, D. L., "Sound Generation by Turbulence and Surfaces in Arbitrary Motion," *Philosophical Transactions of Royal Society*, **A 264**, 1969, pp. 321 - 342.

Gaunaurd, G. C. and Eisler, T. J., "Contributions to the Theory of Acoustic Radiation by Moving Multipole Sources," *U. S. Naval Ordnance Laboratory Report*, TN9578, 1972.

Goldstein, M. E., *Aeroacoustics*, McGraw-Hill International Book Company, New York, 1976, Chap. 3, pp. 116.

Hawkings, D. L., "Noise Generation by Transonic Open Rotors," *Westland Research Paper* 599, 1979.

Jackson, J., *Classical Electrodynamics*, John Wiley & Sons, Inc., New York, 1975, 2nd Ed., Chap. 14, pp. 654 - 679.

Jones, D. S., *The Theory of Electromagnetism*, The MacMillan Company, New York, 1964, Chap. 1, pp. 35 - 42.

Khromov, V. A., "Generalization of Kirchhoff's Theorem for the Case of a Surface Moving in an Arbitrary Way," *Soviet Physics – Acoustics*, Vol. **9**, No. 1, 1963, pp. 68 – 71.

Kirchhoff, G., "Toward a Theory of Light Rays," *Annalen der Physik und Chemie*, **18**, 1883, pp. 663 – 695.

Leppington, F. G. and Levine, H., "The Sound Field of a Pulsating Sphere in Unsteady Rectilinear Motion," *Proceedings of the Royal Society of London*, **A 412**, 1987, pp. 199 – 221.

Lighthill, M. J., "On Sound Generated Aerodynamically. I. General Theory," *Proceedings of the Royal Society of London*, **A 221**, 1952, pp. 564 - 587.

Lowson, M. V., "The Sound Field for Singularities in Motion," *Proceedings of the Royal Society of London*, **A 286**, 1965, pp. 559 – 572.

Lyrintzis, A. S. and George, A. R., "Transonic Blade-Vortex Interactions: the Far Field," *American Helicopter Society National Specialists' Meeting on Aerodynamics and Aeroacoustics*, 25 – 27, February, Arlington, Texas, 1987.

Morfey, C. L. and Tanna, H. K., "Sound Radiation from a Point Force in Circular Motion," *Journal of Sound and Vibration*, **15**(3), 1971, pp. 325 – 351.

Morgans, W. R., "The Kirchhoff Formula Extended to a Moving Surface," *Philosophical Magazine*, **9**, 1930, pp. 141 – 161.

Morse, P. M. and Ingard, K. U., *Theoretical Acoustics*, Princeton University Press, Princeton, New Jersey, 1968.

Munro, D. H., *The Production of Sound by Moving Objects*, Ph. D. Dissertation, Massachusetts Institute of Technology, 1980.

Pierce, A. D., *Acoustics: An Introduction to Its Physical Principles and Applications*, McGraw-Hill, Inc. New York, 1981.

Selby, S. M., *Standard Mathematical Tables* CRR Press, Inc. 1973, 23rd Ed., pp. 557 – 558.

Stratton, J. A., *Electromagnetic Theory*, McGraw-Hill, New York, 1941, pp. 424 – 428.

Tanna, H. K. and Morfey, C. L., "Sound Radiation from Point Sources in Circular Motion," *Journal of Sound and Vibration*, **16**(3), 1971, pp. 337 – 348.

Wu, X.- F. and Akay, A.,"Extended Kirchhoff Integral Theorem for Sound Radiation from Vibrating Surfaces in Motion," (Submitted to *The Journal of the Acoustical Society of America*), 1989.

TRANSIENT FINITE ELEMENT ANALYSIS OF THE EXTERIOR STRUCTURAL ACOUSTICS PROBLEM

P. M. Pinsky and N. N. Aboud
Department of Civil Engineering
Stanford University
Stanford, California

ABSTRACT

Considerable progress has been made in the development of numerical methods for the time-harmonic exterior structural acoustics problem involving solution of the coupled Helmholtz equation. In contrast, numerical solution procedures for the transient case have not been studied so extensively. In this paper a finite element formulation is proposed for solution of the time-dependent coupled wave equation over an infinite fluid domain. The formulation is based on a finite computational fluid domain surrounding the structure and incorporates a sequence of boundary operators on the fluid truncation boundary. These operators are designed to minimize reflection of outgoing waves and are based on an asymptotic expansion of the exact solution for the time-dependent problem. In the fluid domain, a mixed two-field finite element approximation, based on a specialization of the Hu-Washizu principle for elasticity, is proposed and employs pressure and displacement potential as independent fields. Since radiation dissipation renders the coupled system nonconservative, a variational formalism based on the Morse and Feshbach concept of a "mirror-image" adjoint system is used. The variational formalism also accommodates viscoelastic dissipation in the structure (or its coatings) and this is considered in the paper. Very accurate results for model problems involving a single layer of fluid elements have been obtained and are discussed in detail.

1. INTRODUCTION

Numerical procedures for the exterior structural acoustics problem involving the interaction of structures submerged in infinite fluid domains may be classified according to the nature of the time dependence as either time-harmonic or transient. In the time-harmonic case, the coupled Helmholtz equation is solved subject to the Sommerfeld boundary condition. Boundary integral solutions have been successfully employed in the development of numerical techniques, such as the boundary element method, and there is now a very large literature in this area. It is perhaps worth noting that despite the seeming success of such techniques, they are usually severly limited in their applicability to high wave number problems and further research on techniques for overcoming these limitations would seem to be important, since those already developed are either problem dependent, or computationally intensive. Other techniques have also been devised for the time-harmonic problem, and a broad review of these can be found in [1].

In the transient case, one approach has been to solve the boundary integral equation in either the Laplace [2] or the Fourier domain [3] and then use the inverse transform to obtain the time domain solutions. This approach seems expensive in that it requires a large number of boundary solutions in the transformed domain for any reasonably accurate sweep of a realistic bandwidth. Furthermore, this indirect approach is limited in its applicability to linear problems and usually inherits the difficulties of the harmonic boundary integral method (e.g. nonuniqueness of Fourier domain solutions). Another approach is based on the direct time integration of Kirchhoff's retarded potential equation [4,5,6]. One possible drawback to this memory-intensive approach is the expense entailed in storing a large pool of historical data, and that might explain why no actual computational results for three-dimensional fluid interaction with non-rigid structures could be found documented in the literature. Another method that attempts to represent the solution in the exterior exactly has recently been proposed by Givoli [7] under the name of "Dirichlet-to-Neumann" (DtN) method. As in the coupled boundary and finite element method, the DtN method divides the fluid into an interior finite element region and an exterior domain represented by an "exact" boundary condition over the artificial boundary (assumed spherical or elliptical). The DtN method mitigates some of the difficulties associated with the classical boundary integral method (such as asymmetry and kernel singularity), but unfortunately the resulting scheme is non-local in space and time, with the current solution being obtained by convolution over time of previous solutions. This presently appears to be a practical impediment since its use would require modifications on standard integration schemes as well as a very substantial storage effort.

Because of the objections raised above, there is a continued interest in the development of approximate radiation boundary conditions that are computationally less cumbersome. Of these approximate boundary conditions, very few are applicable to both time-harmonic and direct transient analysis. Similarly, many are restricted to special geometries and may have other mathematical restrictions. The simplest approximate radiation boundary conditions are the plane-wave and the curved-wave dampers. The performance of these is known to deteriorate as the position of the artificial boundary approaches the source of perturbation, especially in the low-frequency range, where the virtual mass (or added mass) approximation is more appropriate. The doubly asymptotic approximation (DAA) of Geers [8] on the other hand, is asymptotically exact in the high-frequency and low-frequency limits and is local in time, though not in space. A family of DAAs has been derived by the method of matched asymptotic expansions in [9,10], though actual implementation seems to be still limited to the first member (DAA_1) and simplified forms of the second member (DAA_2^m and DAA_2^c) of that family only [11]. Under the category of radiation dampers,† Bayliss and Turkel[12], and Bayliss, Gunzberger, and Turkel[13] have introduced a sequence of boundary operators which are based on an asymptotic expansion in powers of $(1/r)$ of the exact time-dependent solution. The operators in this sequence are of progressively higher order (and are progressively more difficult to implement) but provide increasing accuracy. In this paper, a direct approach to the transient exterior structural acoustics problem is considered. Finite elements are used in the truncated fluid domain and the above noted boundary operators, recast in a form appropriate for finite element application, are imposed on the artificial boundary. Our numerical results obtained using these boundary operators are encouraging. For one "difficult" problem of high wave number ($ka = 4$), very accurate results have been obtained by use of a single layer of fluid elements surrounding the structure. Boundary element methods based on Helmholtz integral solutions would be expected to perform very poorly in this wave number regime.

In the fluid domain, a mixed two-field finite element approximation employing both the pressure and displacement potential as independent unknowns is used. The variational statement of the governing time-dependent fluid equations is developed from a specialization of the well known three-field Hu-Washizu variational principle for elasticity[14]. Unlike classical elasticity, the exterior structural acoustics problem is rendered nonconservative by the radiation dissipation on the fluid truncation boundary. In order to properly account for this dissipation and retain the inherent "symmetry" of a variational principle, use is made of the concept of an adjoint system and associated variational formalism introduced by Morse and Feshbach [15]. The use of this formalism also allows the consideration of viscoelastic dissipation in the structure or in its coatings and a discussion is included in the paper.

The boundary operators proposed for use in the finite element formulation involve high order temporal and spatial derivatives, the exact order depending on the particular operator selected from the general sequence. In principle, higher order operators are difficult to implement since they involve representing high order spatial derivatives in the finite element approximation as well as the appearance of high order time derivatives in the semi-discrete (time-continuous) equations. The latter condition possibly requiring special temporal integration schemes. On the other hand, the use of the higher order operators will, in principle, allow the fluid truncation boundary to be close to the structure. Some of the implementational issues are explored in this paper and an attempt is made to draw some preliminary conclusions about use of the boundary operator sequence. The results of the numerical tests, given in Section 6, appear very encouraging.

† The term damper is used to indicate a radiation boundary condition that is local in both space and time, i.e. the boundary condition is a differential equation rather than an integro-differential one.

2. VARIATIONAL FORMULATION FOR THE STRUCTURE-FLUID SYSTEM

2.1- Introduction

In this paper, a finite element formulation is proposed for the time-dependent exterior structural acoustics problem involving a viscoelastic structure submerged in an infinite acoustic fluid. As noted in the Introduction, the dissipative sources in this system are the radiation boundary and viscoelastic material behavior of the structure. In practice, this system might describe an elastic shell structure with viscoelastic compliant coatings interacting with an infinite exterior acoustic fluid.

A variational principle, which provides a basis for time-dependent finite element approximation of the structure and fluid, is developed as the main result of the present section. The principle utilizes a special formalism for dissipative systems as described below. This paper does not discuss in detail any specific finite element structural models, such as the multi-layer elastic shell-viscoelastic coatings model of Pinsky and Kim [16,17], although the formulation described is sufficiently general for their direct inclusion.

2.2- Viscoelastic Structure Lagrangian

For present purposes and without loss of generality, it suffices to introduce viscoelastic constitutive equations for both the deviatoric and dilatational response of the structure based on a single generalized Kelvin element as follows:

$$\boldsymbol{\sigma} = \boldsymbol{J}_1 : \boldsymbol{\varepsilon} + \boldsymbol{J}_2 : \dot{\boldsymbol{\varepsilon}} \tag{2.1.a}$$

$$\boldsymbol{J}_1 = \frac{1}{3}(3\kappa - 2\mu)\mathbf{1} \otimes \mathbf{1} + 2\mu\, \mathbf{I\!I} \tag{2.1.b}$$

$$\boldsymbol{J}_2 = \frac{1}{3}(3\vartheta - 2\eta)\mathbf{1} \otimes \mathbf{1} + 2\eta\, \mathbf{I\!I} \tag{2.1.c}$$

where, $\boldsymbol{\sigma}$ and $\boldsymbol{\varepsilon}$ are the stress and strain tensors, respectively, $\mathbf{1}$ and $\mathbf{I\!I}$ are the second-order and fourth-order unit tensors, respectively, μ and κ are the elastic shear and bulk moduli respectively, and η and ϑ are the Kelvin element deviatoric and dilatational viscosity coefficients, respectively. Although certain applications may require more sophisticated models, Equation (2.1) introduces the essential feature of viscoelastic dissipation and also gives rise to constant stiffness and damping matrices in the finite element representation of the structure. It may be noted that the viscoelastic model (2.1) reverts to linear elasticity in the limiting case of zero viscosity coefficients.

A special variational formalism for dissipative systems has been introduced by Morse and Feshbach [15], and has since been used by Gladwell [18] and Craggs [19,20] for damped acousto-structural problems. The procedure involves the use of a "mirror-image" adjoint system in which energy accrues as it is being dissipated in the physical system. The extended system is thus conservative and the equations of motion can be written in Hamilton's canonical form. The first variation of the resulting functional with respect to the adjoint variables (denoted by an asterisk in this paper) gives rise to the Euler-Lagrange equations governing the physical problem, while the variation with respect to the original variables yields those governing the adjoint problem.

Using the Morse and Feshbach formalism, a Lagrangian for a viscoelastic structure with constitutive equation (2.1) is given by $\mathcal{L}_s = \mathcal{T}_s - \mathcal{V}_s$, in which $\mathcal{T}_s$ is the kinetic energy and $\mathcal{V}_s$ is the potential energy, having the form:

$$\mathcal{T}_s = \frac{1}{2}\int_{\Omega_s} \rho_s (\dot{\boldsymbol{u}} \cdot \dot{\boldsymbol{u}}^*) \, d\Omega \qquad (2.2.a)$$

$$\begin{aligned} \mathcal{V}_s = & \frac{1}{2}\int_{\Omega_s} (\nabla \boldsymbol{u} : \boldsymbol{J}_1 : \nabla \boldsymbol{u}^*) \, d\Omega \\ & + \frac{1}{4}\int_{\Omega_s} (\nabla \boldsymbol{u}^* : \boldsymbol{J}_2 : \nabla \dot{\boldsymbol{u}} - \nabla \boldsymbol{u} : \boldsymbol{J}_2 : \nabla \dot{\boldsymbol{u}}^*) \, d\Omega \\ & - \frac{1}{2}\int_{\Gamma_\sigma} (\bar{\boldsymbol{t}} \cdot \boldsymbol{u}^* + \bar{\boldsymbol{t}}^* \cdot \boldsymbol{u}) \, d\Gamma \end{aligned} \qquad (2.2.b)$$

where $\boldsymbol{u}$ is the structure displacement vector, $\bar{\boldsymbol{t}}$ is the external traction vector applied on Γ_σ, ρ_s is the structure mass density, and Ω_s is the structure domain with partial boundary Γ_σ on which the traction vector $\bar{\boldsymbol{t}}$ is prescribed. The Euler-Lagrange equations found by the variation of $\int_{t_0}^{t_1} \mathcal{L}_s dt$ with respect to $\boldsymbol{u}^*$ correspond to the equation of motion and traction boundary condition for the physical structure *in vacuo* and are expressed in terms of the physical displacement $\boldsymbol{u}$. A loading term arising from the surrounding fluid will appear naturally from the coupling procedure described below.

2.3- Two-Field Fluid Lagrangian

We now turn our attention to the acoustic fluid. Several variational formulations for the governing equations of a compressible, irrotational fluid have been proposed as a basis for finite element approximation. These include single field and mixed (multiple) field representations as well as the use of scalar and vectorial fields. See Pinsky and Abboud [21] for a general discussion. In this paper, the use of a mixed two-field scalar-scalar formulation is advocated, following [21,22]. Related two-field variational formulations for the interior structural acoustics problem involving structures containing a finite interior fluid domain have been given by Morand and Ohayon [23,24] and Ohayon [25] for the time-harmonic case. Geradin, Robert, and Huck [26], Olson and Bathe [27], and Liu and Uras [28] have also proposed related formulations for the transient case.

In summary, it may be stated that the principal advantages of the mixed two-field scalar-scalar variational formulation for the fluid is the reduction in the number of unknowns after discretization when compared to a fully or partially vectorial approach, the *a priori* satisfaction of the irrotationality condition, and the resulting symmetry of the coupled system of equations, which is not found when a single scalar field is employed [21].

Under the classical acoustic approximation, momentum balance, continuity and the equation of state are given, respectively, by:

$$p + \rho_o \ddot{\phi} = 0 \qquad \text{in } \Omega_f \qquad (2.3)$$

$$\rho + \rho_o \nabla^2 \phi = 0 \qquad \text{in } \Omega_f \qquad (2.4)$$

$$p - c^2 \rho = 0 \qquad \text{in } \Omega_f \qquad (2.5)$$

where ϕ, p, and ρ are small perturbations in the displacement potential, pressure, and fluid density from the reference state of zero displacement potential, hydrostatic pressure $p_0 \gg p$, and constant density $\rho_o \gg \rho$. The velocity of acoustic wave propagation c is assumed constant. It is noted that elimination of the density ρ in (2.3) – (2.5) results in:

$$p + \rho_o \ddot{\phi} = 0 \qquad \text{in } \Omega_f \qquad (2.6)$$

$$p + \rho_o c^2 \nabla^2 \phi = 0 \qquad \text{in } \Omega_f \qquad (2.7)$$

while elimination of both ρ and p, or ρ and ϕ yields the wave equation in the remaining variable:

$$\nabla^2 p - (1/c^2)\ddot{p} = 0 \qquad \text{in } \Omega_f \qquad (2.8)$$

$$\nabla^2 \phi - (1/c^2)\ddot{\phi} = 0 \qquad \text{in } \Omega_f \qquad (2.9)$$

For the time being it is convenient to ignore the fluid boundary conditions which will be incorporated in Sections 3 and 4.

A mixed three-field variational principle for an acoustic fluid with Euler-Lagrange equations (2.3) – (2.5) may be found by analogy with the Hu-Washizu principle for elasticity [14] and is given by:

$$\mathcal{J}_f = \int_{t_0}^{t_1} \mathcal{L}_f \, dt \qquad (2.10.a)$$

$$\mathcal{L}_f = \left[\frac{1}{2}\rho_o \int_{\Omega_f} \nabla\dot{\phi} \cdot \nabla\dot{\phi} \, d\Omega\right] - \left[\frac{1}{2}\int_{\Omega_f} \left(\frac{\rho}{\rho_o}\right)(\rho_o c^2)\left(\frac{\rho}{\rho_o}\right) d\Omega - \int_{\Omega_f} p\left(\frac{\rho}{\rho_o} + \nabla^2\phi\right) d\Omega\right] \qquad (2.10.b)$$

where ϕ, p and ρ are *independent* fields with variations that vanish at times t_0 and t_1.

Enforcement of the equation of state (2.5) in (2.10.b) eliminates the density ρ and yields a two-field Lagrangian $\mathcal{L}_f = \mathcal{T}_f - \mathcal{V}_f$, where the kinetic energy $\mathcal{T}_f$ and the potential energy $\mathcal{V}_f$ are given by:

$$\mathcal{T}_f = \frac{1}{2}\rho_o \int_{\Omega_f} \nabla\dot{\phi} \cdot \nabla\dot{\phi} \, d\Omega \qquad (2.11.a)$$

$$\mathcal{V}_f = -\frac{1}{2\rho_o c^2}\int_{\Omega_f} p^2 \, d\Omega - \int_{\Omega_f} p \nabla^2 \phi \, d\Omega \qquad (2.11.b)$$

As noted in Pinsky and Abboud [21] and Abboud and Pinsky[22], it is possible to obtain two forms of the fluid Lagrangian $\mathcal{L}_f$ useful for computation. Here, only one of these forms is considered. Performing a Legendre transformation on$\mathcal{T}_f$ in (2.11.a) and using the divergence theorem gives:

$$\mathcal{T}_f = -\frac{1}{2}\rho_o \int_{\Omega_f} \nabla\dot{\phi} \cdot \nabla\dot{\phi} \, d\Omega - \rho_o \int_{\Omega_f} \dot{\phi}\nabla^2\dot{\phi} \, d\Omega + \rho_o \int_{\Gamma_f} \dot{\phi}\nabla\dot{\phi} \cdot \nu \, d\Gamma$$

where ν is the unit outward normal vector to the fluid domain. Using the continuity equation (2.7) to eliminate $\nabla^2\phi$ in $\mathcal{V}_f$ and $\nabla^2\dot{\phi}$ in $\mathcal{T}_f$, we obtain $\mathcal{L}_f = \mathcal{T}_f - \mathcal{V}_f$, where

$$\mathcal{T}_f = -\frac{1}{2}\rho_o \int_{\Omega_f} \nabla\dot{\phi} \cdot \nabla\dot{\phi} \, d\Omega + \frac{1}{c^2}\int_{\Omega_f} p\dot{\phi} \, d\Omega + \rho_o \int_{\Gamma_f} \dot{\phi}\nabla\dot{\phi} \cdot \nu \, d\Gamma \qquad (2.12.a)$$

$$\mathcal{V}_f = \frac{1}{2\rho_o c^2}\int_{\Omega_f} p^2 \, d\Omega \qquad (2.12.b)$$

Anticipating the introduction of dissipative effects, the Morse and Feshbach formalism for nonconservative systems is used to rewrite the fluid Lagrangian as:

$$\mathcal{L}_f = -\frac{1}{2}\rho_o \int_{\Omega_f} \left(\nabla\dot{\phi}\cdot\nabla\dot{\phi}^*\right) d\Omega + \frac{1}{2c^2}\int_{\Omega_f}\left(p\dot{\phi}^* + p^*\dot{\phi}\right) d\Omega + \frac{1}{2}\rho_o\int_{\Gamma_f}\left(\dot{\phi}\nabla\dot{\phi}^* + \dot{\phi}^*\nabla\dot{\phi}\right)\cdot\nu\, d\Gamma - \frac{1}{2\rho_o c^2}\int_{\Omega_f}(pp^*)\, d\Omega \tag{2.13}$$

It may be verified that the first variation of $\mathcal{J}_f = \int_{t_0}^{t_1}\mathcal{L}_f dt$ with respect to ϕ^* and p^*, subject to the condition that the variations $\delta\phi^*$ and δp^* vanish at times t_0 and t_1, gives the momentum balance and and continuity equations in the fluid domain Ω_f. The next section develops a sequence of radiation boundary operators that are then incorporated into the fluid Lagrangian. Section 4 will address the structure-fluid interface conditions.

3. RADIATION BOUNDARY CONDITIONS

3.1- Introduction

Although it is technically feasible to solve the time-dependent exterior structural acoustics problem by use of boundary integral techniques, such an approach is considered to be prohibitively expensive for general three-dimensional fluid-structure interaction problems. Here, a direct approach to the transient problem is considered. Finite element and finite difference methods must be based on a bounded computational domain and involve the introduction of a truncation or artificial boundary inclosing a finite computational fluid domain. Special boundary conditions must be used on the truncation boundary and should produce no reflections of outgoing waves. A number of possible boundary conditions for the transient case have been discussed in the literature and have been surveyed by Zienkiewicz *et. al.* [1].

To date, the most accurate boundary conditions applicable to transient finite element analysis are based on either matching analytical solutions or on the use of radiation dampers. Recently, Givoli [7] has proposed an "exact" boundary condition for the transient problem, but unfortunately the resulting scheme is nonlocal in space and time, with the current solution being obtained by convolution over time of previous solutions. This presently appears to be a practical impediment since its use would require modifications on standard integration schemes as well as a very substantial storage effort. Under the category of radiation dampers, Bayliss and Turkel[12], and Bayliss, Gunzberger, and Turkel[13] have introduced a sequence of boundary operators which are based on an asymptotic expansion of the exact time-dependent solution. The operators in this sequence are of progressively higher order and provide increasing accuracy. With the exception of the doubly asymptotic approximation [8-11] and the Smith algorithm [29], the above noted boundary treatments are the only ones known to us for the transient problem.

3.2- Boundary Conditions for the Physical Problem

For the variational formalism expressed by (2.13), it is necessary to obtain the radiation boundary conditions for the physical problem (waves outgoing) and the adjoint problem (waves incoming). We start with the physical problem. The wave equation and Sommerfeld condition requiring waves to be outgoing are:

$$\nabla^2\phi - \frac{1}{c^2}\ddot{\phi} = 0 \tag{3.1}$$

$$\lim_{r\to\infty} r\left(\frac{\partial\phi}{\partial r} + \frac{1}{c}\dot{\phi}\right) = 0 \tag{3.2}$$

An asymptotic expansion of the solution [12,13,30], based on a separation of variables, can be developed in spherical coordinates as:

$$\phi(t,r,\theta,\varphi) = \sum_{i=1}^{\infty}\frac{f_i(r-ct,\theta,\varphi)}{r^i} \tag{3.3}$$

where r, θ and φ are the radius, circumferential angle and polar angle, respectively. Following Bayliss and Turkel[12], a sequence of operators is introduced and defined by:

$$B_m = \prod_{l=1}^{m}\left(\frac{\partial}{\partial r} + \frac{1}{c}\frac{\partial}{\partial t} + \frac{2l-1}{r}\right) \tag{3.4}$$

Observe that B_m involves radial and temporal partial derivatives up to order m. When B_m is applied to the asymptotic solution given by (3.3), it is observed that[12]:

$$B_m(\phi) = O(1/r^{2m+1}) \tag{3.5}$$

which indicates that B_m annihilates the first m terms of the expansion (3.3). Thus boundary conditions on Γ_∞ can be stated to any desired degree of accuracy in $(1/r)$ as:

$$B_m(\phi) = 0 \tag{3.6}$$

As noted by Bayliss and Turkel[12], this sequence provides increasing accuracy with m which, however, is also a measure of the difficulty of implementation.

3.3- Boundary Conditions for the Adjoint Problem

The corresponding sequence of adjoint boundary operators is developed for the wave equation with the boundary condition requiring waves to be convergent (*i.e.* incoming):

$$\nabla^2\phi^* - \frac{1}{c^2}\ddot{\phi}^* = 0 \tag{3.7}$$

$$\lim_{r\to\infty} r\left(\frac{\partial\phi^*}{\partial r} - \frac{1}{c}\dot{\phi}^*\right) = 0 \tag{3.8}$$

an asymptotic expansion of the solution, based on a separation of variables, can be developed and it takes the form:

$$\phi^*(t,r,\theta,\varphi) = \sum_{i=1}^{\infty}\frac{f_i(r+ct,\theta,\varphi)}{r^i} \tag{3.9}$$

Then the sequence of adjoint boundary operators is defined as:

$$B_m^* = \prod_{l=1}^{m}\left(\frac{\partial}{\partial r} - \frac{1}{c}\frac{\partial}{\partial t} + \frac{2l-1}{r}\right) \tag{3.10}$$

In this case, it can also be shown that

$$B_m^*(\phi^*) = O(1/r^{2m+1}) \tag{3.11}$$

As for the physical problem, B_m^* annihilates the first m terms in the above expansion and a family of adjoint boundary conditions is proposed as:

$$B_m^*(\phi^*) = 0 \tag{3.12}$$

3.4- Finite Element Boundary Conditions

The boundary conditions (3.6) and (3.12) have to be built into the finite element formulation and this is considered in Section 4 where they are treated essentially as Neumann conditions. In order to accomplish this, it is convenient to rephrase the radiation boundary conditions. Consider first the physical boundary condition (3.6) which is rewritten as:

$$B_m(\phi) = \frac{m!}{r^{m-1}}\left(\frac{\partial \phi}{\partial r} - S_m(\phi)\right) = 0 \qquad \text{on } \Gamma_\infty \tag{3.13}$$

Letting $\boldsymbol{n}$ denote the outward unit normal to the truncation boundary Γ_∞, which is now restricted to be a spherical surface enclosing the structure and the finite element fluid domain, we now write

$$\nabla\phi \cdot \boldsymbol{n} = S_m(\phi) \qquad \text{on } \Gamma_\infty \tag{3.14.a}$$

Similarly, for the adjoint boundary condition (3.12) we have

$$\nabla\phi^* \cdot \boldsymbol{n} = S_m^*(\phi^*) \qquad \text{on } \Gamma_\infty \tag{3.14.b}$$

For m equal to 1 and 2, the operators S_m and S_m^* are identified as:

$$S_1(\phi) = -\frac{1}{r}\phi - \frac{1}{c}\dot{\phi} \tag{3.15.a}$$

$$S_1^*(\phi^*) = -\frac{1}{r}\phi^* + \frac{1}{c}\dot{\phi}^* \tag{3.15.b}$$

$$S_2(\phi) = -\frac{r}{c}\frac{\partial\dot{\phi}}{\partial r} - \frac{1}{2r}(2-D)\phi - \frac{2}{c}\dot{\phi} - \frac{r}{c^2}\ddot{\phi} \tag{3.16.a}$$

$$S_2^*(\phi^*) = +\frac{r}{c}\frac{\partial\dot{\phi}^*}{\partial r} - \frac{1}{2r}(2-D)\phi^* + \frac{2}{c}\dot{\phi}^* - \frac{r}{c^2}\ddot{\phi}^* \tag{3.16.b}$$

where D, the Beltrami operator, is given by

$$D = \frac{1}{\sin\varphi}\frac{\partial}{\partial\varphi}\left(\sin\varphi\frac{\partial}{\partial\varphi}\right) + \frac{1}{\sin^2\varphi}\frac{\partial^2}{\partial\theta^2} \tag{3.17}$$

Note that the B_m and B_m^* operators will always involve m^{th} order temporal and radial partial derivatives. The high order radial partial derivatives may not be accurately represented by finite element discretization and will be a potential source of error. However, the mth radial partial derivative can always be expressed in terms of mth order and lower temporal and tangential spatial partial derivatives and first order radial derivatives by recursive use of the wave equation in spherical coordinates:

$$\frac{\partial^2\phi}{\partial r^2} + \frac{2}{r}\frac{\partial\phi}{\partial r} + \frac{1}{r^2}D\phi - \frac{1}{c^2}\ddot{\phi} = 0 \tag{3.18}$$

This approach has been used in obtaining S_2 above (and S_m in (3.13)) which contains second order temporal and tangential spatial derivatives (in D), but no radial partial derivatives of order greater than one. It will be shown later that in the finite element formulation the tangential derivatives can also be reduced to first order.

4. STRUCTURE-FLUID COUPLED SYSTEM

The Lagrangian for the coupled problem is obtained by adding the Lagrangians for the solid and the fluid domains, and imposing the appropriate constraints at the structure-fluid interface Γ_i and at the truncation boundary Γ_∞. The resulting Lagrangian is the main result of this section and is used in Section 5 to obtain the semi-discrete finite element equations for the coupled system.

The structure-fluid interface Γ_i requires a kinematic boundary condition that provides for compatibility of the normal displacement at the interface and a dynamic boundary condition that enforces the equilibrium of the hydrodynamic forces:

$$\boldsymbol{u}\cdot\boldsymbol{n} = \nabla\phi\cdot\boldsymbol{n} \quad ; \quad \boldsymbol{u}^*\cdot\boldsymbol{n} = \nabla\phi^*\cdot\boldsymbol{n} \tag{4.1}$$

$$\boldsymbol{\sigma}\cdot\boldsymbol{n} = -p\boldsymbol{n} = \rho_o\dot{\phi}\boldsymbol{n} \quad ; \quad \boldsymbol{\sigma}^*\cdot\boldsymbol{n} = -p^*\boldsymbol{n} = \rho_o\dot{\phi}^*\boldsymbol{n} \tag{4.2}$$

where $\boldsymbol{n} = -\nu$ is the unit normal vector on Γ_i pointing into the fluid domain Ω_f. The boundary conditions on the truncation boundary Γ_∞ are given by (3.14).

The Lagrangian for the total system is found by adding the structure and fluid Lagrangians (2.2) and (2.13) and imposing the boundary conditions (3.14) and (4.1) by the use of Lagrange multipliers. The Lagrange multipliers may then be identified and the resulting Lagrangian is given by:

$$\begin{aligned}
\mathcal{L} =& \frac{1}{2}\int_{\Omega_s} \rho_s(\dot{\boldsymbol{u}}\cdot\dot{\boldsymbol{u}}^*)\, d\Omega \\
&-\frac{1}{2}\int_{\Omega_s} (\nabla\boldsymbol{u} : \boldsymbol{J}_1 : \nabla\boldsymbol{u}^*)\, d\Omega \\
&-\frac{1}{4}\int_{\Omega_s} (\nabla\boldsymbol{u}^* : \boldsymbol{J}_2 : \nabla\dot{\boldsymbol{u}} - \nabla\boldsymbol{u} : \boldsymbol{J}_2 : \nabla\dot{\boldsymbol{u}}^*)\, d\Omega \\
&+\frac{1}{2}\int_{\Gamma_\sigma} (\bar{\boldsymbol{t}}\cdot\boldsymbol{u}^* + \bar{\boldsymbol{t}}^*\cdot\boldsymbol{u})\, d\Gamma \\
&-\frac{1}{2}\int_{\Omega_f} \rho_o\left(\nabla\dot{\phi}\cdot\nabla\dot{\phi}^*\right) d\Omega \\
&+\frac{1}{2}\int_{\Omega_f} \frac{1}{c^2}\left(p\dot{\phi}^* + p^*\dot{\phi}\right) d\Omega \\
&-\frac{1}{2}\int_{\Omega_f} \frac{1}{\rho_o c^2}(pp^*)\, d\Omega \\
&-\frac{1}{2}\int_{\Gamma_i} \rho_o\left(\dot{\phi}\dot{\boldsymbol{u}}^* + \dot{\phi}^*\dot{\boldsymbol{u}}\right)\cdot\boldsymbol{n}\, d\Gamma \\
&+\frac{1}{4}\int_{\Gamma_\infty} \rho_o\left(\dot{\phi}S_m^*(\dot{\phi}^*) + \dot{\phi}^*S_m(\dot{\phi})\right) d\Gamma
\end{aligned} \tag{4.3}$$

This Lagrangian is expressed in terms of the three independent fields $\boldsymbol{u}$, p and ϕ as well as their adjoint counterparts. The Euler-Lagrange equations of the functional $\mathcal{J} = \int_{t_0}^{t_1}\mathcal{L}dt$, found from the conditions of stationarity, are the governing equations of the coupled system, as follows:

$$\begin{array}{lllr}
\nabla\cdot\boldsymbol{\sigma} - \rho_s\ddot{\boldsymbol{u}} = 0; & \nabla\cdot\boldsymbol{\sigma}^* - \rho_s\ddot{\boldsymbol{u}}^* = 0 & \text{in } \Omega_s & (4.4)\\
\boldsymbol{\sigma}\cdot\boldsymbol{n} - \bar{\boldsymbol{t}} = 0; & \boldsymbol{\sigma}^*\cdot\boldsymbol{n} - \bar{\boldsymbol{t}}^* = 0 & \text{on } \Gamma_\sigma & (4.5)\\
\boldsymbol{\sigma}\cdot\boldsymbol{n} - \rho_o\dot{\phi}\boldsymbol{n} = 0; & \boldsymbol{\sigma}^*\cdot\boldsymbol{n} - \rho_o\dot{\phi}^*\boldsymbol{n} = 0 & \text{on } \Gamma_i & (4.6)\\
\ddot{\boldsymbol{u}}\cdot\boldsymbol{n} - \nabla\ddot{\phi}\cdot\boldsymbol{n} = 0; & \ddot{\boldsymbol{u}}^*\cdot\boldsymbol{n} - \nabla\ddot{\phi}^*\cdot\boldsymbol{n} = 0 & \text{on } \Gamma_i & (4.7)\\
p + \rho_o\ddot{\phi} = 0; & p^* + \rho_o\ddot{\phi}^* = 0 & \text{in } \Omega_f & (4.8)\\
\ddot{p} + \rho_o c^2\nabla^2\ddot{\phi} = 0; & \ddot{p}^* + \rho_o c^2\nabla^2\ddot{\phi}^* = 0 & \text{in } \Omega_f & (4.9)\\
\nabla\ddot{\phi}\cdot\boldsymbol{n} - S_m(\ddot{\phi}) = 0; & \nabla\ddot{\phi}^*\cdot\boldsymbol{n} - S_m^*(\ddot{\phi}^*) = 0 & \text{on } \Gamma_\infty & (4.10)
\end{array}$$

where equations in the first column are those for the physical system found from variation of the adjoint variables and equations in the second column are for the adjoint system found from variation of the physical variables.

Remark 1: The variation of $\boldsymbol{u}$, p and $\partial^i\phi/\partial t^i$ and their adjoint counterparts $\boldsymbol{u}^*$, p^* and $\partial^i\phi^*/\partial t^i$ must vanish at times t_0 and t_1, for integer values of i from 0 to m.

Remark 2: The above formulation requires the finite element approximations on $\boldsymbol{u}, p$, and ϕ to be $\boldsymbol{u} \in C^0(\Omega_s)$, $\phi \in C^0(\Omega_f)$, and $p \in L_2(\Omega_f)$, and likewise for the corresponding adjoint fields.

Remark 3: The continuity equation (2.4) and the coupling boundary conditions (3.14) and (4.1) are easily obtained from the Euler-Lagrange equations (4.9), (4.10), and (4.7) respectively. The constants of integration are all determined to be zero from the initial condition of rest at time t_0.

5. SEMI-DISCRETE FINITE ELEMENT EQUATIONS

Standard finite element interpolations are intoduced for the independent system variables:

$$\begin{aligned} \boldsymbol{u}(\boldsymbol{x},t) &= \boldsymbol{N}_u(\boldsymbol{x})\boldsymbol{d}(t) && \in && C^0\,(\Omega_s) \\ p(\boldsymbol{x},t) &= \boldsymbol{N}_p(\boldsymbol{x})\boldsymbol{p}(t) && \in && L_2\,(\Omega_f) \\ \phi(\boldsymbol{x},t) &= \boldsymbol{N}_\phi(\boldsymbol{x})\boldsymbol{\phi}(t) && \in && C^0\,(\Omega_f) \end{aligned}$$

where $\boldsymbol{N}_u$, $\boldsymbol{N}_p$, and $\boldsymbol{N}_\phi$ contain the finite element global shape functions, and where $\boldsymbol{d}$, $\boldsymbol{p}$, $\boldsymbol{\phi}$ are the solution vectors. The Galerkin finite element approximation to the adjoint fields is based on the same shape functions as those used for the physical fields.

Consider the contribution from the boundary Γ_∞ to the variation of $\mathcal{J}$ with respect to ϕ^*. It may be shown that:

$$\delta\left[\frac{1}{4}\int_{t_0}^{t_1}\int_{\Gamma_\infty}\rho_\circ\left(\dot{\phi}S_m^*(\dot{\phi}^*)+\dot{\phi}^*S_m(\dot{\phi})\right)\,d\Gamma dt\right] = -\frac{1}{2}\int_{t_0}^{t_1}\int_{\Gamma_\infty}\rho_\circ\left(\delta\phi^*S_m(\ddot{\phi})\right)\,d\Gamma dt \qquad (5.1)$$

Consider S_2 given by (3.16.a). The presence of the Beltrami operator D in S_2 implies a requirement of $C^1(\Gamma_\infty)$ continuity of the finite element approximation in the tangential directions. This may be reduced to a C^0 continuity condition by the use of the following identity (integration by parts) which holds on a closed spherical surface, such as Γ_∞:

$$\int_{\Gamma_\infty}\delta\phi^* D\ddot{\phi}\;d\Gamma = -\int_{\Gamma_\infty}\left(\frac{\partial\delta\phi^*}{\partial\varphi}\frac{\partial\ddot{\phi}}{\partial\varphi}+\frac{1}{\sin^2\varphi}\frac{\partial\delta\phi^*}{\partial\theta}\frac{\partial\ddot{\phi}}{\partial\theta}\right)\;d\Gamma \qquad (5.2)$$

We emphasize that for the boundary conditions B_1 and B_2, the finite element approximations need only be C^0. We also note that the result (5.1) has been employed in deriving the Euler-Lagrange equations (4.10).

Now the discretization of $\delta\mathcal{J}=0$ gives rise to the following matrices:

$$\boldsymbol{K}_1 = \boldsymbol{K}_1^T = \mathop{\mathbf{A}}_{e=1}^{n_{el}}\int_{\Omega_s^e}(\nabla\boldsymbol{N}_u)^T\,\tilde{\boldsymbol{J}}_1\,(\nabla\boldsymbol{N}_u)\;d\Omega^e \qquad \text{-structure stiffness-} \quad (5.3)$$

$$\boldsymbol{K}_2 = \boldsymbol{K}_2^T = \mathop{\mathbf{A}}_{e=1}^{n_{el}}\int_{\Omega_s^e}(\nabla\boldsymbol{N}_u)^T\,\tilde{\boldsymbol{J}}_2\,(\nabla\boldsymbol{N}_u)\;d\Omega^e \qquad \text{-structure damping-} \quad (5.4)$$

$$\boldsymbol{M} = \boldsymbol{M}^T = \mathop{\mathbf{A}}_{e=1}^{n_{el}}\int_{\Omega_s^e}\rho_s\boldsymbol{N}_u^T\boldsymbol{N}_u\;d\Omega^e \qquad \text{-structure mass-} \quad (5.5)$$

$$\boldsymbol{Q} = \boldsymbol{Q}^T = \mathop{\mathbf{A}}_{e=1}^{n_{el}}\int_{\Omega_f^e}\frac{1}{\rho_\circ c^2}\boldsymbol{N}_p^T\boldsymbol{N}_p\;d\Omega^e \qquad \text{-fluid stiffness-} \quad (5.6)$$

$$\boldsymbol{E} = \boldsymbol{E}^T = \mathop{\mathbf{A}}_{e=1}^{n_{el}}\int_{\Omega_f^e}\rho_\circ(\nabla\boldsymbol{N}_\phi)^T(\nabla\boldsymbol{N}_\phi)\;d\Omega^e \qquad \text{-fluid mass-} \quad (5.7)$$

$$\boldsymbol{f} = \mathop{\mathbf{A}}_{e=1}^{n_{el}}\int_{\Gamma_\sigma^e}\boldsymbol{N}_u^T\,\bar{\boldsymbol{t}}\;d\Gamma^e \qquad \text{-force vector-} \quad (5.8)$$

$$\boldsymbol{C}_1 = \mathop{\mathbf{A}}_{e=1}^{n_{el}}\int_{\Gamma_i^e}-\rho_\circ\boldsymbol{N}_u^T\boldsymbol{n}\boldsymbol{N}_\phi\;d\Gamma^e \qquad \text{-coupling matrix-} \quad (5.9)$$

$$\boldsymbol{D}_1 = \mathop{\mathbf{A}}_{e=1}^{n_{el}}\int_{\Omega_f^e}\frac{1}{c^2}\boldsymbol{N}_p^T\boldsymbol{N}_\phi\;d\Omega^e \qquad \text{-coupling matrix-} \quad (5.10)$$

$${}^2\boldsymbol{B} = {}^2\boldsymbol{B}^T = \mathop{\mathbf{A}}_{e=1}^{n_{el}}\int_{\Gamma_\infty^e}\frac{\rho_\circ}{r}\left[H_1(m)\boldsymbol{N}_\phi^T\boldsymbol{N}_\phi + \frac{H_2(m)}{2}\left(\boldsymbol{N}_{\phi,\varphi}^T\boldsymbol{N}_{\phi,\varphi}+\frac{1}{\sin^2\varphi}\boldsymbol{N}_{\phi,\theta}^T\boldsymbol{N}_{\phi,\theta}\right)\right]\;d\Gamma^e \qquad \text{-radiation damping-} \quad (5.11)$$

$${}^3\boldsymbol{B}_s = {}^3\boldsymbol{B}_s^T = \mathop{\mathbf{A}}_{e=1}^{n_{el}}\int_{\Gamma_\infty^e}\frac{\rho_\circ}{c}(1+H_2(m))\boldsymbol{N}_\phi^T\boldsymbol{N}_\phi\;d\Gamma^e \qquad \text{-radiation damping-} \quad (5.12)$$

$${}^3\boldsymbol{B}_u = \mathop{\mathbf{A}}_{e=1}^{n_{el}}\int_{\Gamma_\infty^e}\frac{\rho_\circ r}{c}H_2(m)\boldsymbol{N}_\phi^T\boldsymbol{N}_{\phi,r}\;d\Gamma^e \qquad \text{-radiation damping-} \quad (5.13)$$

$${}^4\boldsymbol{B} = {}^4\boldsymbol{B}^T = \mathop{\mathbf{A}}_{e=1}^{n_{el}}\int_{\Gamma_\infty^e}\frac{\rho_\circ r}{c^2}H_2(m)\boldsymbol{N}_\phi^T\boldsymbol{N}_\phi\;d\Gamma^e \qquad \text{-radiation damping-} \quad (5.14)$$

where $\mathop{\mathbf{A}}_{e=1}^{n_{el}}$ denotes the element assembly operator, $\tilde{\boldsymbol{J}}_1$ and $\tilde{\boldsymbol{J}}_2$ are the matrix counterparts of the tensors $\boldsymbol{J}_1$ and $\boldsymbol{J}_2$ in (2.1) and $H_i(m)=1$ if m is greater than or equal to i, and is otherwise zero. The system of equations, corresponding to the stationarity of $\mathcal{J}_1$ with respect to the adjoint variables, takes the form:

$$\begin{bmatrix}\boldsymbol{K}_1 & 0 & 0\\ 0 & \boldsymbol{Q} & 0\\ 0 & 0 & 0\end{bmatrix}\begin{Bmatrix}\boldsymbol{d}\\ \boldsymbol{p}\\ \boldsymbol{\phi}\end{Bmatrix} + \begin{bmatrix}\boldsymbol{K}_2 & 0 & 0\\ 0 & 0 & 0\\ 0 & 0 & 0\end{bmatrix}\begin{Bmatrix}\dot{\boldsymbol{d}}\\ \dot{\boldsymbol{p}}\\ \dot{\boldsymbol{\phi}}\end{Bmatrix} + \begin{bmatrix}\boldsymbol{M} & 0 & \boldsymbol{C}_1\\ 0 & 0 & \boldsymbol{D}_1\\ \boldsymbol{C}_1^T & \boldsymbol{D}_1^T & -(\boldsymbol{E}+{}^2\boldsymbol{B})\end{bmatrix}\begin{Bmatrix}\ddot{\boldsymbol{d}}\\ \ddot{\boldsymbol{p}}\\ \ddot{\boldsymbol{\phi}}\end{Bmatrix} + \begin{bmatrix}0 & 0 & 0\\ 0 & 0 & 0\\ 0 & 0 & -({}^3\boldsymbol{B}_s+{}^3\boldsymbol{B}_u)\end{bmatrix}\begin{Bmatrix}\dddot{\boldsymbol{d}}\\ \dddot{\boldsymbol{p}}\\ \dddot{\boldsymbol{\phi}}\end{Bmatrix} + \begin{bmatrix}0 & 0 & 0\\ 0 & 0 & 0\\ 0 & 0 & -{}^4\boldsymbol{B}\end{bmatrix}\begin{Bmatrix}\ddddot{\boldsymbol{d}}\\ \ddddot{\boldsymbol{p}}\\ \ddddot{\boldsymbol{\phi}}\end{Bmatrix} = \begin{Bmatrix}\boldsymbol{f}\\ 0\\ 0\end{Bmatrix} \qquad (5.15)$$

It is worth noting the following:

(i) With the exception for ${}^3\boldsymbol{B}_u$, the system (5.15) presents a symmetric and very sparse structure.

(ii) The coupling of fields occurs in the extended mass matrix. The formulation is thus of the "mass-coupling" type.

(iii) The elimination of dissipative terms yields the system proposed by Geradin, Robert, and Huck [26] for the transient response of an elastic enclosure filled with a compressible fluid.

The system of equations corresponding to the discretization of a functional with a single field pressure fluid Lagrangian can be obtained by specialization of (5.15). As already noted, p and p^* need only be integrable over Ω_f and accordingly, a finite element approximation which is discontinuous across interelement boundaries can be considered for those fields. A static condensation, at the element level, of the pressure parameters is thus possible. The weak form of the momentum balance equation corresponding to the stationarity of $\mathcal{J}$ with respect to p^* yields:

$$\boldsymbol{p}^e = -(\boldsymbol{Q}^e)^{-1}(\boldsymbol{D}_1^e)\ddot{\boldsymbol{\phi}}^e \tag{5.16}$$

and the matrix system (5.15) reduces to:

$$\begin{bmatrix} \boldsymbol{K}_1 & 0 \\ 0 & 0 \end{bmatrix}\begin{Bmatrix} \boldsymbol{d} \\ \boldsymbol{\phi} \end{Bmatrix} + \begin{bmatrix} \boldsymbol{K}_2 & 0 \\ 0 & 0 \end{bmatrix}\begin{Bmatrix} \dot{\boldsymbol{d}} \\ \dot{\boldsymbol{\phi}} \end{Bmatrix} + \begin{bmatrix} \boldsymbol{M} & \boldsymbol{C}_1 \\ \boldsymbol{C}_1^T & -(\boldsymbol{E}+{}^2\boldsymbol{B}) \end{bmatrix}\begin{Bmatrix} \ddot{\boldsymbol{d}} \\ \ddot{\boldsymbol{\phi}} \end{Bmatrix}$$
$$-\begin{bmatrix} 0 & 0 \\ 0 & ({}^3\boldsymbol{B}_s+{}^3\boldsymbol{B}_u) \end{bmatrix}\begin{Bmatrix} \dddot{\boldsymbol{d}} \\ \dddot{\boldsymbol{\phi}} \end{Bmatrix} - \begin{bmatrix} 0 & 0 \\ 0 & (\boldsymbol{D}_1^T\boldsymbol{Q}^{-1}\boldsymbol{D}_1+{}^4\boldsymbol{B}) \end{bmatrix}\begin{Bmatrix} \ddddot{\boldsymbol{d}} \\ \ddddot{\boldsymbol{\phi}} \end{Bmatrix}$$
$$=\begin{Bmatrix} \boldsymbol{f} \\ 0 \end{Bmatrix} \tag{5.17}$$

where $(\boldsymbol{D}_1^T)(\boldsymbol{Q})^{-1}(\boldsymbol{D}_1)$ is actually formed as:

$$(\boldsymbol{D}_1^T)(\boldsymbol{Q})^{-1}(\boldsymbol{D}_1) = \mathop{\boldsymbol{\mathsf{A}}}_{e=1}^{n_{el}}(\boldsymbol{D}_1^e)^T(\boldsymbol{Q}^e)^{-1}(\boldsymbol{D}_1^e) \tag{5.18}$$

The resulting system, although symmetrically coupled, involves five time orders. To reduce the number of time orders, we invoke the limitation principle for mixed finite elements[31,32] which states that choosing $\boldsymbol{N}_p^e = \boldsymbol{N}_\phi^e$ is equivalent to a pointwise enforcement of the momentum balance equation (2.3). In that case, the reduced system (5.17) degenerates, with an obvious loss of symmetry in the coupling, to the following system corresponding to the discretization of a functional with a single-field (p) fluid Lagrangian:

$$\begin{bmatrix} \boldsymbol{K}_1 & \boldsymbol{C}_1 \\ 0 & (\boldsymbol{E}+{}^2\boldsymbol{B})/\rho_o \end{bmatrix}\begin{Bmatrix} \boldsymbol{d} \\ \boldsymbol{p} \end{Bmatrix} + \begin{bmatrix} \boldsymbol{K}_2 & 0 \\ 0 & ({}^3\boldsymbol{B}_s+{}^3\boldsymbol{B}_u)/\rho_o \end{bmatrix}\begin{Bmatrix} \dot{\boldsymbol{d}} \\ \dot{\boldsymbol{p}} \end{Bmatrix}$$
$$+\begin{bmatrix} \boldsymbol{M} & 0 \\ \boldsymbol{C}_1^T & (\rho_o\boldsymbol{Q}+{}^4\boldsymbol{B}/\rho_o) \end{bmatrix}\begin{Bmatrix} \ddot{\boldsymbol{d}} \\ \ddot{\boldsymbol{p}} \end{Bmatrix} = \begin{Bmatrix} \boldsymbol{f} \\ 0 \end{Bmatrix} \tag{5.19}$$

In the numerical examples given in the next section, (5.19) is employed as a basis for the calculations. The equations were integrated in time using the trapezoidal rule[33]

6. NUMERICAL EXAMPLES

The problems investigated are primarily designed to study the performance of the boundary operators, rather than that of the coupled formulation as a whole. Consequently, the computations will be made using the single-field pressure formulation(5.19) for a massless structure submerged in an unbounded acoustic fluid domain.

6.1- Uniformly Pulsating Sphere

Consider the problem of a submerged massless sphere, of radius a, pulsating harmonically with a uniform radial acceleration $\ddot{u} = \ddot{u}_o \sin(\omega t + \psi)$. The phase angle is chosen such that $\psi = \tan^{-1}(ka)$ where the wave number k is defined as the ratio of the exciting frequency ω to the wave speed c. The boundary value problem for the acoustic fluid can be formulated, with either a Neumann or a Dirichlet boundary condition, as follows:

$$\frac{d^2p}{dr^2} + \frac{2}{r}\frac{dp}{dr} - \frac{1}{c^2}\ddot{p} = 0 \text{ for } r \in]a, +\infty[\tag{6.1}$$

$$\frac{dp}{dr} = -\rho_o\ddot{u} \quad \text{at} \quad r = a \tag{6.2}$$

(or) $$p = \frac{\rho_o a\ddot{u}_o}{\sqrt{1+(ka)^2}}\sin\omega t \text{ at } r = a \tag{6.3}$$

In either case, the steady-state time-dependent pressure solution is given by:

$$p(r,t) = -\frac{\rho_o a^2\ddot{u}_o}{\sqrt{1+(ka)^2}}\frac{1}{r}\sin k(r-a-ct) \tag{6.4}$$

It is seen that the pressure solution is one-dimensional, in that it depends on the single coordinate r. The $(1/r)$ dependence of the pressure amplitude indicates a wave front which is exactly spherical. In terms of the expansion (4.3), f_1 is the only non-zero term, and therefore the boundary condition $B_1(p) = 0$ is exact. Any higher order boundary condition, $B_2(p) = 0$ for example, is then obviously also exact.

A transient finite element solution of the above problem was obtained, using the Dirichlet boundary condition (6.3). The system starts from rest at $t = 0$ and is allowed to reach the steady-state harmonic solution with time. The computations were performed for a sphere, of radius $a = 1$, pulsating with a radial acceleration $\ddot{u} = \sqrt{2}\sin(\omega t + \pi/4)$. The acoustic fluid was chosen to have a density $\rho_o = 1$ and a wave speed $c = 1$. Due to the spherical symmetry of the problem, it suffices to discretize a single solid sector subtending angles of 1 rad. at the center of the sphere. Three-dimensional 8-node linear and 27-node quadratic acoustic elements were employed. The mesh discretization was kept constant with an element dimension of unity along the radial direction. The finite element mesh was not varied since the purpose of these numerical examples is to evaluate the performance of the proposed dampers as affected by their order m and their distance from the origin r_∞, rather than by mesh refinement.

The progression of the solution with time is depicted in Fig. 1, for a 9 element mesh and a time-step $\Delta t = 0.5$. The artificial boundary Γ_∞ is kept at $r_\infty = 10$, the boundary operators B_1 and B_2 are applied, and the finite element transient solution is compared to the exact solution (6.4). It is seen, in Fig. 1(a), that the condition of rest ahead of the wave front is very well captured. We also note that solutions obtained with different radiation boundary conditions are identical since the dampers are inactive as long as the wave has not reached the truncation boundary. The first perturbation reaches Γ_∞ at time $t = 9$ and the dampers are activated from then on. Figures 1(b) to 1(d) demonstrate the effectiveness of the proposed dampers in inhibiting spurious reflections from Γ_∞, especially when the quadratic acoustic element is used.

The effects of reducing the distance r_∞ to the artificial boundary are illustrated in Fig. 2. We recall that such effects should not theoretically exist for the B_1 and B_2 dampers since they are exact for the present problem. Effectively, the results show that the solution remains virtually unaffected as the truncation boundary is brought closer to the disturbance source; even a single 27-node element with $r_\infty = 2$ predicts the solution accurately (see Fig 2(c)). This is certainly not the case for the classical plane-wave damper (PW), which exhibits significant reflection with quadratic elements, as can be seen in Fig 2(b).

We note a slight decrease in accuracy in the above results when the B_2 boundary condition is used instead of B_1. This is due to finite element errors being amplified by the additional terms involved in the B_2 boundary condition, without the benefit of a better match with the exact solution since B_1 is already exact.

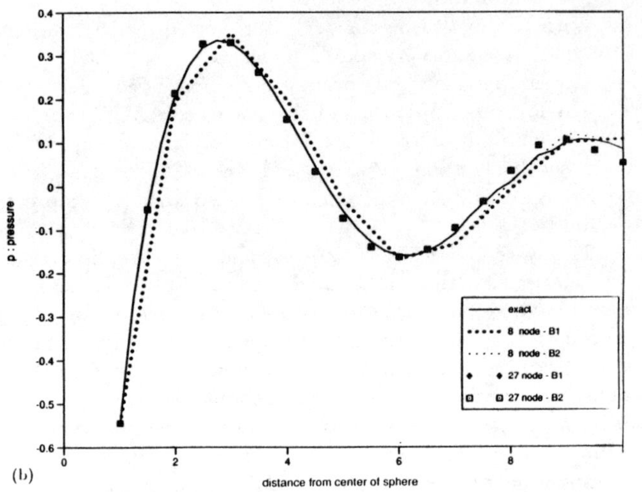

(b)

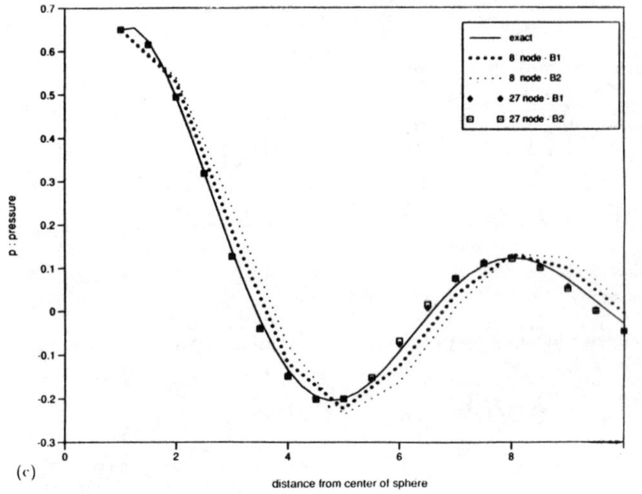

(c)

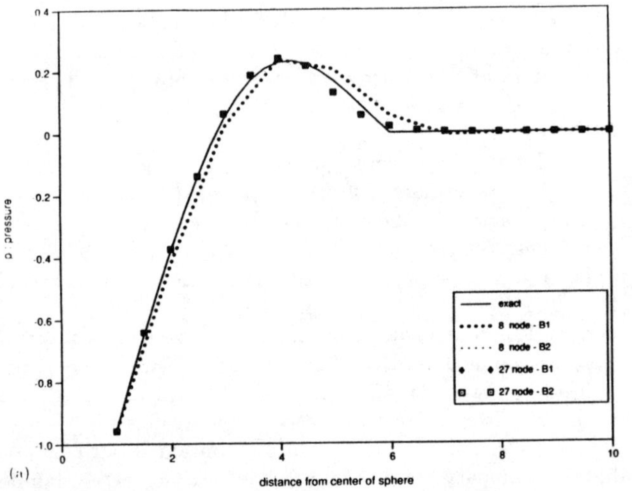

(a)

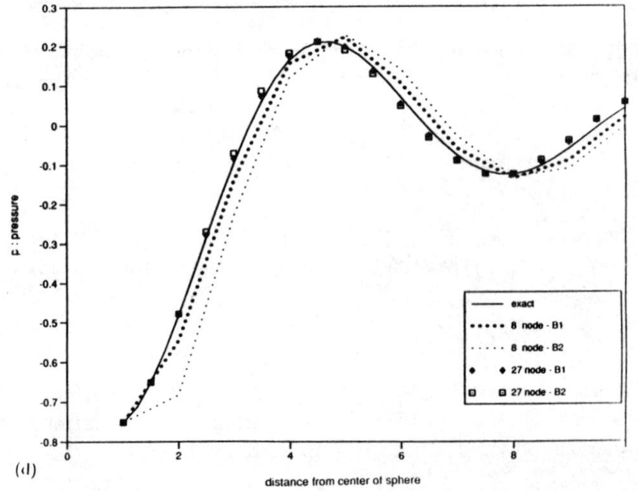

(d)

Fig. 1. Uniformly pulsating sphere. Pressure solution at times:
(a) $t = 5$,
(b) $t = 10$,
(c) $t = 15$,
(d) $t = 18$.

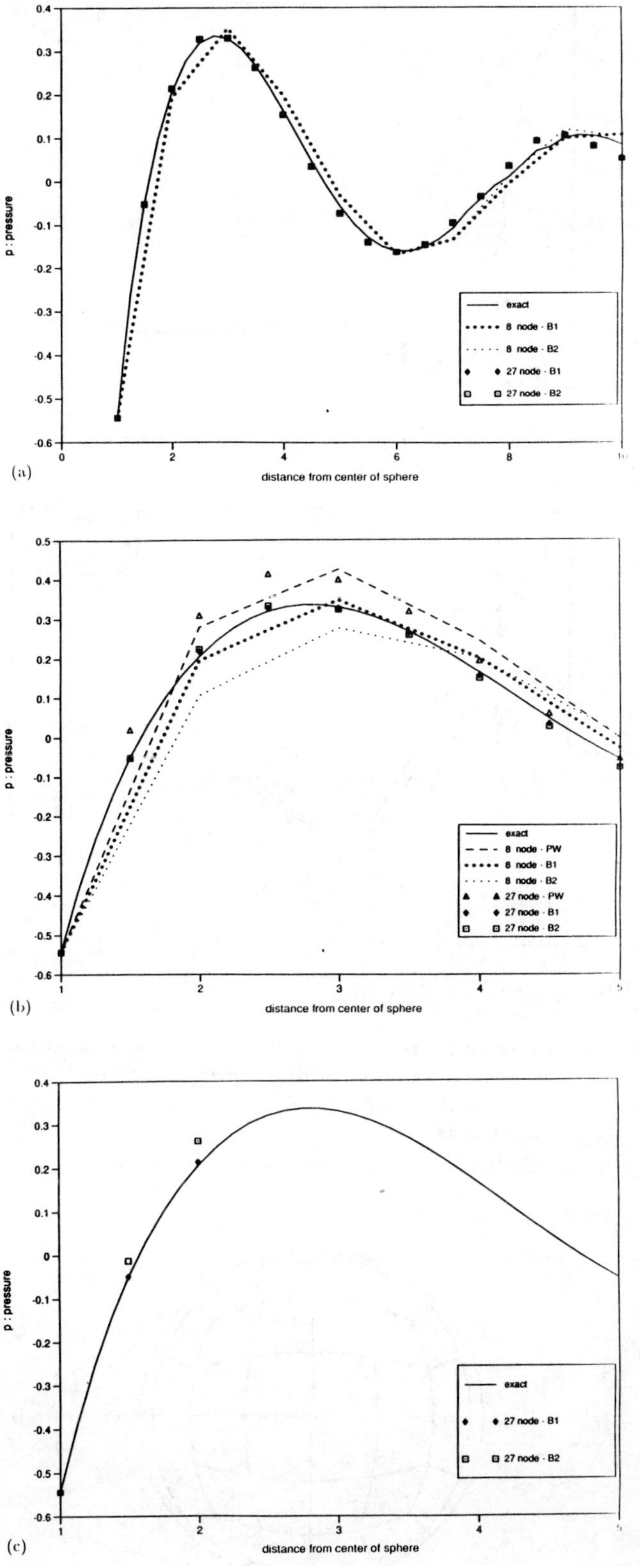

Fig. 2. Uniformly pulsating sphere. Pressure solution at time $t = 10$:
(a) 9 elements, $r_\infty = 10$;
(b) 4 elements, $r_\infty = 5$;
(c) 1 element , $r_\infty = 2$.

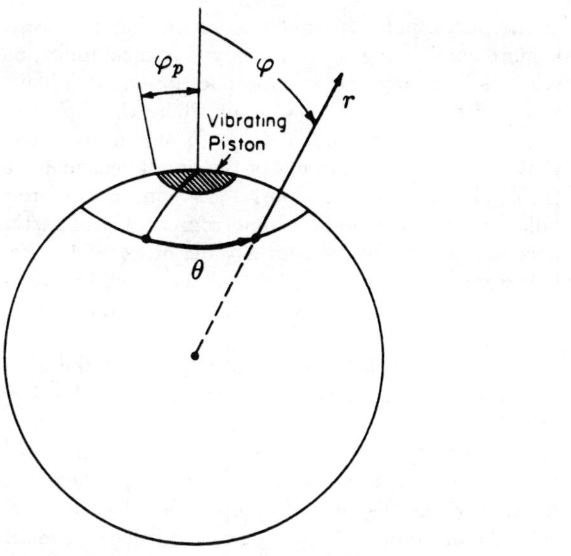

Fig. 3. Circular piston on a spherical baffle.

6.2- Circular Piston on a Spherical Baffle

We examine a second problem where the radiating waves do not present a spherical front, and therefore the B_1 boundary condition is no longer exact.

Consider the problem of a circular piston subtending a polar angle $0 \leq \varphi \leq \varphi_p$ on a submerged massless and rigid sphere of radius a (see Fig. 3). The piston vibrates harmonically with a uniform radial acceleration $\ddot{u} = \ddot{u}_o \sin(\omega t)$. The boundary value problem for the acoustic fluid can be stated as:

$$\frac{\partial^2 p}{\partial r^2} + \frac{2}{r}\frac{\partial p}{\partial r} + \frac{1}{r^2}Dp - \frac{1}{c^2}\ddot{p} = 0 \qquad \text{for } r \in]a, +\infty[\qquad (6.5)$$

$$\partial p/\partial r = -\rho_o \ddot{u} \qquad \text{for } r = a \text{ and } 0 \leq \varphi \leq \varphi_p \qquad (6.6)$$

$$\partial p/\partial r = 0 \qquad \text{for } r = a \text{ and } \varphi_p < \varphi < \pi \qquad (6.7)$$

The steady-state time-dependent pressure solution is then given by:

$$p(r, \varphi, t) = \frac{\rho_o \ddot{u}_o}{2k}\left[F_i \cos\omega t - F_r \sin\omega t\right] \qquad (6.8)$$

$$F_r + iF_i = \sum_{n=0}^{\infty} \frac{h_n(kr)}{h_n'(ka)}\left[P_{n-1}(\cos\varphi_p) - P_{n+1}(\cos\varphi_p)\right] P_n(\cos\varphi) \qquad (6.9)$$

where the functions P_n and h_n are the Legendre polynomials and the spherical Hankel functions of the first kind respectively. Details of the derivation of this solution can be found in [15,34].

This problem presents some interesting features in terms of the wave patterns it generates. As the wave, generated at the piston, travels along longitudes, it is attenuated by a geometrical spreading loss until it reaches the equator. Beyond the equator ($\varphi > \pi/2$), the trace length of the wave front decreases, thus producing a geometric focusing effect which can be important at the antipode($\varphi = \pi$). As the diameter of the spherical baffle increases when compared to a wavelength, the pressure field generated departs more and more from the single point source pattern. When the diameter is of the same order or larger than a wavelength, the spherical baffle gives rise to a "shadow" region below the equator. In this "shadow" zone but away from the antipode, the pressure amplitude is significantly lower than the one at the piston pole. It should be noted that the present problem is quite challenging to both analytic and computational solution schemes. The ray-acoustic method, for example, fails to discern any waves in the "shadow" zone where the diffracted field prevails. The wave-harmonic series (6.9) converges very slowly in that region for large values of the nondimensional wave number (ka). The creeping-wave series [34] converges more rapidly but fails near the antipode. Some computational methods do not necessarily fair any better: a value of $ka = 4$, as we shall use below, puts this problem beyond the reach of standard boundary integral techniques which fail by non-uniqueness in this ka regime.

A transient finite element solution of the above problem is obtained using a time increment $\Delta t = 0.5$ for the trapezoidal rule integration algorithm. The system starts from rest at $t = 0$ and is driven towards the steady-state solution with time. The axisymmetric nature of the problem allows the reduction of the discretized computational acoustic domain to a solid sector delimited by any two longitudes. A relatively coarse mesh consisting of a single layer of three-dimensional quadratic acoustic elements was utilized to discretize the computational domain defined by:$1 \leq r \leq 1.5$ and $0 \leq \theta \leq \pi/2$. A total of 26 isoparametric 27-node elements were used. The trace of the three-dimensional fluid finite element discretization on the baffle surface is displayed in Fig. 4. Actual computations were performed using only a quadrant of the hemisphere shown. The computations were performed for a spherical baffle, of radius $a = 1$, with a large piston ($\varphi_p = 22.5°$) pulsating harmonically with a radial acceleration $\ddot{u} = \sin t$. The acoustic fluid considered has a density $\rho_0 = 1$ and a wave velocity $c = 0.25$. The computational boundary Γ_∞ is brought very close to the disturbance source such that $r_\infty = 1.5$.

The pressure solution on the baffle surface, at the truncation boundary, and on a surface midway between the two, is depicted in Fig. 5 at two times in the steady-state regime. It is seen that the solution is accurately captured throughout the domain, even in the "shadow" zone where the diffracted field prevails. The error in the computed solutions includes errors associated with the boundary operators as well as spatial and temporal discretization.

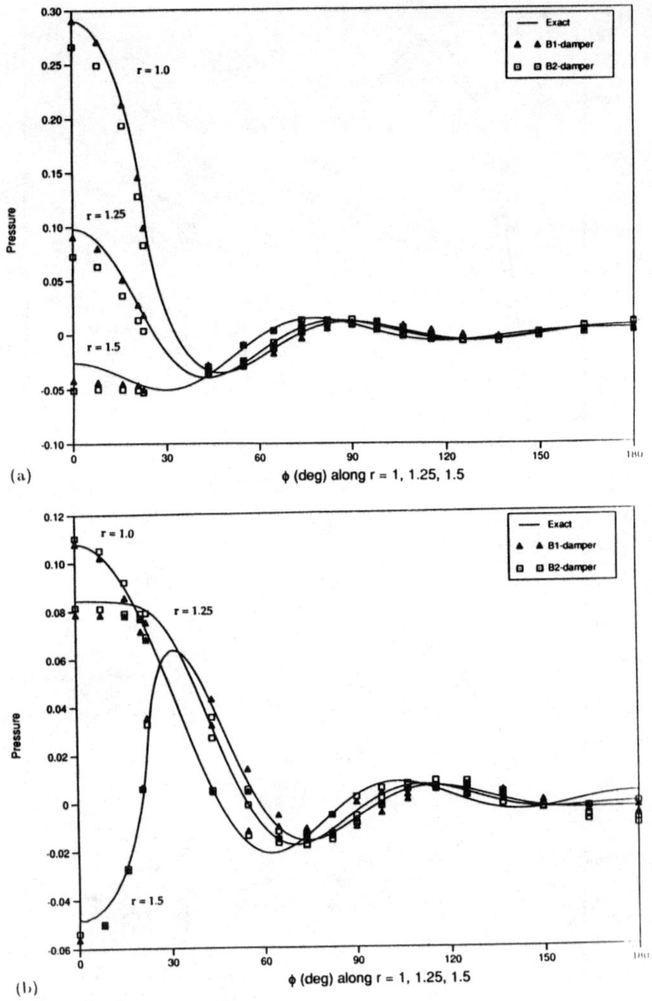

Fig. 5. Pressure distribution on the baffle surface, truncation boundary, and intermediate surface at times:
(a) $t = 46$,
(b) $t = 48$.

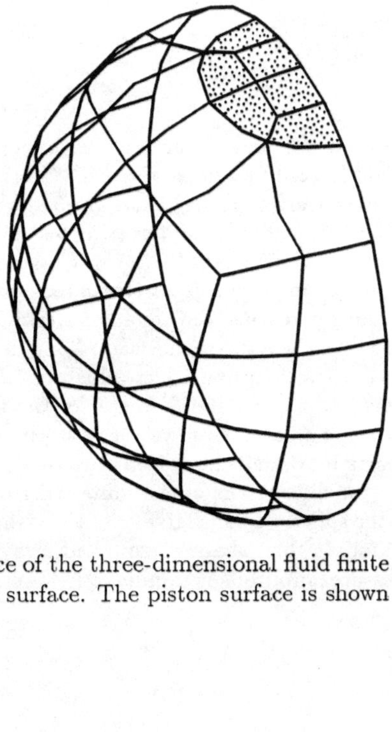

Fig. 4. Trace of the three-dimensional fluid finite element mesh on the baffle surface. The piston surface is shown speckled.

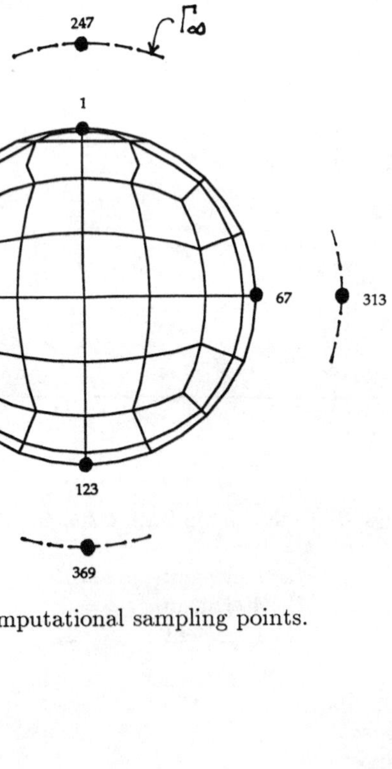

Fig. 6. Location of computational sampling points.

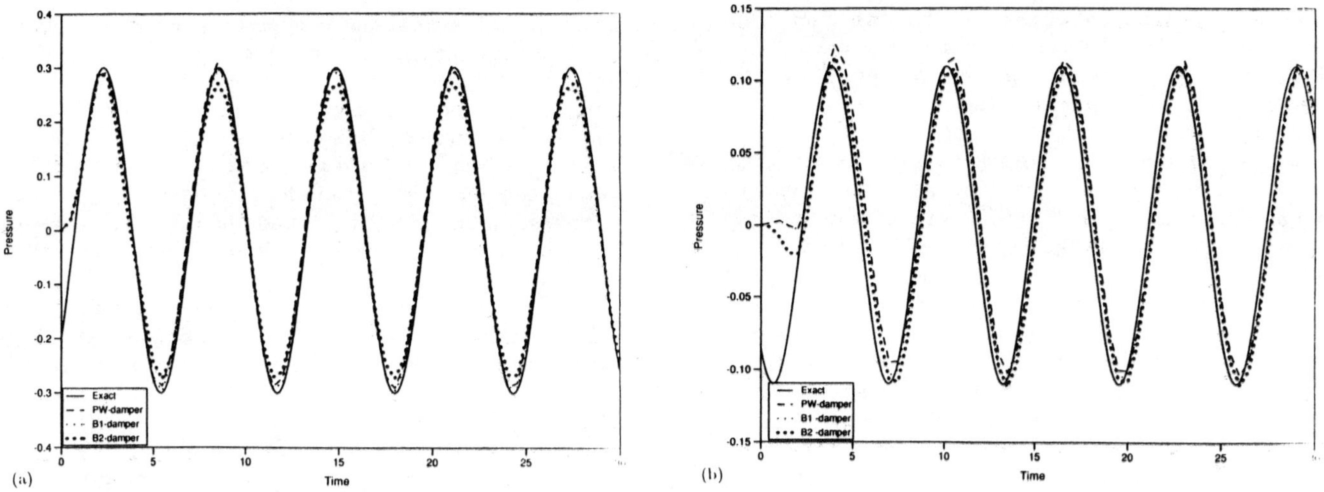

Fig. 7. Pressure solution at the pole: (a) on piston surface, (b) on truncation boundary.

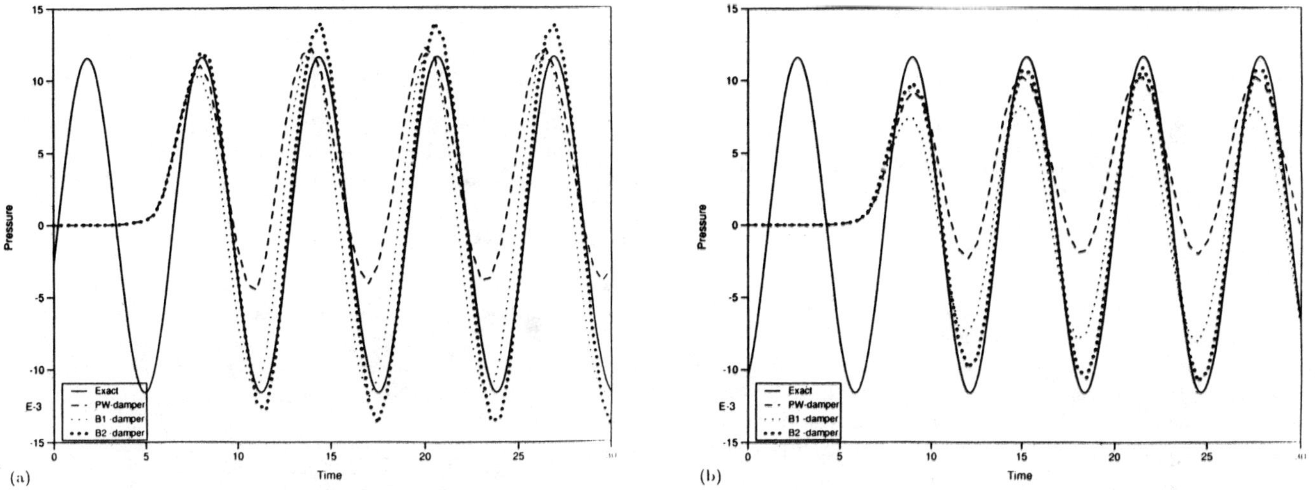

Fig. 8. Pressure solution at the equator: (a) on piston surface, (b) on truncation boundary.

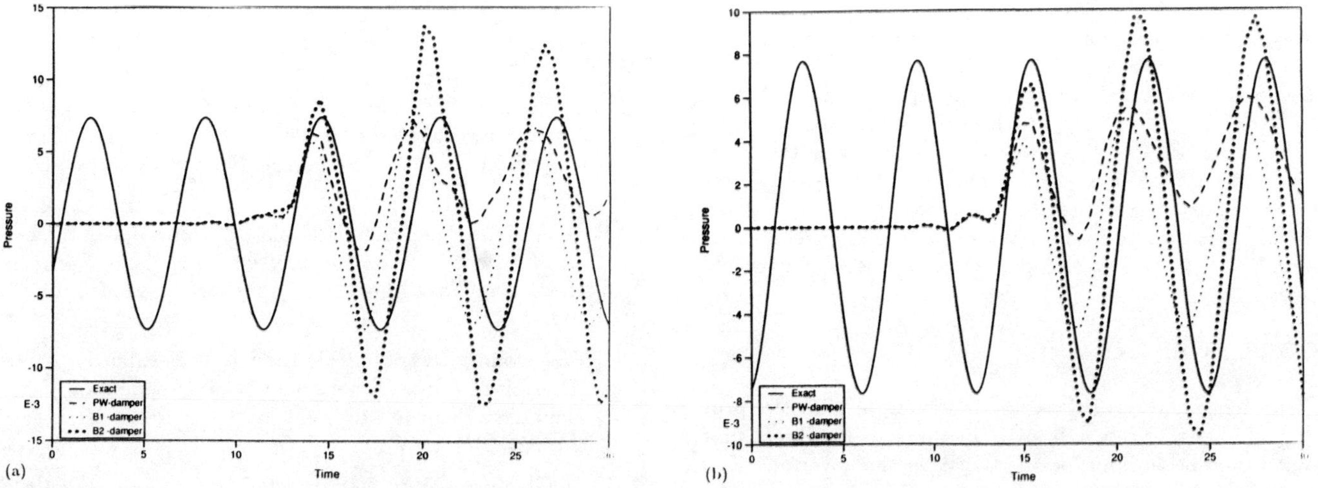

Fig. 9. Pressure solution at the antipole: (a) on piston surface, (b) on truncation boundary.

The curves show computed solutions at times when many spurious reflections between the truncation boundary and baffle could have occurred. The accuracy and lack of spurious reflection demonstrates the effectiveness of the boundary operators, for this problem.

We next consider the progression with time of the transient solution at six representative points in the fluid domain. These points are indicated in Fig. 6, and the results for these points are shown in Fig. 7-9. In the region near the piston, the pressure amplitude achieves its greatest values. Numerical results are shown for the points 1 and 247 which are located at the pole on the piston surface and truncation boundary, respectively, as indicated in Fig. 6. Results for both B_1 and B_2 dampers are shown in Fig. 7. On the piston surface, at point 1, the computed pressure amplitude at steady-state conditions is within 1.3% of the exact solution for the B_1 damper and 9.0% for the B_2 damper. On the truncation boundary, at point 247, the computed steady-state pressure amplitude has an error of 4.0% for B_1 and 1.8% for B_2. Both dampers produce negligible phase errors in this region.

Fig. 8 displays the results at the equator for point 67 on the baffle surface and point 313 on the truncation boundary. We note that the pressure amplitude at these points has been attenuated down to a few percent of that at the piston surface. Both dampers capture the pressure amplitude solution fairly well. On the other hand, the pressure phase solution shows very small errors. The computed solution on the baffle surface, at point 67, has a phase error of 7.7% for the B_1 boundary condition and 1.3% for the B_2 damper. On the truncation boundary, at point 313, the phase error is 4.7% for B_1 and 0.6% for B_2.

In the "shadow" region below the equator, the numerical solution captures the "creeping-wave" solution accurately. In this region, pressure amplitudes are significantly lower than those prevailing at the pole, and therefore a much better measure of the accuracy of the numerical solution is the phase error. The B_2 boundary condition yields "shadow-zone" solutions with phase errors below 8%. The B_1 boundary operator, though less accurate than B_2, is still reasonably accurate with respect to phase angle errors.

The solution at the antipode, for point 123 on Γ_i and for point 369 on Γ_∞, is shown in Fig. 9. We note that, though the focusing effect is not pronounced in the present example, the proposed finite element formulation seems to have the capability of detecting such a large gradient at the antipode. It is seen that the B_2 boundary condition manages to capture the exact phase solution with a 90% accuracy in this complex region.

The superiority of the proposed dampers over the classical plane-wave damper is clearly manifested. The plane-wave damper not only exhibits severe amplitude, phase, and dispersion errors, but also performs very poorly in inhibiting spurious wave reflection from Γ_∞, as indicated by the lack of symmetry in the computed solution.

7. CONCLUSIONS

Transient analysis of the exterior structural acoustics problem is a challenging research area. In this paper a direct finite element formulation is proposed, based on the governing time-dependent equations via a mixed variational formalism. The formulation employs a finite computational fluid domain and therefore depends critically on the accuracy of the radiation boundary conditions employed on the truncation boundary. Other sources of error include spatial and temporal discretization errors.

A sequence of boundary operators described by Bayliss and Turkel [12], which provide increasing orders of accuracy, have been incorporated into the general coupled finite element formulation. For two members of this sequence, implementational details have been described and some numerical results are reported. The numerical results for two model problems are encouraging. In one problem having a high wave number regime, very accurate results have been obtained with a single layer of fluid elements.

It is interesting to note that the numerical results were generally consistent with the theoretical prediction that the B_2 operator is more accurate than the B_1 operator. However, is is not apparent from the numerical results that the difference is highly significant. Generally, B_1 gave less amplitude error and B_2 gave less phase error. We plan to consider the implementation of even higher order operators in future research. We believe that an optimal solution scheme for the higher order boundary operators will employ a partitioned solution scheme in which the boundary and domain contributions will be treated with separate and different time integration schemes.

ACKNOWLEDGMENTS

The material in this paper is based upon work supported by the Office of Naval Research under Grant No. N00014-84-K-0715, for which the authors are grateful.

REFERENCES

[1] O.C. Zienkiewicz, P. Betess, T.C. Chiam and C. Emson, "Numerical Methods for Unbounded Field Problems and a New Infinite Element Formulation," in A.J. Kalinowski (ed.), *Computational Methods for Infinite Domain Media-Structure Interaction, AMD Vol. 46*, ASME, 1981, pp 115-148.

[2] G.D. Manolis and D.E. Beskos, "Dynamic Stress Concentration Studies by Boundary Integrals and Laplace Transform," *Int. J. Num. Meth. Engng*, **17**, 244-259 (1981).

[3] G.D. Manolis, "A Comparative Study on Three Boundary Element Method Approaches to Problems in Elastodynamics," *Int. J. Num. Meth. Engng*, **19**, 73-91 (1983).

[4] P.H.L. Groenenboom and B.V. Neratoom, "The Application of Boundary Elements to Steady and Unsteady Potential Fluid Flow Problems in Two and Three Dimensions," in C.A. Brebbia (ed.), *Boundary Element Methods*, Springer Verlag, Berlin, 1981.

[5] W.J. Mansur and C.A. Brebbia, "Formulation of the Boundary Element Method for Transient Problems Governed by the Scalar Wave Equation," *Appl. Math. Modelling*, **6**, 307-311 (1982).

[6] J.L. Dohner, R. Shoureshi, and R.J. Bernhard, "Transient Analysis of Three-Dimensional Wave Propagation Using the Boundary Element Method," *Int. J. Numer. Methods Eng.*, **24**, 621-634 (1987).

[7] D. Givoli, *A Finite Element Method for Large Domain Problems*, Ph.D. Thesis, Stanford University, May 1988.

[8] T. L. Geers, "Doubly Asymptotic Approximations for Transient Motions of Submerged Structures," *J. Acoust. Soc. Am.*, **64**, 1500-1508 (1978).

[9] C. A. Felippa, "Top-Down Derivation of Doubly Asymptotic Approximations for Fluid-Structure Interaction Analysis," in R. Shaw *et al.* (ed.), *Proc. 2nd Int. Symp. Innovative Numerical Analysis for the Applied Enginnering Sciences*, 79-88, University Press of Virginia, Charlottesville (1980).

[10] C. A. Felippa, "A Family of Early-Time Approximations for Fluid-Structure Interaction," *J. Appl. Mech.*, **47**, 703-708 (1980).

[11] T. L. Geers and C. A. Felippa, "Doubly Asymptotic Approximations for Vibration Analysis of Submerged Structures," *J. Acoust. Soc. Am.*, **73**(4), 1152-1159 (1983).

[12] A. Bayliss and E. Turkel, "Radiation Boundary Conditions for Wave-like Equations," *Comm. Pure Appl. Math.*, **XXXIII**, 707-725 (1980).

[13] A. Bayliss, M. Gunzberger, and E. Turkel, "Boundary Conditions for the Numerical Solution of Elliptic Equations in Exterior Regions," *ICASE Report No. 80-1*, NASA, Langley, 1978

[14] K. Washizu, *Variational Methods in Elasticity and Plasticity*, 3rd. ed., Pergamon Press, Oxford, 1982.

[15] P.M. Morse and H. Feshbach, *Methods of Theoretical Physics*, McGraw-Hill, New York, 1953. (cf. Chap. 3 for variational formalism, Chap. 11 for numerical example).

[16] P.M. Pinsky and K.O. Kim, "A Multi-Director Formulation for Elastic-Viscoelastic Layered Shells," *Int. J. Num. Meth. Engng.*, **23**(12), 2213-2244 (1986).

[17] P.M. Pinsky and K.O. Kim, "A Multi-Director Formulation for Nonlinear Elastic-Viscoelastic Layered Shells," *Comput. & Struct.*, **24**(6), 901-913 (1986).

[18] G.M.L. Gladwell, "A Variational Formulation of Damped Acousto-Structural Vibration Problems," *J. Sound Vib.*, 4(2), 172-186 (1966).

[19] A. Graggs, "The Transient Response of a Coupled Plate-Acoustic System Using Plate and Acoustic Finite Elements," *J. Sound Vib.*, **15**, 509-528 (1971).

[20] A. Graggs, "An Acoustic Finite Element for Studying Boundary Flexibility and Sound Transmission between Irregular Enclosures," *J. Sound Vib.*, **30**(3), 343-357 (1973).

[21] P.M. Pinsky and N.N. Abboud, "Two Mixed Variational Principles for Exterior Fluid-Structure Interaction Problems," *Comput. Struct.*, (to appear in 1989).

[22] N.N. Abboud and P.M. Pinsky, "A Mixed Variational Principle for the Structure-Exterior Fluid Interaction Problem," in W.L. Keith, E.M. Uram, and A.J. Kalinowski (ed.), *Acoustic Phenomena and Interaction in Shear Flows over Compliant and Vibrating Surfaces, Proc. 1988 Int. Symp. on Flow-Induced Vibration and Noise, Vol. 6*, ASME, pp 137-148.

[23] H. Morand and R. Ohayon, "Substructure Variational Analysis of the Vibrations of Coupled Fluid-Structure Systems. Finite Element Results," *Int. J. Num. Meth. Engng.*, **14**(5), 741-755 (1979).

[24] H. Morand and R. Ohayon, "Variational Formulations for the Elasto-Acoustic Vibration Problem: Finite Element Results," *Proc. 2nd. Int. Symp. F.E.M. Flow Problems*, St. Margharita, Italy, 1976, pp 785-796.

[25] R. Ohayon, "Fluid-Structure Modal Analysis. New Symmetric Continuum-Based Formulations. Finite Element Applications," in G.N. Pande and J. Middleton (ed.), *Transient/Dynamic Analysis and Constitutive Laws for Engineering Materials, Proc. Int. Conf. Num. Meth. Engng.: Theory and Applications, NUMETA '87, Vol. 2*, Swansea, U.K., Jul. 1987. (cf. T46/1).

[26] M. Geradin, G. Robert, and A. Huck, "Eigenvalue Analysis and Transient Response of Fluid-Structure Interaction Problems," *Engng. Comput.*, **1**(2), 151-160, Jun. 1984.

[27] L.G. Olson and K.J. Bathe, "Analysis of Fluid-Structure Interactions. A Direct Symmetric Coupled Formulation Based on the Fluid Velocity Potential," *Comput. & Struct.*, **21**(1/2), 21-32 (1985).

[28] W.K. Liu and R.A. Uras, "Variational Approach to Fluid-Structure Interaction with Sloshing," *Nucl. Engrg. Des.*, **106**, 69-85 (1988).

[29] W. D. Smith, "A Non-reflecting Plane Boundary for Wave Propagation Problems," *J. Comput. Phys.*, **15**, 493-503 (1974).

[30] F.G. Friedlander, "On the Radiation Field of Pulse Solutions of the Wave Equation," *Proc. Roy. Soc. London*, **A.269**, 53-65 (1962).

[31] B. Fraejis de Veubeke, "Displacement and Equilibrium Models in the Finite Element Method," in O.C. Zienkiewicz and G.S. Holister (ed.), *Stress Analysis*, Wiley, London, U.K., 1965. (cf. Chap. 9).

[32] H. Stolarski and T. Belytschko, "Limitation Principles for Mixed Finite Elements Based on the Hu-Washizu Variational Formulation," in R.L. Spilker and K.W. Reed (ed.), *Hybrid and Finite Element Methods, AMD Vol. 73*, ASME, 1985, pp 123-132.

[33] T.J.R. Hughes, *The Finite Element Method*, Prentice-Hall, 1987.

[34] M.C. Junger and D. Feit, *Sound, Structures, and Their Interaction*, 2nd. ed., MIT Press, 1986 (cf. Chap. 6 & 12).

AN IMPROVED COMPUTATIONAL METHOD FOR PREDICTING THE TRANSIENT ACOUSTIC FIELD FROM IMPACT MACHINERY

D. H. Chun
Becton Dickinson Research Center
Computer Aided Engineering Department
Research Triangle Park, North Carolina

T. H. Hodgson
Center for Sound and Vibration
Mechanical and Aerospace Engineering Department
North Carolina State University
Raleigh, North Carolina

Abstract

A recent extension of the Boundary Element Method in order to predict the transient acoustic field radiated from an impacted three-dimensional elastic structure is described. The improved computational method presented for calculating the transient acoustic field combines a hybrid explicit and implicit time integration method in order to reduce the numerical instability inherent in the explicit scheme for near-field element spacing and to improve the overall computational accuracy and efficiency for large scale problems.

Boundary conditions such as velocities on the surface of the structure are first calculated using one of the Finite Element Method structural analysis programs such as SAP IV, ANSYS, NASTRAN, or I-DEAS, which also permit pre- and post- processing capabilities. Verification of the procedure was obtained by use of a simple plate test case as well as the comparison of experimental results from a hand-operated punch press with the predicted numerical method.

1 Introduction

With the advancement of digital computer techniques, numerical methods are being developed to identify noise radiation from the source in order to control the radiated noise. Through the use of the finite-element method (FEM) and/or the boundary element method (BEM), an arbitrarily shaped vibrating body can be characterized. Subsequently, acoustic radiation patterns can be calculated using the well-known Helmholtz wave equation for the steady-state motions introduced by Chen and Schweikert[1], Chertock[2], Kleinman and Roach[3], for instance. Although computational techniques for steady-state responses can produce non-unique solutions, these methods have been well documented in the literature as described by Copley[4], Schenck[5], Koopman and Benner[6].

Transient noise radiation problems have inherent difficulties in the application of conventional analysis, both in numerical and experimental techniques. The main difficulties occur because noise generating mechanisms for the transient impact case are quite different from the steady-state case. In addition, the numerical analysis requires large amounts of computer memory, disk storage and CPU time in order to implement the time and frequency domain analyses.

The recent application of numerical techniques to transient noise radiation problems has been accelerated with the introduction of the BEM by Banerjee et al[7] and Brebbia[8]. Subsequently, numerical formulations for some two- and three- dimensional transient problems have been developed by Groenenboom[9], Manolis[10], Mansur and Brebbia[11,12], Rizzo et al [13,14,15]. In general, there are two major solution procedures for the transient wave equation, namely the transformed domain solution technique and the direct time integration scheme. The transformed solution procedure using the Laplace transform is analogous to the steady-state solution except the Fourier transform inversion is used to obtain the final solution in the time domain, see for instance Dubner and Abate[16]. Although there are some trade-offs, the direct time integration scheme (or time-stepping method) of the integral equation generally offers better computational efficiency and accuracy because of inherent difficulties in inversion with the transformed domain solution technique for complicated problems.

A numerical technique for solving the three-dimensional transient wave propagation problem in a rectangular cavity using the explicit time integration method was reported by Dohner et al[17]. However, the explicit time integration scheme has some disadvantages due to its use of massive computer storage as well as CPU time in order to obtain a reasonable solution to a large scale problem such as the transient analysis for a punch press. Here an improvement in the numerical technique for the transient noise radiation problem is described by combining the explicit and implicit time integration schemes in order to obtain better numerical stability and efficiency and to apply the method to a more complicated geometry. Extensive numerical computations were performed to understand both transient structural vibrations and sound radiation from a Roper and Whitney No. 34 hand-operated punch press. The wave mechanism and energy flow in this machine were characterized in order to develop solution procedures for noise radiation by an impulsive device. This machine was chosen because it is more accessible than a full-size punch press.

2 Theory

A wide variety of numerical techniques have been studied in the field of acoustic radiation problems in the past decade. Among them, the solution of problems involving a discretized surface integral is commonly referred to as the BIE method. Unlike the FEM, this method only considers the bounding surfaces in the solution procedure for three dimensional field problems involving an infinite domain. Since the boundary (or surface) of the body is only a concern for the field problems, the BIE method reduces the dimension of the domain in the problems by one, while the FEM requires the discretization of the entire domain.

The acoustic radiation from a vibrating body is governed by the wave equation. The general form of the inhomogeneous scalar wave equation for the pressure P at the point (r_o, t_o) and the Neumann type boundary condition (BC) is given by

$$\nabla_o^2 P(r_o,t_o) - \frac{1}{c^2}\frac{\partial^2 P}{\partial t_o^2} = -F(r_o,t_o), \qquad \frac{\partial P}{\partial n} = -\rho\frac{\partial^2 w}{\partial t_o^2} \qquad (2.1)$$

where F describes a general acoustic source and n is an outward normal to the boundary, w is the normal component of the displacement, and ∇_o^2 is the Laplace operator, see Fig.1.

As mentioned earlier, there are two different approaches in solving equation (2.1). One method is to transform the equation into the Laplace domain. This then yields an inhomogeneous Helmholtz equation given by

$$\nabla_o^2 \tilde{P}(r_o) - \frac{s^2}{c^2}\tilde{P}(r_o) = \tilde{F}(r_o) \qquad (2.2)$$

where $\tilde{P}$ and $\tilde{F}$ represent transformed variables. This technique implies that the integral equation is first solved using steady-state solution procedures. Then, the frequency domain solution is transformed back to reconstitute the time domain response using, say, Fourier synthesis, (see ref. 16 which avoids problems associated with inverse Laplace transforming in the complex domain). The other scheme solves the integral equation directly in the time domain using a step-by-step time integration method known as the time-stepping method. The time integration method generally provides more accurate results than the Laplace transform method, since the Laplace transform solution procedure sometimes has difficulties with inversion of the frequency domain solution.

With the usual Green's function, G = $\delta[R/c - (t-t_o)]/4\pi R$, conditions of causality and reciprocity, and the usual integral properties of the Dirac delta function, the pressure field P(r,t) may be described by the following integral equation form

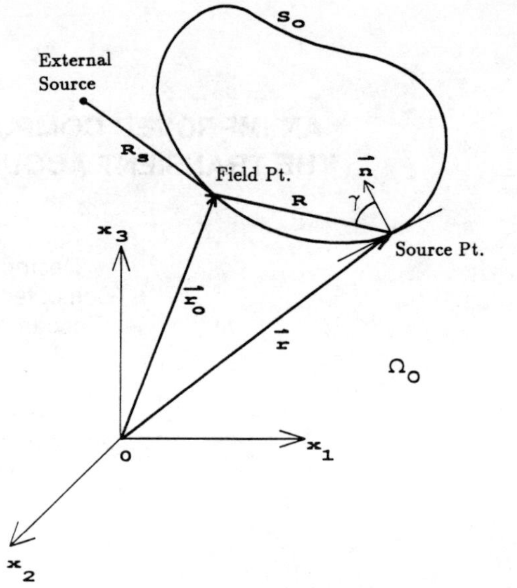

Fig.1 The Coordinate System for a Radiating Surface

$$C_oP(r,t) = \frac{1}{2\pi}\left[\int_{s_o}\frac{1}{R}\left(\frac{\partial P}{\partial n}\right)_{t_{ret}} dS_o + \int_{s_o}\frac{1}{R^2}\frac{\partial R}{\partial n}P_{t_{ret}}\, dS_o + \int_{s_o}\frac{1}{cR}\frac{\partial R}{\partial n}\left(\frac{\partial P}{\partial t_o}\right)_{t_{ret}} dS_o + \int_{\Omega_o}\frac{F(r_o,t-R/c)}{R}\, d\Omega_o\right] + \frac{2}{c^2}\int_{\Omega_o}\left(G\frac{\partial P}{\partial t_o} - P\frac{\partial G}{\partial t_o}\right)\Big|_{t_o=0} d\Omega_o \qquad (2.3)$$

where,

$$\frac{\partial R}{\partial n} = \cos\gamma, \qquad \frac{\partial P}{\partial n} = -\rho\ddot{w}, \qquad t_{ret} = t - R/c$$

where R is the distance between the field and source points and with $C_o = 2$ for field points outside the body, $C_o = 1$ for field points on the surface of the body, $C_o = 0$ for field points inside the body.

3 Numerical Implementation

If the surface S_o is discretized into N planar elements small enough to allow that the variable P and $\partial P/\partial n$ are assumed to be spatially constant on each element but varying with time, then the equation (2.3) becomes for the kth time step t_k

$$C_oP(r_i,t_k) = \sum_{j=1}^{N}\left[\frac{1}{2\pi}\int_{s_j}\frac{1}{R_{ij}}(-\rho\ddot{w}_j)_{t_{ret}} dS_j + \frac{1}{2\pi}\int_{s_j}\frac{1}{R_{ij}^2}\cos\gamma_{ij}P_{t_{ret}} dS_j + \frac{1}{2\pi}\int_{s_j}\frac{1}{cR_{ij}}\cos\gamma_{ij}\left(\frac{\partial P}{\partial t_o}\right)_{t_{ret}} dS_j\right] + \frac{1}{2\pi}\int_{\Omega_o}\frac{F(r_o,t-R/c)}{R}\, d\Omega_o + \frac{2}{c^2}\int_{\Omega_o}\left(G\frac{\partial P}{\partial t_o} - P\frac{\partial G}{\partial t_o}\right)\Big|_{t_o=0} d\Omega_o \qquad (3.1)$$

with i describing the field point and j the source point and R_{ij} = $|r_i - r_j|$. If the source term, F, in equation (3.1) can be represented by a Dirac delta function given by

$$F(r_o, t_o) = f(t_o)\delta(r_o - r_s) \tag{3.2}$$

where r_s is the location of the source. Then the integral for source term in equation (3.1) may be rewritten as

$$\frac{1}{2\pi}\int_{\Omega_o} \frac{F(r_o, t - R/c)}{R} d\Omega_o = \frac{1}{2\pi}\frac{f(t_{R_s})}{R_s} \tag{3.3}$$

where $R_s = |r_i - r_s|$, $t_{R_s} = t_k - R_s/c$. By assuming that the body is initially at rest, the last integral term for the initial condition in equation (3.1) will become zero by virtue of causality.

Thus the pressure field at element i on the surface of the body can be formulated at the kth time step in the form

$$P(r_i, t_k) = \frac{1}{2\pi}\sum_{j=1}^{N}\left[\frac{1}{R_{ij}}(-\rho\ddot{w}_j)_{t_{ret}} + \frac{1}{R_{ij}^2}\cos\gamma_{ij}P_{j_{t_{ret}}}\right.$$

$$\left. + \frac{1}{cR_{ij}}\cos\gamma_{ij}\left(\frac{\partial P_j}{\partial t}\right)_{t_{ret}}\right]S_j + \frac{1}{2\pi}\frac{f(t_{R_s})}{R_s} \tag{3.4}$$

The variables at retarded time in each time step shown in equation (3.4) are evaluated using a cubic spline interpolation in an explicit method, but combined also with an implicit method (see later).

One of the difficulties in computing the equation (3.4) may arise when the source point and the field point coincide on the surface of the body, i.e. i=j. The integrands in equation (3.1) become singular since $R_{ii} = 0$. This problem may be avoided by using the planar element approximation which considers the following limiting process [6,18]. For example, the second integral term in the equation (3.1) may be approximated by

$$\frac{1}{2\pi}\int_{s_i}\frac{1}{R_{ii}^2}\cos\gamma_{ii}P_{t_{ret}}dS_i \cong \frac{1}{2\pi}\lim_{\Delta s_i \to 0}\int_{s_i - \Delta s_i}\frac{1}{R_{ii}^2}\cos\gamma_{ii}P_{t_{ret}}dS_i \tag{3.5}$$

where Δs_i is an increment of area surrounding the source point on element i.

Since the term, $\partial R/\partial n = cos\gamma$, is zero for the planar element due to the orthogonality between the outward normal and the surface of the element, only the first integral term in equation (3.1) needs to be considered whenever the field point and source point coincide. Therefore,

$$\frac{1}{2\pi}\int_{s_j}\frac{1}{R_{ij}}(-\rho\ddot{w}_j)_{t_{ret}}dS_j = -\frac{1}{2\pi}(\rho\ddot{w}_i)_{t_{ret}}\int_{s_i}\frac{1}{R}dS_i\,, \quad i = j \tag{3.6}$$

This integral can be evaluated via two-dimensional Gaussian quadrature formula.

In order to illustrate the procedure in more detail, a general quadrilateral element is considered for the simple case of the singular integral in the form

$$\int_s \frac{1}{R}\,dS = \int_{-1}^{1}\int_{-1}^{1}\frac{1}{R(\xi,\eta)}|J(\xi,\eta)|\,d\xi\,d\eta$$

$$\cong \sum_{j=1}^{\sqrt{n}}\sum_{i=1}^{\sqrt{n}}\frac{1}{R(\xi,\eta)}|J(\xi,\eta)|\,W_jW_i \tag{3.7}$$

where ξ and η are the local coordinates, J is the Jacobian of the transformation, and W_i and W_j are the weighting functions.

Although the 1/R term approaches infinity as R tends to zero, this phenomenon does not occur in practice. This can be illustrated by the damped system response at the resonant frequency, where the peak amplitude occurs with a finite value. Since the system, in reality, always possesses a certain amount of damping which will prevent the response from diverging, the Gaussian quadrature formulation can achieve an accurate solution in practice.

Although the approximation of the variables using planar elements with uniform distributions on each element has produced good results, the use of higher-order interpolating functions representing the variables generally gives a beneficial trade-off in accuracy of the solution for the same number of elements, as described by Seybert et al[19].

4 An Improved Hybrid Explicit-Implicit Method

In the original version of the Boundary Element Transient Acoustic Analysis (BETAA) program, all the variables at retarded time on each time step were either interpolated or extrapolated explicitly as a linear or higher-order combination of the computationally known past values. This method is referred as to the explicit or forward integration method. Though the numerical procedure based on the explicit integration method has produced good results when the time increment, Δt, does not exceed a certain fraction of the smallest period of the structure, there is ample motivation to improve the numerical time integration scheme. Usually most explicit methods are conditionally stable, while the most unconditionally stable methods are implicit, however, there is a trade-off in computing time.

In order to solve a problem such as the punch press with the explicit method for an upper frequency limit of 2 kHz, the time step, Δt, required might be 20 μsec or smaller to obtain an acceptable accuracy in the final solution. The time step (20 μsec) was selected based on the smallest period of the structure of interest as well as the smallest size of the element. The total number of time steps which then result will be so large to be incompatible with mini- or micro-computers as well as for numerical stability in the final solution. Thus a combined explicit-implicit method (or hybrid method) has been developed to improve the computational efficiency and the numerical stability, see Chun[20].

In order to categorize the above methods, further discussion about the role of Δt and the distance R_{ij}, which comes from the geometry of the body, is required. When $\Delta t \leq$ Minimum value of R_{ij}/c, all variables could be obtained by interpolating from the known past values explicitly, since the effect at the field point could not be caused by the source point within the time interval Δt. However, when $\Delta t \geq$ Minimum value of R_{ij}/c, the solution for the unknown variables does not only depend on the previous values but also on the present values, as yet unknown, causing the solution procedure to be iterative.

One of the difficulties, which may arise when employing the explicit-implicit time integration method into the equation (3.1), is due to the inherent inaccuracy of numerical differentiation. The time derivative of the pressure in equation (3.4) may be approximated by the order of interpolating polynomials as follows.

$$\left.\frac{\partial P}{\partial t}\right|_k = \frac{P_k - P_{k-1}}{\Delta t}, \quad (Polynomial\ \ Order = 1)$$

$$\left.\frac{\partial P}{\partial t}\right|_k = \frac{P_{k-2} - 4P_{k-1} + 3P_k}{2\Delta t}, \quad (Polynomial\ \ Order = 2)$$

$$\left.\frac{\partial P}{\partial t}\right|_k = \frac{-2P_{k-3} + 9P_{k-2} - 18P_{k-1} + 11P_k}{6\Delta t},$$

$$(Polynomial\ \ Order = 3)\ (4.1)$$

where k represents the kth time step. Since the full advantage of the higher-order formulas may be realized as Δt becomes progressively smaller, the linear approximation was adopted to further formulate the equation (3.1) in implicit time integration steps.

Assuming that the retarded time, t_{ret}, exists between t_k and t_{k-1}, then the pressure and its time derivative on element j may be approximated in the form

$$P_{j_{t_{ret}}} = (1 - \alpha_{ij})P_j^k + \alpha_{ij}P_j^{k-1}, \quad \left.\frac{\partial P_j}{\partial t}\right|_{t_{ret}} = \frac{P_j^k - P_j^{k-1}}{\Delta t} \quad (4.2)$$

where $\alpha_{ij} = (t_k - t_{ret})/(t_k - t_{k-1})$. Thus whenever $\Delta t \geq R_{ij}/c$, the pressure at a field point on the jth element is influenced by the element i at the kth time step and may be evaluated by

$$P(r_i, t_k) = \frac{1}{2\pi}\left[\frac{1}{R_{ij}}(-\rho\ddot{w}_j)_{t_{ret}} + \frac{1}{R_{ij}^2}\cos\gamma_{ij}\{(1-\alpha_{ij})P_j^k + \right.$$

$$\left.\alpha_{ij}P_j^{k-1}\} + \frac{1}{cR_{ij}}\cos\gamma_{ij}\left(\frac{P_j^k - P_j^{k-1}}{\Delta t}\right)\right] S_j + \frac{1}{2\pi}\frac{f(t_{R_s})}{R_s} \quad (4.3)$$

Notice that the RHS equation includes the values at present time step, yet unknown. Fig. 2 shows the graphical view of the numerical evaluations.

By substituting the equation (4.2) into equation (3.1), the system of linear equations for the combined explicit-implicit method yields the matrix form

$$[A]\cdot\{P\} = \{B\} \quad (4.4)$$

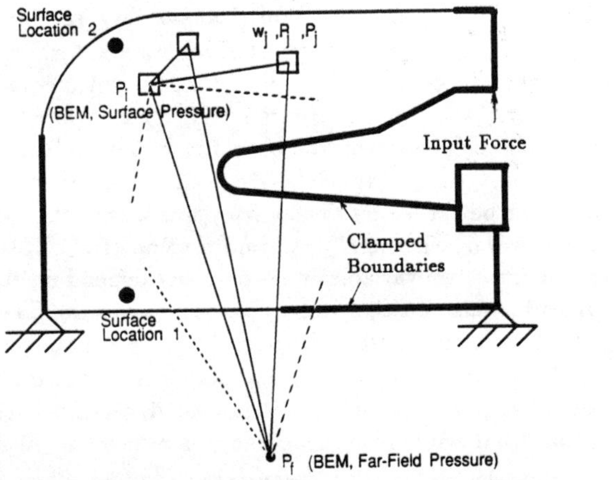

Fig.2 Geometrical Arrangement of the Elements in the Boundary Element Method

where,

$$A_{ij} = -\frac{1}{2\pi}\left[\frac{1}{R_{ij}^2}\cos\gamma_{ij}(1-\alpha_{ij}) + \frac{1}{cR_{ij}}\frac{\cos\gamma_{ij}}{\Delta t}\right] S_j, \quad (i \neq j)$$

$$A_{ij} = 1, \quad (i = j)(4.5)$$

$$B_i^k = \frac{1}{2\pi}\sum_{j=j_{ex}}^{N}\left[\frac{1}{R_{ij}}(-\rho\ddot{w}_j)_{t_{ret}} + \frac{1}{R_{ij}^2}\cos\gamma_{ij}P_{j_{t_{ret}}} + \right.$$

$$\left.\frac{1}{cR_{ij}}\cos\gamma_{ij}\left(\frac{\partial P_j}{\partial t}\right)_{t_{ret}}\right] S_j + \frac{1}{2\pi}\sum_{j=j_{im}}^{N}\left[\frac{1}{R_{ij}}(-\rho\ddot{w}_j)_{t_{ret}} + \right.$$

$$\left.\left\{\frac{1}{R_{ij}^2}\cos\gamma_{ij}\alpha_{ij} - \frac{1}{cR_{ij}}\frac{\cos\gamma_{ij}}{\Delta t}\right\} P_j^{k-1}\right] S_j + \frac{1}{2\pi}\frac{f(t_{r_s})}{R_s} \quad (4.6)$$

Here, in the expression for B_i^k, there are two types of terms namely j_{ex} and j_{im} which represent the element j for explicit and implicit time integration steps respectively. That is for the cases $\Delta t \leq R_{ij}/c$ and $\Delta t \geq R_{ij}/c$, which can be illustrated as follows.

In procedure (a), referring to Fig. 3 below, when t_{ret} exists before the $(k-1)^{th}$ time step when the source and field points are relatively far apart (i.e., $\Delta t \leq R_{ij}/c$), then the retarded time values could be accurately evaluated explicitly by using a four point cubic spline interpolation between the $(k-1)^{th}$ value and below in the case shown here. This method is referred as the explicit integration method, since the variables are evaluated using the known past values up to the $(k-1)^{th}$ time step.

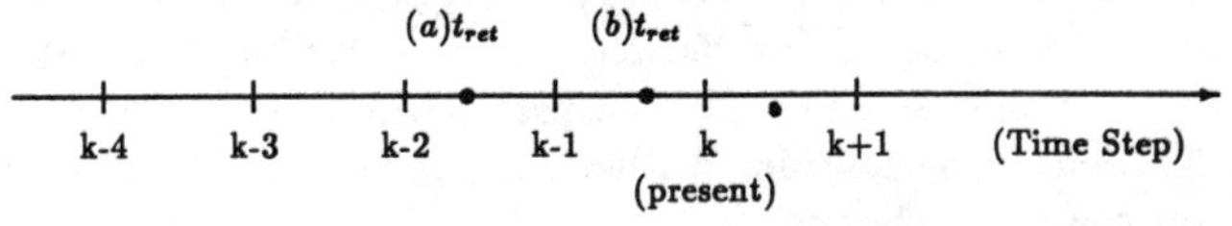

Fig.3 Schematic of the Time Steps for Time Integration Method

In procedure (b), when t_{ret} occurs after the $(k-1)^{th}$ time step when the source and field points are relatively close together (i.e., $\Delta t \geq R_{ij}/c$), then the solution for the unknown variables depends on the present values at the kth time step, as yet unknown, see Fig. 3. This method is referred as the implicit integration method. A simple extrapolation from previous (k-1), (k-2), etc. values would lead to a diverging solution. Instead each unknown k^{th} value now leads to an additional equation and is solved by an iterative procedure in the pressure matrix {P}. A further simplification may be obtained in the present case in that f(t)=0, ie, no external sources are present.

This hybrid time integration method took advantage of the reduced computational time of the explicit scheme as well as the numerical stability of the implicit scheme. A wider range of element sizes for describing the punch press could be used which resulted in a mesh of fewer elements. As mentioned earlier, the implicit method solves the system of linear equations whose solution procedure normally requires most of the computing time. Computing time was further reduced by using the FORTRAN subroutine Sparse Linear Equations Package, SPARSPAK, in BETAA. In this form, the effective system matrix needed to be factorized only once for the entire response calculation.

In order to show the validity of these numerical procedures, a test case was run for a simply-supported square-plate of dimensions $36 \times 36 \times 1/4$(inches) with a pulse-type forcing function of width 0.2 msec applied at the plate center. The plate was discretized by using 144 planar square elements for computing the surface pressure. A time step of 0.2 msec. was selected in the solution procedure. The results are shown in Fig. 4 for the first 5 time steps with the three parameters, i.e. surface velocity, pressure and instantaneous acoustic intensity displayed. As shown in the figure, the "wake" could be observed following the initial sharp wave-front which is similar to the case of the transient membrane.

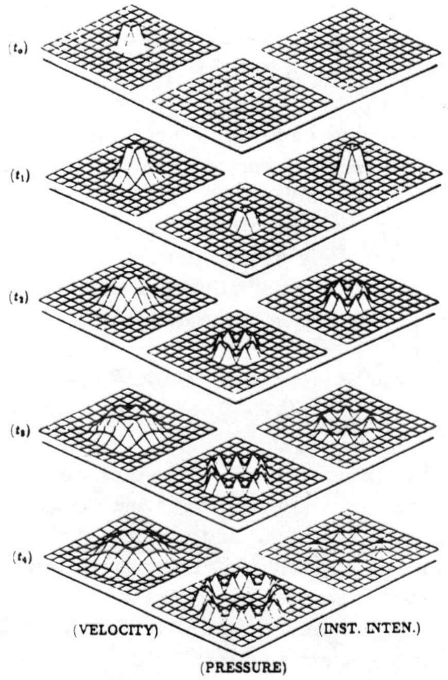

Fig.4 Time Responses for Impacted Simply-Supported Plate at t_o=0.2 msec, $2t_o$, $3t_o$ etc.

It may be noted that for comparison, the explicit integration method uses 24.2 minutes of solution time to obtain the time period of up to 20 msec as compared with the hybrid integration method which uses 17.3 minutes of solution time on the VAX 11/785 computer. This enhancement is due, in part, to the use of sparse matrix solution techniques to solve the system of linear equations.

The input boundary conditions used for the surface accelerations were obtained by differentiating the calculated surface displacements via a cubic spline interpolation. Solid modelling and the structural response of the plate were obtained using the pre- and post-processing FEM structural analysis code, ANSYS. The flow chart for the overall numerical procedure is shown in Fig. 5.

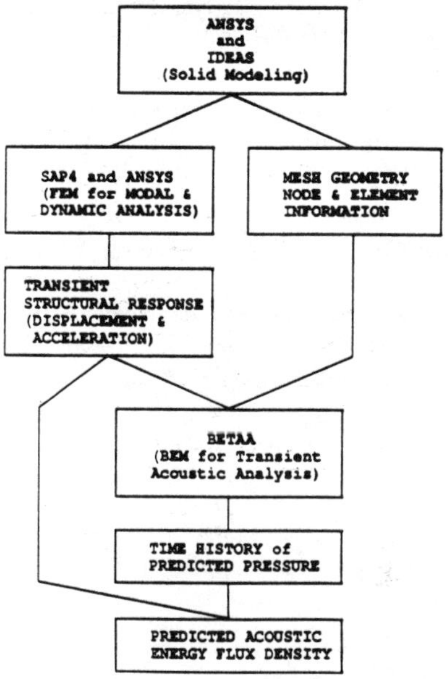

Fig.5 Flow Chart for the Overall Numerical Analysis

5 Application to the Punch Press (Steady-State)

5.1 Modal Analysis

In order to conduct the acoustic analysis, it is necessary to supply the BC's for the wave equation. The BC's in this case are the surface accelerations which could be obtained by differentiating the surface displacement.

ANSYS was first used for the dynamic analysis, since the punch press model was built from this software. ANSYS was used for the preliminary model check using only 132 elements due to a wavefront limit with our current educational version at NCSU. Consequently, SAP IV was used for the dynamic analysis, although one of the disadvantages of SAP IV is the excessive amount of scratch disk space required for large scale problems, especially for problems of the forced dynamic response type. Thus it was decided to modify the SAP IV program to reduce the amount of disk I/O and saving the CPU time by efficient coding methods. It is noted that, for comparison, the CPU time was reduced from 112 seconds to 52 seconds for the plate model using 132 elements on a MicroVAX II. This CPU time reduction was obtained simply by reforming the element library for each of the two element types, and increase the working matrix size, and the modification of the computational flow for the disk I/O. The overall gain from these modifications for the forced dynamic response of the punch press was a reduction in disk space from 45 Mbytes to 5 Mbytes and a reduction in CPU time from 3.5

Hrs to 52 min.

As a preliminary understanding of the dynamic behavior of the punch press, a computational steady-state modal analysis was performed to calculate the eigenvalues and corresponding eigenvectors using the modified SAP IV program. The extracted eigenvalues and eigenvectors were stored for the later forced dynamic analysis by using the "RESTART" capability. The first 10 natural frequencies were 201 Hz, 205 Hz, 360 Hz, 368 Hz, 397 Hz, 590 Hz, 679 Hz, 762 Hz, 820 Hz and 879 Hz. Typical mode shapes (for the fifth mode) are shown in Fig. 6 for motion of side-plates. Other parts such as cross-plates, punch and die area were treated as lumped masses.

5.2 Acoustic Radiation from Punch Press

In order to demonstrate the generalized BEM code for the steady-state analysis, the surface pressure and acoustic intensity patterns were calculated for the 5th mode shape following the previous work by Chun[18].

The pressure field on the surface for the BEM steady-state analysis has the form given by

$$P(r) = \frac{1}{2\pi}\int_{s_o} P(r_o)\frac{e^{ikR}}{R}\left(\frac{1}{R} - ik\right)cos\gamma\, dS_o$$

$$-\frac{i\omega\rho}{2\pi}\int_{s_o} U(r_o)\frac{e^{ikR}}{R}\, dS_o + \frac{Fe^{ikR_s}}{4\pi R_s} \quad (5.1)$$

where k is the wave number, ρ is the density of air, F is the noise source and R_s is the distance between the field point and the noise source location.

The discretization process yields the system of linear equations on the computer in the form of

$$[(1 - F\frac{e^{ikR_s}}{4\pi R_s})\delta_{ij} - D_{ij}]\{P_j\} = [M_{ij}]\{U_{ij}\} \quad (5.2)$$

where,

$$D_{ij} = \frac{k}{2\pi}\int_{s_j}\frac{e^{ikR_{ij}}}{R_{ij}}\left(\frac{1}{R_{ij}} - i\right)cos\gamma_{ij}\, dS_j,\ M_{ij} = -\frac{ik}{2\pi}\int_{s_j}\frac{e^{ikR_{ij}}}{R_{ij}}\, dS_j \quad (5.3)$$

Here M_{ij} and D_{ij} are the so-called monopole and dipole terms respectively.

The BEM program is now used to predict the surface pressure and to calculate the surface acoustic intensity, again typical results (for the fifth mode) are shown in Fig. 6. As the figure indicates, maximum acoustic intensity usually corresponds to the maximum deflection while there is little intensity along the nodal lines. However this phenomena is not always applicable to the transient response which will be discussed in the next chapter.

6 Transient Analysis of the Punch Press

6.1 Input Force Time History

Throughout this investigation, it was confirmed many times that the input force is one of the most important parameters in the prediction of the dynamic response of the punch press.

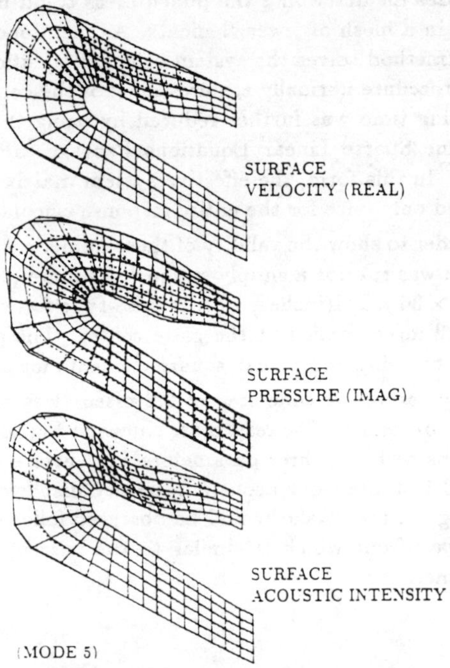

Fig.6 The Steady-State Response for the 5th Mode of the Punch Press

The time history of the forcing function to open the C-frame of the punch press may be represented based on measurements, as shown in Fig. 7. As indicated in this figure, there is a period of time in which the C-clamp of the frame is statically loaded up. Then, as the tool die forms the work piece, the C-frame is released and produces radiated noise as a result of this transient. The time constant of release corresponds to a frequency of the order of 10 kHz such that the frequencies from 10 kHz down to DC are stored as energy in the structure. (This is a subtle point but is a most important physically as a result of the DC level of the time waveform.)

However, the actual typical input force time history has a more complicated shape containing some decaying oscillatory signals, immediately after the material fracture, with relatively high amplitude. It was felt that this residual oscillatory force was caused by an interaction between the punch, die and stock material while the C-frame of the press body is excited. In the analysis, it was important to incorporate these measured details of the force time-history after material fracture. Without this experimental information, incorrect displacements over some areas of the press will be predicted.

6.2 Forced Dynamic Response

Using the above described forcing function, a forced dynamic response analysis by mode superposition was performed using the modified SAP IV by means of the restarting procedure from the steady-state case in the program. It should be noted here that the mode superposition method is used for the FEM structural forced dynamic analysis to obtain the displacements over the press in the time domain. The direct time integration method is applied for calculating the surface and far-field acoustic pressures.

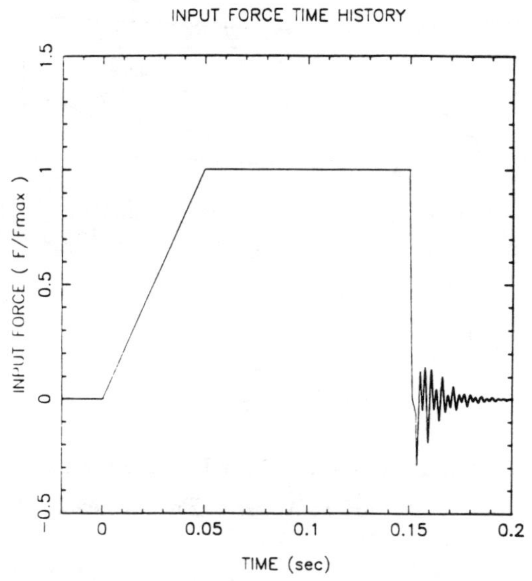

Fig.7 Input Force Time-History of the Punch Press

The advantage of mode superposition is that a variety of forced response analysis can be carried out with relatively small additional CPU time and disk space usage, once the frequencies and mode shapes are obtained. However, the direct integration method requires the solution of the triangularized effective stiffness matrix in each time step, it is a time-consuming task for a large scale problem in general.

The displacement time histories were calculated with a time step of 0.05 msec which corresponds to a sampling rate of 20 kHz. Although this rate is higher than that required by the analysis limit of 2 kHz in the study, it was selected in order to obtain better accuracy for numerical differentiation. For comparison purposes, two locations on the press surface are of interest, namely locations 1 and 2, see Fig. 2, which correspond to the maximum and minimum acoustic energy flux but with similar displacement amplitude.

6.3 Surface Acoustic Energy Flux Density

Using the geometry information obtained from ANSYS and the surface accelerations over the press which were obtained from the dynamic analysis using the modified SAP IV, the surface pressures on the punch press were then calculated using the Boundary Element Transient Acoustic Analysis (BETAA) program via equation (4.4). It should be noted that the use of a planar element further simplifies the solution procedure, because $cos\gamma_{ij} = 0$, when element i and j are on the same plane. This enabled the use of a sparse matrix technique which resulted in a more efficient solution of the system of linear BEM acoustic equations. All the retarded time values were calculated using a four-point cubic spline interpolation and the numerical integrations were carried out based on a 36-point Gaussian quadrature scheme.

The time step used for calculation was 0.05 msec to allow a good accuracy in the solution, see earlier remarks. A value of 0.1 seconds was selected for the time period for the transient responses, since most of the signals decayed away after 0.1 seconds of duration. All computations were done on the Departmental VAX 11/785 computer system and a MicroVAX II. It is noted that the total CPU time to compute 1000 time-steps on the MicroVAX II took 26 hours of pure CPU time and more than several Mbytes of scratch disk space. This indicates that future studies on this type of problem require either the super-computing environment or parallel processor technology.

The predicted surface pressures for locations 1 and 2 as compared with measured results are shown in Fig. 8 and Fig. 9. The acoustic energy density shown in these figures are obtained by the definition of the acoustic energy flux density in the form of

$$E = \int_{t_1}^{t_2} \vec{I} \cdot \vec{n}\, dt \tag{6.1}$$

where the vector $\vec{n}$ defines the orientation of the area considered. And in the frequency domain,

$$E(r,f) = Re[P(r,f) \cdot U^*(r,f)] \tag{6.2}$$

where P and U^* are the pressure and the complex conjugate of the velocity spectra respectively, t_1 and t_2 correspond to the time interval of the transient event, and $\vec{I_i}$ is the instantaneous acoustic intensity defined by

$$\vec{I_i}(r,t) = Re\{p(r,t)\} \cdot Re\{\vec{u}(r,t)\} \tag{6.3}$$

Referring to Fig. 8, which describes location 1, very strong sound radiation occurs at apparently a strong modal line in both displacement, pressure and acoustic energy flux density. Another observation may be made concerning results for location 2 shown in Fig. 9. Even though the amplitude of the deflection is of the order of the maximum as at the location 1, there is little sound radiation from this area modally. However, as will be seen below, these results describe only a part of the overall sound mechanism.

The overall surface acoustic energy density level is shown on a relative basis by contour lines in Fig. 10. These levels were obtained from the acoustic energy spectra for the transient noise by adding all the frequency components from 0 to 2 kHz. As the figure indicates, the bottom corner is the strongest sound radiator while the upper middle portion is in the mid-range. The top corner of the C-frame shows little radiation as indicated in Fig. 9.

It is noted that this energy density level clearly shows where the main noise radiation occurs for the transient event of the punch press. This completes the transient noise source identification by numerical means.

6.4 Far-Field Acoustic Energy Flow

Using the surface pressure informations predicted from BEM, the BETAA program enables the calculation of the pressure field at any place in the domain. In order to investigate the far-field transient noise radiation from the punch press, a field point in the domain was selected at 10 ft away from the punch press sideplate. The pressure field equation for the far-field is obtained from equation (3.1) by substituting $C_o = 2$. It is noted that the evaluation of the pressure at time, t_k, does not require the

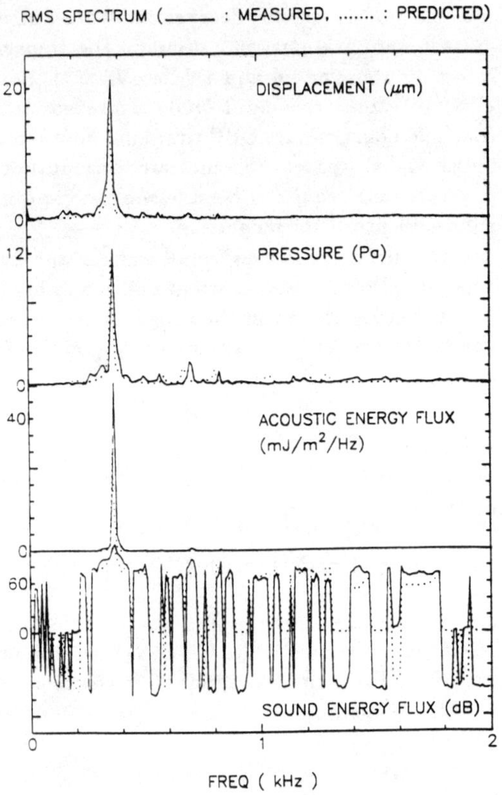

Fig.8 Comparison of the Predicted and Measured Frequency Spectra at Location 1 on the Press

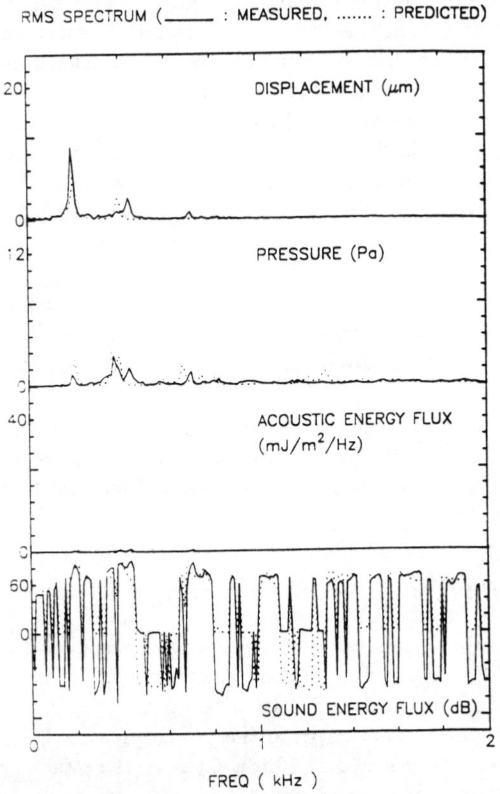

Fig.9 Comparison of the Predicted and Measured Frequency Spectra at Location 2 on the Press

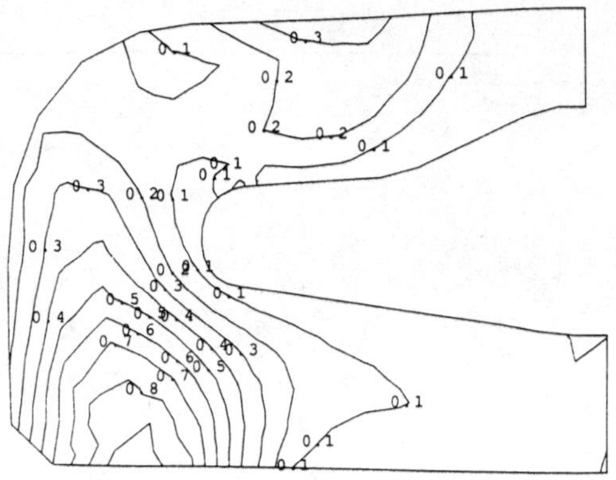

Fig.10 Contours of the Relative Overall Surface Acoustic Energy Level

implicit time integration scheme because the distance between source point and field point, R_{ij}, is generally long enough so that the Green's function effect may not happen within a single time step. Thus, the numerical evaluation of the far-field response for the punch press can be simplified by the equation

$$P_i^k = \frac{1}{4\pi} \sum_{j=1}^{N} \left[\frac{1}{R_{ij}} (-\rho \ddot{w}_j)_{t_{ret}} + \frac{1}{R_{ij}^2} \cos\gamma_{ij} P_{j_{t_{ret}}} + \frac{1}{cR_{ij}} \cos\gamma_{ij} \left(\frac{\partial P_j}{\partial t} \right)_{t_{ret}} \right] S_j \qquad (6.4)$$

The predicted and measured time histories of the far-field pressure and its frequency spectrum are shown in Fig. 11 and Fig. 12 respectively. It is noted that although there is a time-delay effect in the far-field response due to the wave traveling some distance, in this case 10 ft, the figure only shows the triggered portion for the FFT purpose. The figure shows some peaks in the spectrum plot which might be corresponding to the modes.

7 Discussion and Conclusion

Fig. 8 describes the surface location corresponding to the maximum sound radiation and shows the matching of several frequency components among the three illustrated parameters. The experimental results are also plotted on this graph and are typical of the excellent agreement obtained between measured and predicted results at all locations. A comparison at location 2, see Fig. 9, indicates little sound radiation from this surface location even though a maximum displacement amplitude occurred there, see also Fig. 10. The agreement between the results from both the BEM and experimental analyses demonstrates the accuracy of the computational technique as well as the validity of the surface acoustic intensity probe method.

In order to confirm the numerical prediction of the radiated far-field noise, Fig. 12 shows the time history of the measured far-field pressure response triggered at material fracture and the RMS spectrum. By comparing these results with the predicted

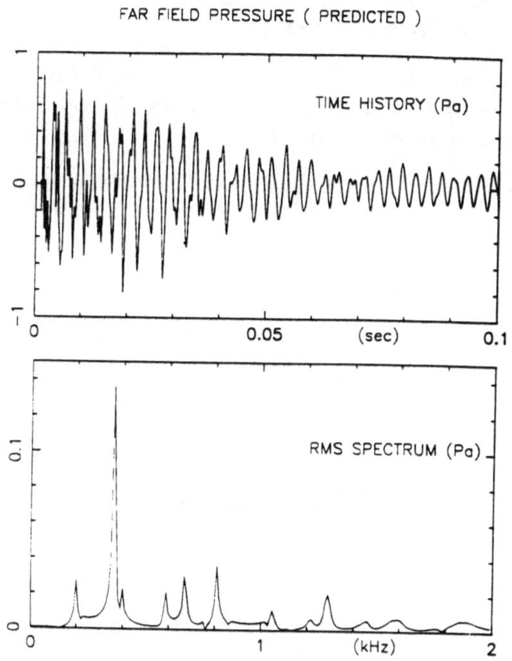

Fig.11 Predicted Far-Field Pressure Time History and Frequency Spectrum at 10 ft Distance from the Punch Press

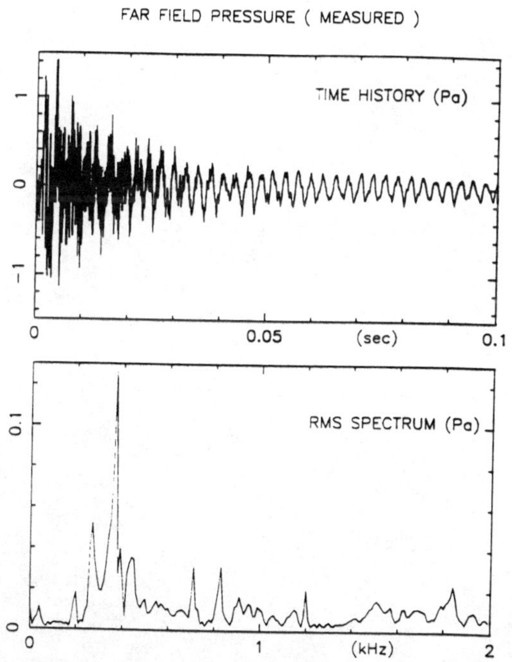

Fig.12 Measured Far-Field Pressure Time History and Frequency Spectrum at 10 ft Distance from the Punch Press

response shown in Fig. 11, it is noted that both figures are in good agreement except for some minor differences shown in the measured response. It was believed that the slight differences might be caused by the resolution limit of predicted frequency response in the numerical technique as well as by excitations of the press supporting frame and the sustaining floor of the anechoic chamber.

Both figures indicate the presence of some strong modal lines. However, these are not the major noise sources as can be seen from the dB plot of sound energy flux in Fig. 8, where a study was performed to confirm the significance of the non-resonant frequency components in transient wave mechanism by simulating the sound energy reduction on the computer. Using the computerized measured data, any modal frequencies within 5 % (13 dB) of the highest peak were set to zero. A reduction of less than 5 dB in overall SPL results. For a steady-state case, this reduction could be of order 15 dB. This numerical study confirms that the overall acoustic energy is not necessarily contained in so-called characteristic modes. Instead, in the time available between the next impact, the millions of progressive waves generated by the impact force are the major sound radiation mechanism in this transient wave problem. These non-resonant waves, though small in amplitude, should not be neglected since there are many of them as compared with the number of the natural modes, see also Coleman and Hodgson[21].

Thus, the conventional modal analysis approach to reduce the overall transient noise level may not be successful. This study demonstrates why punch press noise has not been easily reduced. In order to achieve transient noise reduction in this case, it would appear that all modes in the structure must be suppressed. This suppression may require a large amount of constrained layer damping applied over the whole press surface. At this time, this would appear to be uneconomical.

References

[1] Chen, L.H. and Schweikert, D.G. "Sound Radiation from an Arbitrary Body", J. Acous. Soc. Am. Vol.35, No.10, October 1963.

[2] Chertock, G. "Sound Radiation from Vibrating Surfaces", J. Acous. Soc. Am. Vol.36, No.7, July 1974.

[3] Kleinman, R.E. and Roach, G.F. "Boundary Integral Equations for the Three- Dimensional Helmholtz Equation", STAM Review, Vol.16, No.2, April 1974.

[4] Copley, L.G. "Fundamental Results Concerning Integral Representations in Acoustic Radiation", J. Acous. Soc. Am. Vol.44, No.1, 1968.

[5] Schenck, H.A. "Improved Integral Formulation for Acoustic Radiation Problems", J. Acous. Soc. Am. Vol.44, No.1, 1968.

[6] Koopman, G.H. and Benner, H. "Method for Computing the Sound Power of Machines Based on the Helmholtz Integral", J. Acous. Soc. Am. Vol.71, No.1, 1982.

[7] Banerjee, P.K. and Butterfield, R. 'Developments in Boundary Element Methods I', Applied Science Pub., London, 1979.

[8] Brebbia, C. 'The Boundary Element Method for Engineers', John Wiley, New York, 1978.

[9] Groenenboom, P.H.L. "The Application of Boundary Elements to Steady and Unsteady Potential Flow Problems in Two and Three Dimensional Problems", in Boundary Element Methods (ed. Brebbia, C.A.), Springer Verlag, Berlin, 1981.

[10] Manolis, G.D. "A Comparative Study on Three Boundary Element Method Approaches to Problems in Elastodynamics", Int. J. Num. Method Eng., Vol.19, 1983.

[11] Mansur, W.J. and Brebbia, C.A. "Formulation of the Boundary Element Method for Transient Problems Governed by the Scalar Wave Equation", Appl. Math. Modelling, Vol.6, August 1982.

[12] Mansur, W.J. and Brebbia, C.A. "Numerical Implementation of the Boundary Element Method for Two Dimensional Transient Scalar Wave Propagation Problems", Appl. Math. Modelling, Vol.6, August 1982.

[13] Rizzo, F.J. and Shippy, D.J. "A Method of Solution for Certain Problems of Transient Heat Conduction", AIAA Journal, November 1970.

[14] Rizzo, F.J. and Shippy, D.J. "An Advanced Boundary Integral Equation Method for Three-Dimensional Thermoelasticity", Int. J. Num. Meth. Eng., Vol.11, 1977.

[15] Rizzo, F.J., Shippy, D.J., and Rezayat, M. "A Boundary Integral Equation Method for Radiation and Scattering of Elastic Waves in Three Dimensions", Int. J. Num. Meth. Eng., Vol.21, 1985.

[16] Dubner, H. and Abate J. "Numerical Inversion of Laplace Transforms by Relating Them to the Finite Fourier Cosine Transform", J. Asso. Com. Machinery, Vol.15, No.1, January 1968.

[17] Dohner, J.L., Shoureshi, R., and Bernhard, R.J. "Transient Analysis of Three-Dimensional Wave Propagation Using the Boundary Element Method", Int. J. Num. Meth. Eng., Vol.24, 1987.

[18] Chun, D.H. "Measurement of Surface Acoustic Intensity and Comparison with the Finite-Element Method", M.S. Thesis, Mechanical and Aerospace Engineering Department, North Carolina State University, August 1984.

[19] Seybert, A.F., Soenarko, B., Rizzo, F.J, and Shippy, D.J. "An Advanced Computational Method for Radiation and Scattering of Acoustic Waves in Three Dimensions", J. Acous. Soc. Am. Vol.77, No.2, 1985.

[20] Chun, D.H. "A Computational and Experimental Study of Steady and Transient Acoustic Energy Flow from a Complex Mechanical Structure", Ph.D. Dissertation, Mechanical and Aerospace Engineering Department, North Carolina State University, December 1988.

[21] Coleman, R.B. and Hodgson, T.H. "Acoustic Intensity Measurements on a 350 kN Automatic Punch Press", Inter-Noise 83, Edinburgh, Scotland, 1983.

RADIATION OF SOUND FROM A FINITE LENGTH ORTHOTROPIC CYLINDRICAL SHELL EXCITED BY AN INTERNAL ACOUSTIC PRESSURE

H. Koriyama
Energy Engineering and Design Department
NKK Corporation
Yokohama, Japan

S. Chonan
Department of Mechanical Engineering
Tohoku University
Sendai, Japan

An analysis is presented for the radiation of Sound from a simply supported orthotropic cylindrical shell excited by a sinusoidally varying, uniform internal acoustic pressure. The problem is studied based on a thick shell theory in which the effects of rotatory inertia and shear deformation are taken into account. The air surrounding the shell is described by the Helmholtz linear acoustic equation. The governing equations of the system are solved by applying both the method of modal expansion and the method of complex Fourier transform with respect to the axial space variable. The radiation intensity around the shell is derived and calculated numerically for various values of the physical parameters involved.

1 INTRODUCTION

With the advance of technology, It has become commonly observed that metals such as steel and aluminum are replaced by newly developed lightweight nonmetals. However, such materials have a deficiency in that the sound insulation characteristics are not always satisfactory compared with the metals, which is a cause of the complaint that the noise leaking out of the structure has increased these days. This is true particularly for the lightweight cylindrical shells installed indoors. It is therefore said that the investigation on the acoustic characteristics of such structures has become more important than ever.

Limiting the problem to the acoustic characteristics of cylindrical shells, Koval(1980), Fuller(1986a, 1986b, 1988) and Chonan(1988) studied the transmission of acoustic waves into or from an infinitely long cylindrical shell perfectly exposed in a free field. Further, Cummings et al.(1984), Chonan and Koriyama(1988) and Chonan(1989) investigated the radiation of sound from an infinitely long cylindrical shell which is partially exposed in a free field and excited by a positive traveling internal acoustic wave.

In all those papers it was assumed that the shell wall has no structural discontinuities in the axial direction, hence no reflected waves in the wall. When the shell is of finite length, both positive and negative traveling waves appear on the wall due to the reflection at the ends, and standing structural waves are formed. As a problem related to the finite length cylindrical shell, Harari and Sandman(1976) studied the response of orthotropic laminated cylindrical shells embedded in an acoustic fluid and excited by a time-harmonic internal load. Numerical results and discussion were presented on the displacement response of the shell wall. Junger(1985) investigated the sound field excited by vibrating boundaries in a cylindrical shell. In view of those papers it appears, to the authors' knowledge, that so far detailed study has not been done on the acoustic characteristics of a finite length cylindrical shell. It is in this context that the subject discussed in this paper is that of the sound radiation from a simply supported orthotropic cylindrical shell excited by a sinusoidally varying, uniform internal pressure. The governing equations of the system are solved by applying both the method of modal expansion and the method of complex Fourier transform with respect to the axial space variable. The radiation intensity around the shell is derived and calculated numerically for various values of the physical parameters involved.

2 FORMULATION OF THE PROBLEM

Figure 1 shows a finite length orthotropic cylin-

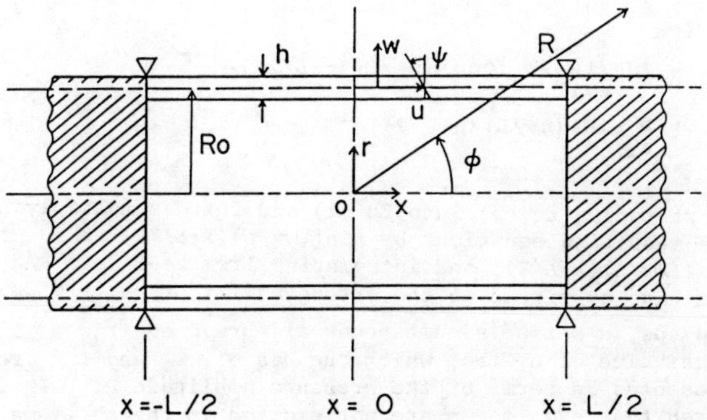

Fig.1 Finite length orthotropic cylindrical shell

drical shell of mean radius R_0, thickness h and length L. The rigid extensions of the vibrating section, at either end, are assumed as shown in the figure to make the theoretical analysis simple. It has been reported that such extensions make virtually no difference to the acoustic radiation(Cummings et al., 1984). The shell is excited by an axisymmetric internal pressure, whereby the θ-component of the displacement vanishes. The wall displacements in the x- and r-directions are denoted by u and w, and the rotation angle in the xr-plane by ψ. In this case, denoting the internal acoustic pressure by $p(x,t)$, one has the equations of motion of the shell as(Chonan, 1984)

$$[a_{ij}][u,\ w,\ \psi]^T=[0,\ b_2,\ 0]^T,\quad i,j=1,2,3, \tag{1}$$

where

$$a_{11}=(E_x h+P_x)\partial^2/\partial x^2-\rho h\partial^2/\partial t^2,$$
$$a_{12}=a_{21}=E_\nu(h/R_0)\partial/\partial x,$$
$$a_{13}=a_{31}=-E_x(h^3/12R_0)\partial^2/\partial x^2+\rho h\partial^2/\partial t^2,$$
$$a_{22}=-(\kappa G_x h+P_x)\partial^2/\partial x^2+E_\theta(1/R_0^2)(h+h^3/12R_0^2)+\rho h\partial^2/\partial t^2,$$
$$a_{23}=a_{32}=\kappa G_x h\partial/\partial x,$$
$$a_{33}=E_x(h^3/12)\partial^2/\partial x^2-\kappa G_x h-\rho h^3/12,$$
$$b_2=(1-h/2R_0)p(x,t).$$

Here, E_x, E_θ, E_ν and G_x are the elastic moduli relating the stresses to strains through $\sigma_{xx}=E_x\varepsilon_{xx}+E_\nu\varepsilon_{\theta\theta}$, $\sigma_{\theta\theta}=E_\nu\varepsilon_{xx}+E_{\theta\theta}\varepsilon_{\theta\theta}$ and $\tau_{xr}=G_x\gamma_{xr}$; P_x is the axial pre-stress in the shell wall, $\kappa(=\pi^2/12)$ is the shear coefficient, ρ is the mass density, and t is time.

In the analysis, the radiation load on the outer surface of the shell is neglected. This seems to be justifiable in view of the fact that Astley and Cummings(1984) found that radiation loading had a negligible effect on the vibrating response of the walls of rectangular ducts to internal sound field,even at resonance. It is further assumed that the internal forcing function is a constant pressure independent of shell motion. Thus the solution to be presented is an uncoupled solution. The pressure within the shell is written as

$$p(x,t)=Pe^{i\omega t}, \tag{2}$$

where P is the pressure amplitude and ω is the angular frequency.

In the following, one will work with the shell simply supported at both ends. The displacements and rotation in this case are expressed in the form

$$\{u,\psi\}=\sum_{n=1}^{\infty}\{u_n,\Psi_n\}\cos[(n\pi/L)(x+L/2)]e^{i\omega t},$$
$$w=\sum_{n=1}^{\infty}W_n\sin[(n\pi/L)(x+L/2)]e^{i\omega t}. \tag{3}$$

Substituting Eq.(3) into Eq.(1) and further multiplying the resulting equations by $\sin[(m\pi/L)(x+L/2)]$ or $\cos[(m\pi/L)(x+L/2)]$ and integrating from $x=-L/2$ to $L/2$ one has, by virtue of the orthogonality property of sin and cos, a set of simultaneous algebraic equations for each value of n, from which one has U_n, W_n and Ψ_n represented in terms of the pressure amplitude P. It is noted that even modes are not excited in the shell wall for the uniform loading under consideration, that is, $U_n=W_n=\Psi_n=0$ for n=2,4,6,....

One next obtains the acoustic field in the air surrounding the shell wall. The pressure p_0 in the air is governed by the Helmholtz equation

$$[\partial^2/\partial r^2+(1/r)\partial/\partial r+\partial^2/\partial x^2]p_0=(1/c_0^2)\partial^2 p_0/\partial t^2, \tag{4}$$

where $c_0(=[K_0/\rho_0]^{1/2})$ is the sound speed, K_0 is the bulk modulus, and ρ_0 is the air density. p_0 is related to the shell vibration through

$$\partial(p_0/\rho_0)/\partial r|_{r=R_0+h/2}=-\partial^2 w_0/\partial t^2,\quad -\infty<x<\infty, \tag{5}$$

where

$$w_0(x)=w \quad |x|<L/2,$$
$$=0 \quad |x|\geq L/2. \tag{6}$$

Here, w is the radial displacement of the shell wall given by Eq.(3).

The solution of Eq.(4) satisfying the boundary condition (5) can be determined by applying the method of complex Fourier transform with respect to x, which is defined by

$$\bar{F}(\xi)=\int_{-\infty}^{+\infty}f(x)e^{i\xi x}dx,\quad f(x)=(1/2\pi i)\int_{-\infty}^{+\infty}\bar{F}(\xi)e^{-i\xi x}d\xi. \tag{7}$$

Transforming Eq.(4) with respect to x and solving the resulted equation give(Cummings et al., 1984)

$$\bar{p}_0=P_0H_0^{(2)}(\xi_r r)e^{i\omega t},\quad \xi_r^2=\xi_0^2-\xi^2=(\omega/c_0)^2-\xi^2 \tag{8}$$

where $H_0(\ )$ is the Hankel function of the second kind, and P_0 is an unknown constant. Substituting Eq.(8) into the transformed boundary condition (5), one has P_0 represented by W_n. Further, substituting P_0 into Eq.(8) and applying the inversion integral, one obtains the acoustic pressure p_0 in the form

$$p_0=\frac{\rho_0\omega^2e^{i\omega t}}{\pi}\int\frac{H_0^{(2)}[(\xi_0^2-\xi^2)^{1/2}r]e^{i\omega t}}{(\xi_0^2-\xi^2)^{1/2}H_0^{(2)\prime}[(\xi_0^2-\xi^2)^{1/2}(R_0+h/2)]}\times\sum W_n\left(\frac{n}{L}\pi\right)\left\{\frac{\cos(\xi L/2)}{(n\pi/L)^2-\xi^2}\right\}d\xi. \tag{9}$$

In general, ducts are installed separated far from the human ear, i.e., $1\ll r$. In this case the Hankel function is approximated by

$$H_0^{(2)}(\xi_r r)=(2/\pi\xi_r r)^{1/2}e^{-i(\xi_r r-\pi/4)} \tag{10}$$

One further introduces the spherical co-ordinates (R,ϕ,θ) in Eq.(9). Here, (R,ϕ,θ) are related to the original cylindrical co-ordinates through

$$r=R\sin\phi,\quad x=R\cos\phi,\quad \theta=\theta. \tag{11}$$

Equation (9) modified by substitution of Eqs.(10) and (11) is here calculated by using the method of steepest descent(Ewing et al., 1957). The result obtained is

$$p_0=\frac{2i\rho_0\omega^2e^{i\omega t}e^{-iR\xi_0}}{\pi R\xi_0\sin\phi H_0^{(2)\prime}[\xi_0(R_0+h/2)]}\times\sum W_n\left(\frac{n}{L}\pi\right)\left\{\frac{\cos(\xi_0 L\cos\phi/2)}{(n\pi/L)^2-(\xi_0\cos\phi)^2}\right\}. \tag{12}$$

The acoustic intensity at point (R,ϕ,θ) is

$$I_0=|p_0|^2/2\rho_0 c_0 \,. \tag{13}$$

The total radiated power from the shell is found by calculating

$$S=\int_0^{2\pi}d\theta\int_0^{\pi}I_0 R^2\sin\phi d\phi. \tag{14}$$

Dividing S by the area of outer shell surface, one obtains the averaged intensity of sound radiated from the shell wall:

$$I_t=S/2\pi(R_0+h/2)L. \tag{15}$$

To represent I_t in a non-dimensional form, one here introduces a constant reference intensity I_i. The intensity of a plane acoustic wave traveling in an infinite rigid duct is used as I_i. Here, I_i is given by

$$I_i=P^2/2\rho_0 c_0 \,, \tag{16}$$

where P is the pressure amplitude. One defines the radiation coefficient R_c as

$$Rc=10\log_{10}(I_i/I_t). \tag{17}$$

Substituting Eqs.(15) and (16) into Eq.(17), one has

$$Rc=10\log_{10}\{(\frac{\pi}{2\rho_0 c_0\omega})^2\frac{(R_0+h/2)L}{J}\}, \tag{18}$$

where

$$J=\int_0^{\pi}\frac{1}{\sin\phi|H_0^{(2)\prime}[\omega(R_0+h/2)\sin\phi/c_0]|^2}\times[\sum(\frac{W_n}{P})\frac{(n\pi/L)\sin^2(n\pi/2)\cos(\omega L\cos\phi/2c_0)}{(n\pi/L)^2\ (\omega\cos\phi/c_0)^2}]^2 d\phi$$

In the following, one will work with Eq.(18) to investigate the acoustic characteristics of the system. R_c is a parameter showing the radiation resistance of the shell wall against the internal disturbance. For example, $R_c=\infty$ means that the shell emanates no sound energy outside the wall. The function J cannot be obtained analytically, hence it was determined by the numerical integration based on the Simpson's 1/3 rule.

3 NUMERICAL RESULTS AND DISCUSSION

Numerical results have been obtained for the orthotropic FRP shell. The reference parameters used are

$$E_x=1.0\times10^{10}\ \mathrm{N/m^2},\ E_\theta=1.0\times10^{10}\ \mathrm{N/m^2},$$
$$E_\nu=0.3\times10^{10}\ \mathrm{N/m^2},\ G_x=0.35\times10^{10}\ \mathrm{N/m^2},$$
$$\rho=1.3\times10^3\mathrm{kg/m^3},\ R_0=0.05\ \mathrm{m},$$
$$h=0.004\ \mathrm{m},\ L=0.7\ \mathrm{m},\ P_x=0\ \mathrm{N/m}. \tag{19}$$

Some of the parameters are varied to study their effects on the acoustic characteristics of shell. The air density and sound velocity are $\rho_0=0.0012\times10^3\mathrm{Kg/m^3}$ and $c_0=340\mathrm{m/s}$, respectively.

Figure 2 shows the R_c curve for two values of the vibrating length of shell L. Results have been obtained by taking the first 15 modes of vibration into ac-

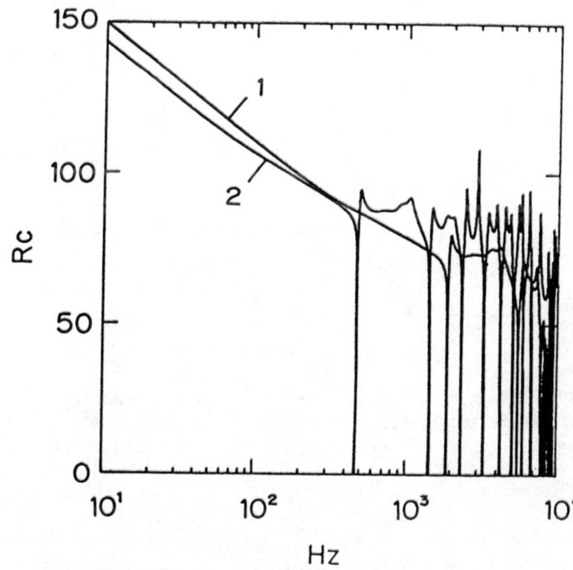

Fig.2 Variation of R_c curve with an increasing radiating length of shell; shell parameters(except L) are as shown in Eq.(19); curve 1, L(m)=0.7; curve 2, =2.8.

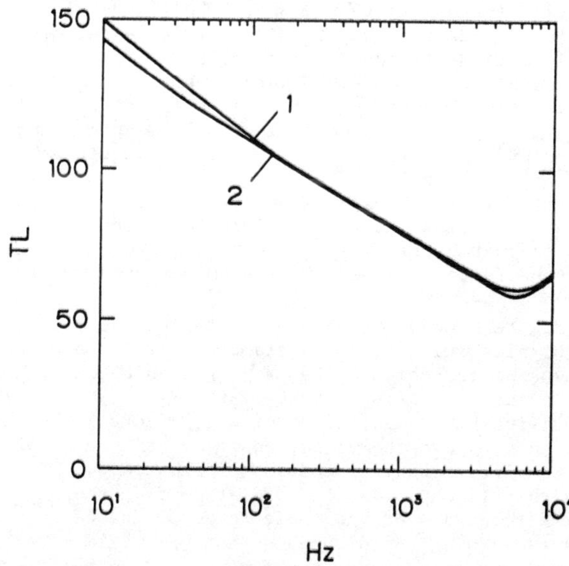

Fig.3 Variation of TL curve for an infinite shell exposed in free field by the length L and excited by a positive traveling internal plane acoustic wave; shell parameters(except L) are as shown in Eq.(19); curve 1, L(m)=0.7; curve 2, =2.8.

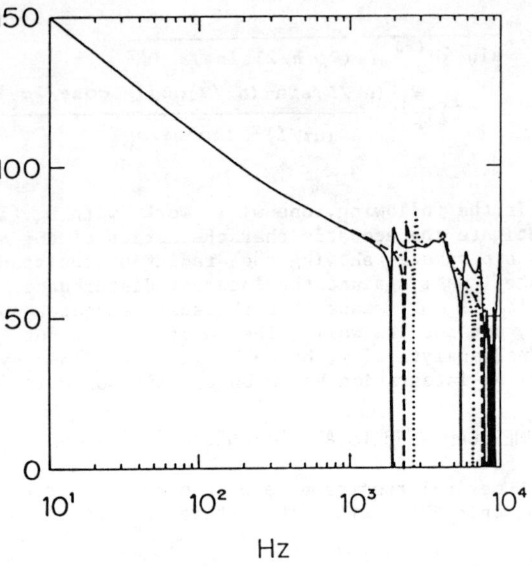

Fig.4 Variation of R_c curve with an increasing E_x ; shell parameters(except E_x) are as shown in Eq.(19); ———, E_x(N/m^2)=1.0×10^{10} ; — — — —, =1.5×10^{10} ; ·······, =2.0×10^{10} .

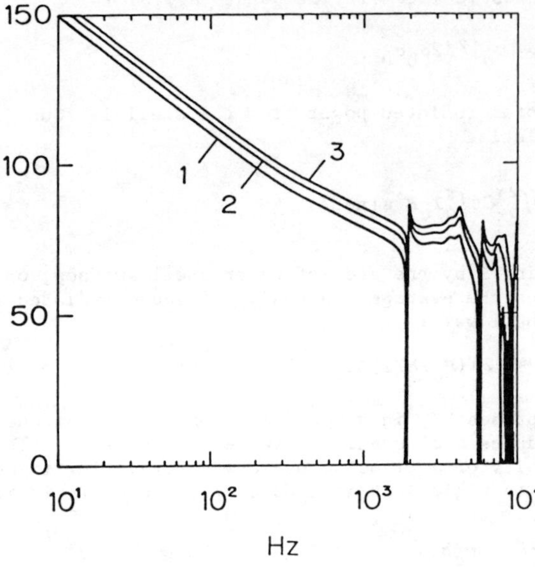

Fig.5 Variation of R_c curve with an increasing E_θ ; shell parameters(except E_θ) are as shown in Eq.(19); curve 1, E_θ(N/m^2)=1.0×10^{10} ; curve 2, =1.2×10^{10} ; curve 3, =1.4×10^{10} ; curve 4, =1.6×10^{10} ; curve 5, =1.8×10^{10} ; curve 6, =2.0×10^{10} .

count, i.e., n=15. R_c takes an infinite value at 0Hz, hence no acoustic radiation from the shell wall. It then decreases monotonically with an increase of the frequency and drops significantly at the first resonance frequency, whereby the sound insulation of the shell wall is degenerated. After the first drop, the R_c curve has repeated drops due to the appearance of the higher structural resonances. It is found from the curves that a long shell(curve 2) has a lower R_c(higher radiation intensity) than a short shell(curve 1) in the range of frequency lower than the first resonance(f_1= 350Hz), while this tendency is reversed for the frequencies higher than f_1. Figure 3 shows, as a comparison, the transmission loss TL of an infinite shell partially exposed in free field by the length L and excited by a positive traveling plane acoustic wave(Chonan and Koriyama, 1988), where TL is defined by $TL=10\log_{10}(I_i/I_t)$; the expressions of I_i and I_t are the same as Eqs. (15) and (16). It is found that the acoustic characteristics of the system are much affected by the factor whether the system has structural resonances or not.

Figure 4 shows the R_c curves taking the axial modulus of elasticity E_x as a parameter. The resonance frequencies for which R_c has a deep drop become greater with an increase of E_x. However the R_c value is rather insensitive to the variation of E_x. Figure 5, on the other hand, shows the variation of R_c as a function of E_θ, the circumferential modulus of elasticity. It is found that in this case the radiation from the shell wall is reduced over the whole range of frequency with an increase of E_θ. This means that the maximum reinforced direction of FRP should be placed in the circumferential direction when one needs to reduce the sound radiation from the shell wall.

The modulus of elasticity E_ν is the parameter prescribing the amount of the normal stress appeared in annex to the extension in the other direction. It cannot be modified artificially, however it may be of interest to see to what extent the radiation is affected by E_ν. Figure 6 is a plot of the R_c curve with E_ν taken as a parameter. It is seen that R_c is decreased uniformly and the radiation is enhanced with an increasing E_ν.

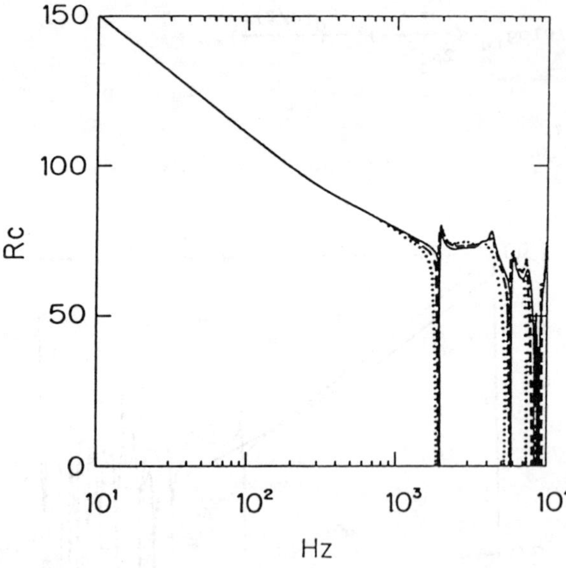

Fig.6 Variation of R_c curve with an increasing E_ν ; shell parameters(except E_ν) are as shown in Eq.(19); ———, E_ν(N/m^2)=0.2×10^{10} ; — — — —, =0.35×10^{10} ; ·······, =0.50×10^{10} .

Figure 7 shows the effect of G_x on the value of R_c. G_x is a parameter that controls the stiffness of wall, but it is found that R_c plays no important role on the acoustic characteristics of shell.

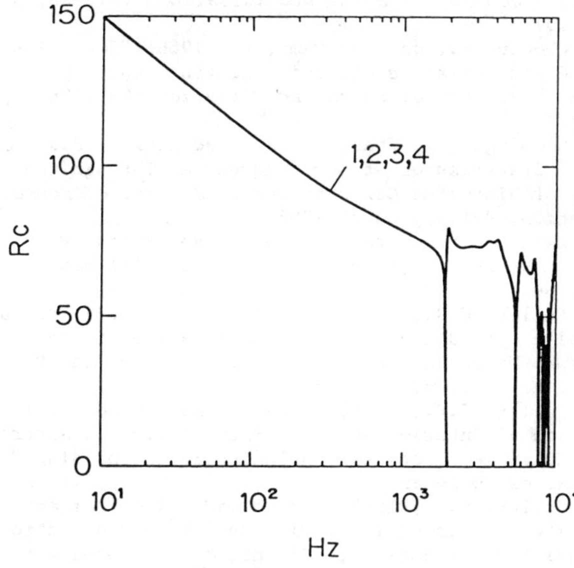

Fig.7 Variation of R_C curve with an increase of G_x; shell parameters(except G_x) are as shown in Eq.(19); curve 1, G_x(N/m^2)=0.35×10^{10}; curve 2, =0.50×10^{10}; curve 3, =0.65×10^{10}; curve 4, =0.80×10^{10}.

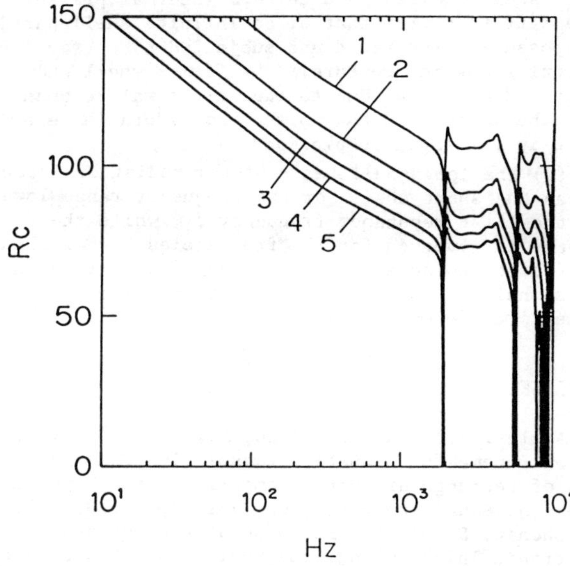

Fig.9 Variation of R_C curve with an increasing R_0; shell parameters(except R_0) are as shown in Eq.(19); curve 1, R_0(m)=0.01; curve 2, =0.02; curve 3, =0.03; curve 4, =0.04; curve 5, =0.05.

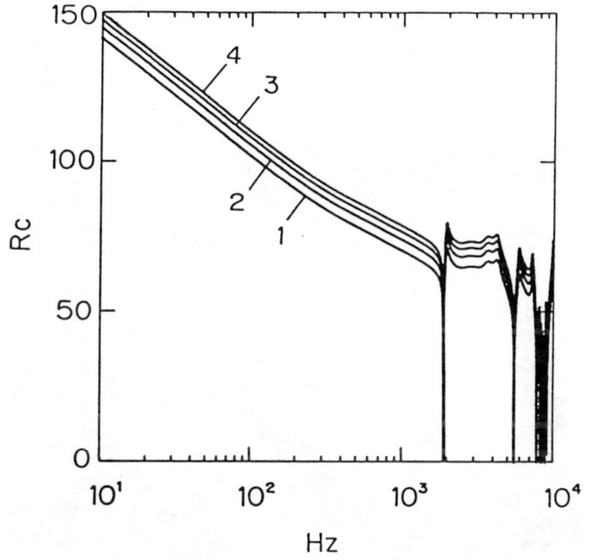

Fig.8 Variation of R_C curve with an increasing h; shell parameters(except h) are as shown in Eq.(19); curve 1, h(m)=0.002; curve 2, =0.003; curve 3, =0.004; curve 4, =0.005.

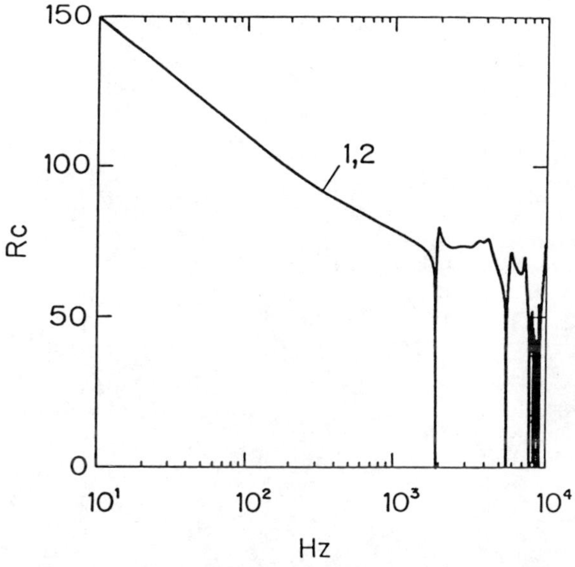

Fig.10 Variation of R_C curve with an increasing P_x; shell parameters(except P_x) are as shown in Eq.(19); curve 1, P_x/E_xh=1.0×10^{-3}; curve 2, =-1.0×10^{-3}.

Figure 10 shows the R_C curves when the shell is stressed in the axial direction. Two cases in which the shell wall is subjected to the tensile and compressive yield stresses are presented. It is seen that the resonance frequencies and the acoustic characteristics are insensitive to the variation of the pre-load. This is due to the fact that for the geometrical and physical parameters under consideration, the shell stiffness in the radial direction is sufficient to suppress the effect of the axial load.

4 CONCLUSIONS

A theory has been developed for the radiation of sound from a simply supported orthotropic cylindrical shell excited by a sinusoidally varying, uniform internal acoustic pressure. The radiation coefficient of the shell wall, R_C, has been derived and calculated in detail for the FRP shell. The results obtained can be summarized as follows.

(1) The radiation characteristics of a finite

length shell excited by a uniform internal pressure is quite different from that of an infinite shell partially exposed in free field and subjected to a traveling internal acoustic pressure. The former shell shows repeated drops in R_C due to the structural resonances while the latter one has no such considerable deterioration in the sound insulation.

(2) The long shell has a higher radiation intensity than the short shell for the frequency range lower than the first resonance frequency f_1, while the tendency is reversed for the frequencies higher than f_1.

(3) The sound insulation of shell wall is enhanced if the shell is constructed with its maximum stiffness in the circumferential direction.

REFERENCES

Astley, R.J., and Cummings, A., 1984, "A Finite Element Scheme for Acoustic Transmission Through the Walls of Rectangular Ducts: Comparison With Experiment," *Journal of Sound and Vibration*, Vol.92, pp.387-409.

Chonan, S., 1984, "Response of a Prestressed, Orthotropic Thick Cylindrical Shell Subjected to a Pressure Pulse," *Journal of Sound and Vibration*, Vol. 93, pp.31-38.

Chonan, S., 1988, "Low Frequency Soundproof Characteristics of Orthotropic Two-Layered Cylindrical Shells," *Journal of Sound and Vibration*, Vol.126, pp. 407-415.

Chonan, S., and Koriyama, H., 1988, "Sound Insulation Characteristics of Finite Length Orthotropic Shells," *Journal of Sound and Vibration*, Vol.126, pp. 525-532.

Cummings, A., Chang, I.-J., and Astley, R.J., 1984, "Sound Transmission at Low Frequencies Through the Walls of Distorted Circular Duct," *Journal of Sound and Vibration*, Vol.97, pp.261-286.

Ewing, W.M., Jardetzky, W.S., and Press, F., 1957, Elastic Waves in Layered Media, McGraw-Hill Book Co., Inc., New York, Toronto, London, pp.365-369.

Fuller, C.R., 1986a, "Radiation of Sound From an Infinite Cylindrical Elastic Shell Excited by an Internal Monopole Source," *Journal of Sound and Vibration*, Vol.109, pp.259-275.

Fuller, C.R., 1986b, "Analytical Method for Investigation of Interior Noise Characteristics in Aircraft With Multiple Propellers Including Synchrophasing," *Journal of Sound and Vibration*, Vol.109, pp.141-156.

Fuller, C.R., 1988, "Sound Radiation From an Infinite Elastic Cylinder With Dual-Wave Propagation - Intensity Distributions," *Journal of Sound and Vibration*, Vol.122, pp.479-490.

Koval, L.R., 1980, "Sound Transmission Into a Laminated Composite Cylindrical Shell," *Journal of Sound and Vibration*, Vol.71, pp.523-530.

A NUMERICAL STUDY OF THE COUPLED ELASTIC/ACOUSTIC WAVE PROPAGATION

A. F. Seybert and T. W. Wu
Department of Mechanical Engineering
University of Kentucky
Lexington, Kentucky

ABSTRACT

In this paper, the coupled acoustic/elastic boundary value problem is solved by an integral equation method. Both radiation and scattering of sound from an elastic solid or shell are examined. The elastic shell can be either a thick or thin shell. The coupled boundary value problem is reformulated into a pair of boundary integral equations, one for the acoustic domain (the host fluid) and the other for the elastic body. The integral equations are coupled by appropriate continuity conditions at the interface between the elastic body and the fluid. For an elastic shell, time harmonic loading in the form of either traction or displacement boundary conditions is prescribed on the inner surface of the body. The integral equations are then reduced to an algebraic form using quadratic isoparametric elements and solved simultaneously to determine the unknown pressure and displacements on the interface as well as the unknown displacements or tractions on the inner surface. The numerical results are compared to an analytic solution for a simple radiation problem. If the elastic shell is extremely thin, a coupled FEM/BEM technique is also suggested.

INTRODUCTION

The radiation and scattering of sound from an elastic body have been of practical importance in underwater acoustics. When the elastic body is submerged in a heavy fluid, such as sea water, the vibration characteristic of the body will be affected by the surrounding sound pressure. On the other hand, since the sound pressure itself also depends on the vibration of the elastic body, the whole system becomes a coupled elastic/acoustic problem. There have been many papers published using a veriety of analytical methods to examine the scattering of sound from hollow and solid elastic cylinders and spheres [1-5]. However, when the elastic body has an arbitrary shape, methods amenable to numerical implementation are usually more desirable than approximate analytical methods. In this category we include the T-matrix method [6-9], integral equation methods and hybrid methods such as the finite element/integral equation method [10].

The T-matrix method is attractive for obtaining farfield data if the shape of the scattering obstacle does not deviate substantially from the basis (e.g., spherical, spheroidal) used . In addition, the T-matrix method cannot be used to obtain solution data near the surface of the scatterer, either in the fluid or within the elastic body, depending, again, on the basis used.

Integral equation methods have been used for radiation and scattering from bodies with known velocity or impedance distribution [11-15], but have not been used for elastic bodies. However a prototype problem for acoustic scattering from elastic bodies is the scattering of sound from fluid obstacles [16]. In Ref. 16 a pair of integral equations, one for the interior of the fluid obstacle and one for the host fluid and both valid at the interface, were written and solved numercally.

Integral equation methods are often criticized due to their nonuniqueness problems. Nonuniqueness occurs at certain characteristic frequencies that depend on the shape of the body, the boundary conditions imposed and the type of integral equation used [17]. However, recent evidence [18] indicates that nonuniqueness need not be a major concern when using integral equation methods if one employs the CHIEF method [11] and interior point checking [18]. At high frequencies where there may be clustering of the characteristic frequencies, one may substitute one of several asymptotic approximations [19] in place of an integral equation.

In this paper we use a direct integral equation method to solve problems in acoustic radiation and scattering from arbitrarily shaped elastic bodies, both solids and shells. The coupled acoustic/elastic boundary value problem is reformulated into a pair of integral equations, one for the interior (elastic) domain and one for the exterior (acoustic) domain. Continuity conditions on the surface are used to couple the integral equations. The resulting integral equations are reduced to an algebraic form using isoparametric elements and solved to determine all the unknown boundary data. Subsequently, quadrature is

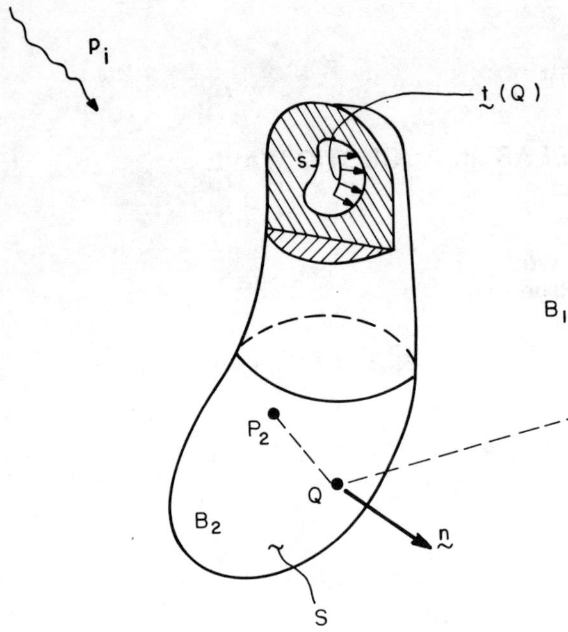

Fig. 1. Nomenclature for radiation or scattering of acoustic waves from an elastic body.

used to obtain the field solution within the body or in the host fluid. In its numerical form the integral equation method is referred to as the boundary element method (BEM).

The BEM model is tested for a several radiation and scattering problems in which the elastic body is either a hollow or solid sphere of various material compositions. When the elastic body reduces to a thin shell, an alternative method to model the elastic part is the finite element method (FEM). The BEM is still used to model the acoustic part. In this paper, we also compare the BEM solution with a coupled FEM/BEM solution for a thin shell problem.

FORMULATION

For acoustic radiation or scattering from an elastic body, we must solve a boundary value problem involving coupling between the elastic body B_2, bounded by the surface S, and the surrounding host fluid B_1, see Figure 1. For scattering problems, we denote the prescribed incident wave as p_i. We consider only time-harmonic wave motion although more generalized wave motion may be investigated using the Laplace transform and Fourier synthesis. The elastic body is assumed to be homogeneous, isotropic and linearly elastic, and the fluid is assumed to be inviscid. Body forces are assumed zero. Under these conditions the wave equation in the elastic body is [20]

$$(c_1^2 - c_2^2)\, \underset{\sim}{\nabla} (\underset{\sim}{\nabla} \cdot \underset{\sim}{u}) + c_2^2\, \nabla^2 \underset{\sim}{u} + \omega^2 \underset{\sim}{u} = \underset{\sim}{0}\ , \quad P_2 \in B_2, \tag{1}$$

where c_1 and c_2 are the wave speeds of the P and S waves, respectively, u is the vector displacement field at a point P_2 and ω is the angular frequency of oscillation. The corresponding equation in the fluid is the Helmholtz equation:

$$\nabla^2 p + (\omega/c)^2 p = 0 \quad , \quad P_1 \in B_1, \tag{2}$$

where $p = p_s + p_i$ is the total acoustic pressure at a point P_1, p_s is the scattered acoustic pressure and c is the speed of sound in the fluid.

Boundary conditions may be specified along the surfaces s and S. Along s, where B_1 and B_2 are not in contact, we may impose surface tractions, surface displacements, or a linear combination of the two, corresponding to a well-posed boundary value problem. The appropriate boundary conditions at the interface S are:

1. The normal displacement is continuous and related to the normal gradient of the acoustic pressure through the momentum equation

$$u_n = \frac{1}{\rho_o \omega^2} \frac{\partial p}{\partial n} \quad , \tag{3a}$$

where ρ_0 is the mean density of the fluid.

2. The normal stress is continuous and equal to the acoustic pressure

$$t_n = -p \quad , \tag{3b}$$

where the normal n is directed into B_1.

3. The elastic shear stresses (both components) vanish

$$t_s = 0 \quad . \tag{3c}$$

In addition to these conditions, p (or p_s for scattering) must satisfy the Sommerfeld radiation condition.

To reformulate the above problem in terms of integral equations valid at the boundary S + s, we use the boundary integral equation for the exterior acoustics problem [e.g., Ref. 13] and a companion interior integral equation [21] for the elastic body. The resulting boundary integral equation formulation for our coupled problem is:

$$\underset{\sim}{C}^T(P_2)\underset{\sim}{u}(P_2) = \int_{S+s} [\ \underset{\sim}{U}^T(P_2,Q)\underset{\sim}{t}(Q) - \underset{\sim}{T}^T(P_2,Q)\underset{\sim}{u}(Q)]dS(Q), \tag{4}$$

which is valid for P_2 in B_2 or on S + s and

$$C(P_1)p(P_1) = \int_S [p(Q)\frac{\partial \psi}{\partial n}(P_1,Q) - \frac{\partial p}{\partial n}(Q)\psi(P_1,Q)]dS(Q) + 4\pi p_i(P_1) \tag{5}$$

which is valid for P_1 in B_1 or on S.

In Eq. (4) U and T are the Stokes' displacement and traction tensors, respectively, which describe the fields at Q due to a time-harmonic point force at P_2 of frequency ω. The tensor $C(P_2)$ is the identity tensor I for P_2 in B_2 or a real function of the geometry of S + s in the vicinity of P_2 for P_2 on S + s. If there is a unique tangent plane at P_2 on S + s (e.g., no edge of corner, such as when S is the surface of a cube) then $C(P_2)$ is I/2.

In Eq. (5) $\psi(P_1,Q)$ is the free-space Green function which describes the acoustic pressure at a point Q due to a point acoustic source at P_1. The coefficient $C(P_1)$ is 4π for P_1 in B_1 and 2π for P_1 on S, providing there is a unique tangent plane at P_1. It is noted that the Sommerfeld radiation condition is taken care of in Eq. (5).

Because the Stokes' solution is singular at P_2, Eq. (4) will become singular for P_2 on S + s. However, the resulting boundary integral equation converges in the Cauchy principal value sense. Special care has to be used in integrating Eq. (4) numerically. Similarly,

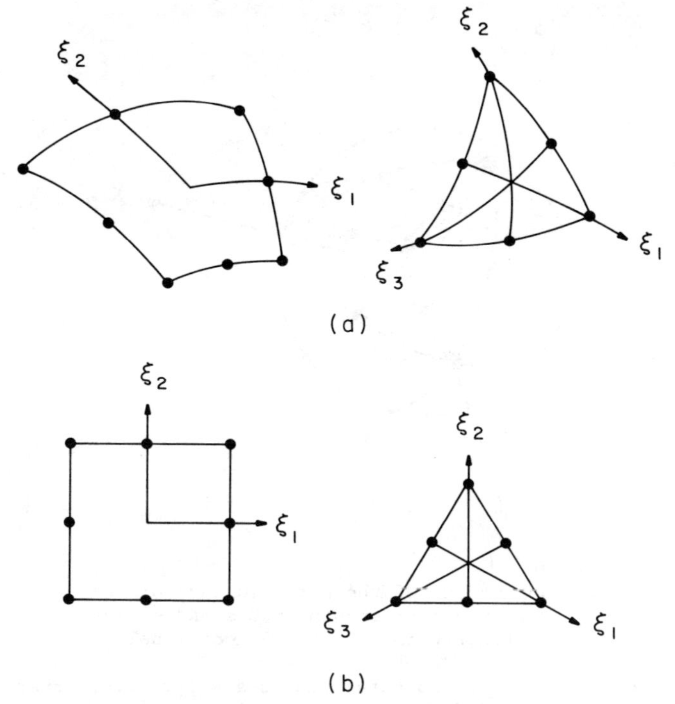

Fig. 2. (a) Curvilinear quadrilateral and triangular isoparametric elements, (b) corresponding parent elements.

Eq. (5) will become singular for P_1 on S. However, the resulting boundary integral equation converges in the normal sense because the free-space Green function is only weakly singular.

In Eq. (4) the unkowns are the tractions and the displacements on the boundary S and either the tractions or the displacements on s. In Eq. (5) the unknowns are the pressure and the normal derivative of the pressure on S. Using the conditions at the interface, Eqs. (3), one set of unknowns on S may be eliminated from Eqs. (4) and (5). The remaining two unknowns on S and the unknown on s may be found by solving Eqs (4) and (5) simultaneously.

NUMERICAL IMPLEMENTATION

A numerical solution to the boundary integral equations (Eqs. (4) and (5)) can be achieved by discretizing the boundary into a number of surface elements. In this paper, we use isoparametric quadratic elements. The global Cartesian coordinates X_i (i=1, 2, 3) of any points of an element are assumed to be related to the nodal coordinates $X_{i\alpha}$ by

$$X_i\ (\xi) = \sum_\alpha N_\alpha(\xi)\ X_{i\alpha}\ ,\ \alpha = 1,\ 2,\ \ldots,6 \text{ or } 8 \qquad (6)$$

in which $N_\alpha(\xi)$ are second order shape functions of the local coordinates $(\xi)=(\xi_1,\ \xi_2)$. Equation (6) is an implicit transformation in which a surface element is mapped into a plane square or a plane equilateral triangle, see Fig. 2. Note that for triangular elements, $\xi_3 = 1 - \xi_1 - \xi_2$.

Next, we use the same quadratic shape functions to interpolate the boundary variables. For the elastodynamic boundary integral equation, Eq. (4), we assume that on each element

$$u_i(\xi) = \sum_\alpha N_\alpha(\xi)u_{i\alpha} \qquad , \qquad (7)$$

$$t_i(\xi) = \sum_\alpha N_\alpha(\xi)t_{i\alpha} \qquad , \qquad (8)$$

where $u_{i\alpha}$ and $t_{i\alpha}$ are the nodal values of the boundary displacements and tractions, respectively. Substituting Eqs. (6)-(8) into Eq. (4) and using appropriate numerical integration rules for quadrilateral elements and triangular elements, we obtain

$$[H]\{u\} = [\tilde{G}]\{t\} \qquad , \qquad (9)$$

where H and $\tilde{G}$ are the assembled coefficient matrices resulting from the numerical integration.

Similarly, for the Helmholtz integral equation, Eq. (5), we assume that the sound pressure on each element can be represented by

$$p(\xi) = \sum_\alpha N_\alpha(\xi)p_\alpha \qquad , \qquad (10)$$

where p_α are the nodal values of pressure. Since the normal derivative of pressure is related to the normal displacement by Eq. (3a), we rewrite Eq. (5) as

$$C(P_1)p(P_1) = \int_S [-\frac{\partial\psi}{\partial n}\ p - \psi\rho_o\omega^2(\underset{\sim}{u}\bullet\underset{\sim}{n})]\ dS + 4\pi p_i(P_1) \qquad (11)$$

and we interpolate the displacement components u_i of each element according to Eq. (7). Substituting Eqs. (6), (7) and (10) into Eq. (11) and carrying out the numerical integration, we get

$$[T]\{u\} = [D]\{p\} + 4\pi\{p_i\} \qquad , \qquad (12)$$

where T and D are the assembled coefficient matrices.

The coupled elastic/acoustic system of equations can be obtained by simply combining Eqs. (9) and (12) and using appropriate interface conditions. Recall that the tractions on the elastic/acoustic interface can be represented by

$$t_i = -pn_i \qquad (13)$$

Hence part of the traction unknowns in Eq. (9) can be transformed into pressure. We rewrite Eq. (9) as

$$[H]\{u\} = [G]\{t\} - [F]\{p\} \qquad (14)$$

where G and F are coefficient matrices obtained by modifying $\tilde{G}$ of Eq. (9). Finally, we combine Eqs. (12) and (14) and, assuming tractions are prescribed on the inner solid surface, we get

$$\begin{bmatrix} H_{11} & F_1 & H_{12} \\ T & -D & 0 \\ H_{21} & F_2 & H_{22} \end{bmatrix} \begin{bmatrix} u_1 \\ p \\ u_2 \end{bmatrix} = \begin{bmatrix} G_{12}t_2 \\ 4\pi p_i \\ G_{22}t_2 \end{bmatrix} \qquad (15)$$

where the subscripts 1 and 2 denote the contribution from the elastic/acoustic interface and the inner solid surface, respectively. On the other hand, if displacements rather than tractions are prescribed on the inner surface, Eq. (15) becomes

$$\begin{bmatrix} H_{11} & F_1 & -G_{12} \\ T & -D & 0 \\ H_{21} & F_2 & -G_{22} \end{bmatrix} \begin{bmatrix} u_1 \\ p \\ t_2 \end{bmatrix} = \begin{bmatrix} -H_{12}u_2 \\ 4\pi p_i \\ -H_{22}u_2 \end{bmatrix} . \quad (16)$$

The linear system of equations, Eq. (15) or (16), can be solved for the nodal values of displacements and acoustic pressure on the elastic/acoustic interface as well as the displacements (or tractions) on the inner solid surface.

Although theoretically there is no limitation on the closeness between the outer surface S and the inner surface s, the BEM may require a large number of integration points and elements if the elastic body reduces to an extermely thin shell. An alternative way to model an elastic thin shell is the finite element method (FEM). If we use the FEM to model the elastic thin shell and the BEM to model the acoustic field, the resulting matrix representation is

$$\begin{bmatrix} K - \omega^2 M & L \\ T & -D \end{bmatrix} \begin{bmatrix} u \\ p \end{bmatrix} = \begin{bmatrix} f \\ 4\pi p_i \end{bmatrix} \quad (17)$$

where K is the stiffness matrix, M is the mass matrix, f is the forcing vector due to the applied internal loading, and L takes care of the loading from the surrounding sound pressure. Note that in Eq. (17) both u_1 and u_2 reduce to the middle surface displacement u in thin shell theory and all the rotation angles are assumed to be "slave" degrees of freedom. Also note that the Helmholtz integral equation (Eq. (5)) and the continuity conditions (Eqs. (3)) are satisfied on the middle surface rather than the outer surface S.

EXAMPLE PROBLEMS

We tested the BEM method by solving several radiation and scattering problems for hollow and solid elastic spheres of various compositions. Table I shows the material properties used for the test cases in this study.

In all of the test cases, the host fluid was sea water; the density and speed of sound used was 1026 kg/m³ and 1500 m/s, respectively.

Because of a symmetry option in our BEM code, only one half of the sphere was modeled. The initial BEM model (see Fig. 3) consisted of 12 triangular elements on each of the inner and outer hemispheres, a total of 58 nodes. A second model was created by dividing each triangle in half to produce a mesh of 24 triangular elements on each hemisphere and 114 total nodes.

A. Radiation and scattering from a hollow or solid sphere

The first problem that we consider is the radiation from a hollow steel sphere in sea water with inner

Table I. Material properties of elastic bodies used in current study (E is Young's modulus and ν is Poisson's ratio).

Material	E (Pa)	ν	ρ_0(kg/m³)	c_1(m/s)	c_2(m/s)
steel	2.07×10^{11}	0.3	7810	5973	3193
aluminum	7.10×10^{10}	0.33	2700	6242	3144
hard rubber	2.30×10^{9}	0.4	1100	2117	864

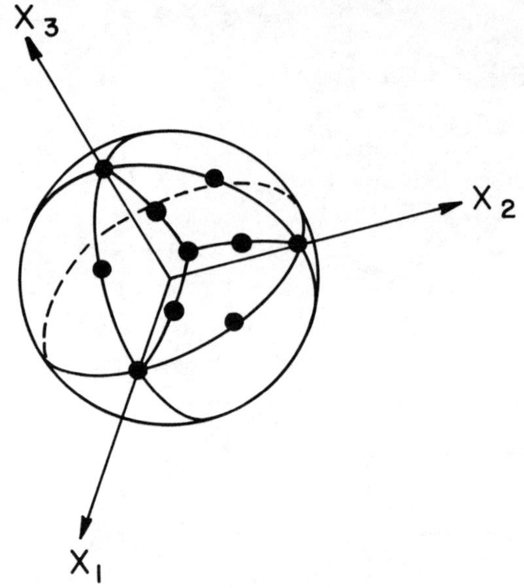

Fig. 3. One octant of the outer surface of the hollow sphere showing nodes and element configuration for the 58 node model.

radius b = 0.5 m and outer radius a = 1 m. The inner surface is loaded with a uniform time-harmonic pressure of amplitude P_0. As this problem has an exact analytical solution [22] it is a good test problem with which to compare our BEM results.

Figure 4 shows the sound pressure at a distance of 3a from the center of the hollow sphere plotted versus ka where k is the wave number in the fluid. The data in Fig. 4 and Table II show the rapid convergence that may be expected with the BEM. The numerical solution, however, is not as good at the resonance frequency (ka=8) as it is at other frequencies. A better numerical result at the resonance frequency can be obtained by a refined BEM mesh. Our numerical study shows that the error at ka=8 will be reduced to 0.7% if a mesh of 386 nodes is used.

Next, we consider a test case consisting of the scattering of an incident plane wave of amplitude P_i by the hollow sphere described above. As the inside of

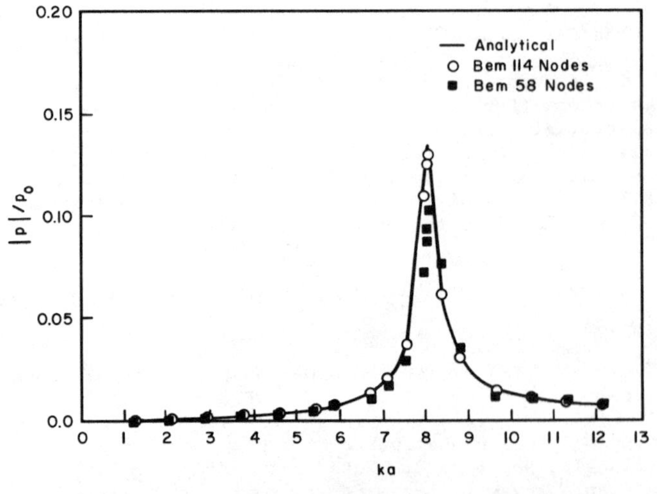

Fig. 4. Normalized sound pressure at a distance of 3a from the center of a hollow steel sphere in sea water excited by an uniform internal pressure P_0.

Table II. Normalized sound pressure $|p|/P_0$ on the surface of a hollow steel sphere with uniform internal pressure P_0

ka	Analytical Solution	BEM 58 nodes Solution	Error	BEM 114 nodes Solution	Error
2	3.2373×10^{-3}	3.0689×10^{-3}	5.2%	3.1948×10^{-3}	1.3%
4	9.0293×10^{-3}	8.3093×10^{-3}	8.0%	8.8544×10^{-3}	1.9%
8	3.9760×10^{-1}	2.5102×10^{-1}	36.9%	3.6160×10^{-1}	9.1%
10	3.9073×10^{-2}	3.6392×10^{-2}	6.9%	3.8527×10^{-2}	1.4%

the sphere is evacuated, the inner surface has a zero traction. All of the scattering data were obtained with the 114 node BEM model discussed above. Figure 5 shows the polar plot of the ratio of the scattered pressure to the incident pressure at a distance of 3a from the center of the sphere for a sphere of various composition when ka = 1. The plane wave is incident on the hollow sphere from the right in Fig. 5. The results for the rigid sphere are also shown in Fig. 5 for comparison.

The effect of material composition on the scattered pressure may be seen quite clearly in Fig. 5. The amplitude of the back-scattered pressure is small when the sphere is either aluminum or rubber as compared to the steel sphere. In the forward-scattering direction, the effects of geometrical diffraction and transmission through the sphere become superposed with the rubber sphere exhibiting the largest forward-scattered pressure. In Fig. 6 we compare the scattering from a hollow and a solid rubber sphere for ka = 1.

The effect of the frequency of the incident wave on the scattering of sound by a hollow rubber sphere is

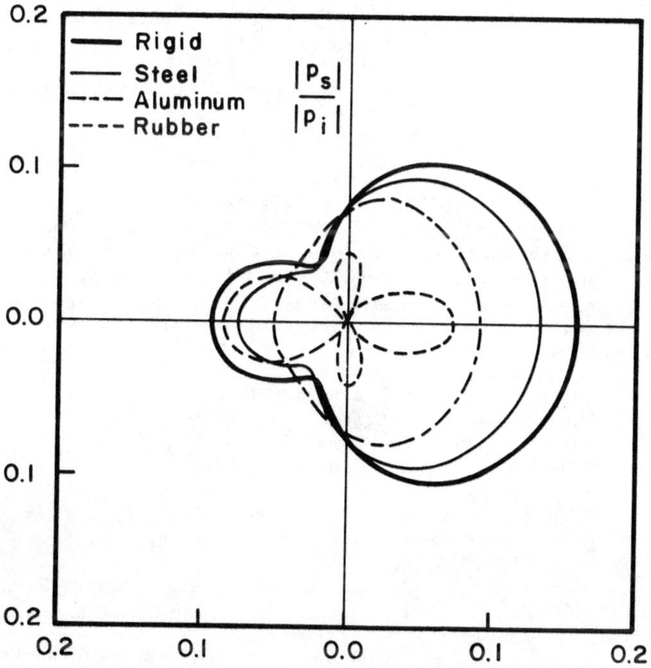

Fig. 5. Polar plot of the ratio of the scattered to the incident pressure at a distance of 3a from the center of a hollow sphere of various compositions in sea water when ka = 1.

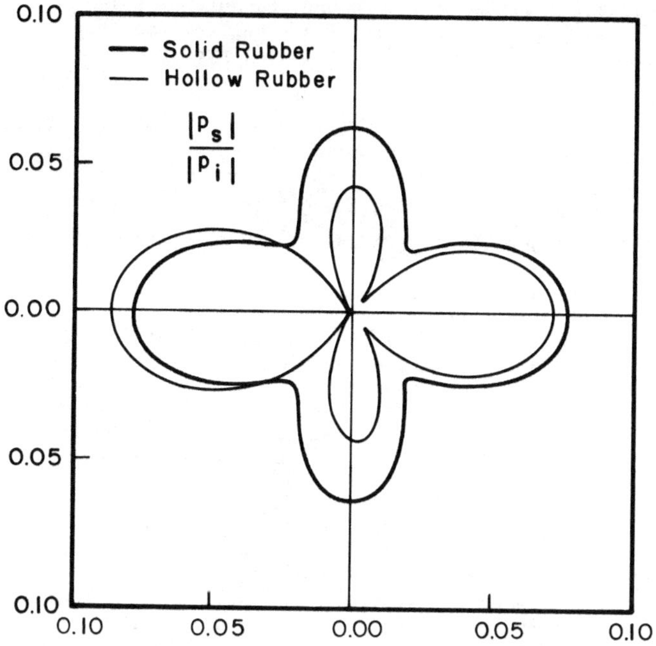

Fig. 6. Polar plot of the ratio of the scattered to the incident pressure at a distance of 3a from the center of a hollow and solid rubber sphere in sea water when ka = 1.

shown in Fig. 7. Figure 7 shows the polar plot of ratio of the back-scattered pressure to the incident pressure as a function of ka on the surface of the sphere. Clearly distinguishable in Fig. 7 is the superposition of the resonant behavior and the background scattering of the sphere.

B. Radiation and scattering from a thin shell

For elastic thin shells, the BEM is still applicable if some care is taken in choosing the number of integration points. Here we consider the radiation from a spherical steel shell in sea water with inner radius b = 0.975 m and outer radius a = 1.025 m. Hence, the thickness of the thin shell is 0.05 m. If the inner surface is loaded with a uniform time-harmonic pressure of amplitude P_0, the sound pressure

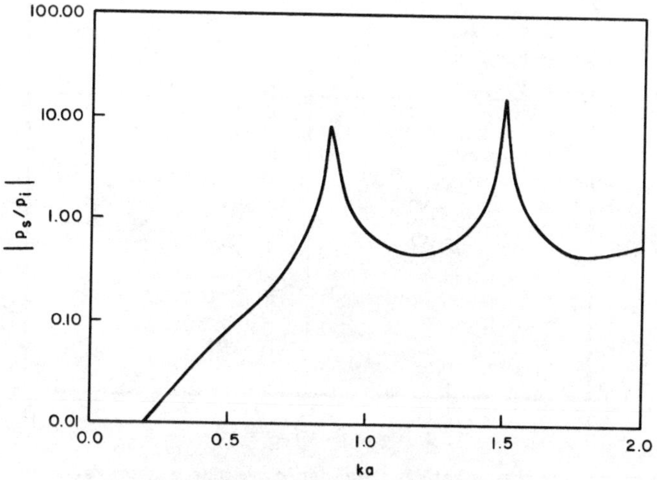

Fig. 7. Ratio of the back-scattered pressure to the incident pressure on the surface of a hollow rubber sphere in sea water as a function of ka.

Table III. Normalized sound pressure $|p|/P_0$ on the surface of a steel thin shell with uniform internal pressure P_0

ka	Analytical Solution	BEM 58 nodes Solution	BEM 58 nodes Error	BEM 114 nodes Solution	BEM 114 nodes Error
1.025	5.6961×10^{-2}	4.6409×10^{-2}	18.5%	5.3787×10^{-2}	5.6%
2.05	1.6080×10^{-1}	1.3048×10^{-1}	18.9%	1.5167×10^{-1}	5.7%

solution on the surface of the spherical thin shell is shown in Table III.

From Table III, we can see that the BEM model still gives a fairly accurate solution with 114 nodes. If we keep refining the mesh to 386 nodes, the error will be reduced to 0.1% for ka = 1.025. An alternative way to solve the thin shell problem is to use the coupled FEM/BEM method (Eq. (17)). With only 45 FEM nodes used on the thin shell, the error of the coupled FEM/BEM solution at ka = 1.025 is 3.6%.

Finally, we use the same FEM/BEM model (45 nodes) to test a scattering problem. The spherical steel shell has a radius of 1 m and thickness 0.05 m. The directivity of the scattered pressure at a distance r = 3 m from the center of the spherical thin shell is shown in Fig. 8. Also shown in Fig. 8 is the BEM solution with 114 nodes. Recall that the coupled FEM/BEM technique evaluates the Helmholtz integral equation and matches the continuity conditions at the middle surface rather than the real outer surface. Therefore, in order to make a fair comparison, the BEM result in Fig. 8 was obtained by using a = 1 m and b = 0.95 m. Since the shell is not extremely thin, slight difference may be observed when we compare a thin shell model (FEM/BEM) with a real 3-D solid model (BEM). Nevertheless, both solutions in Fig. 8 exhibit a similar behavior.

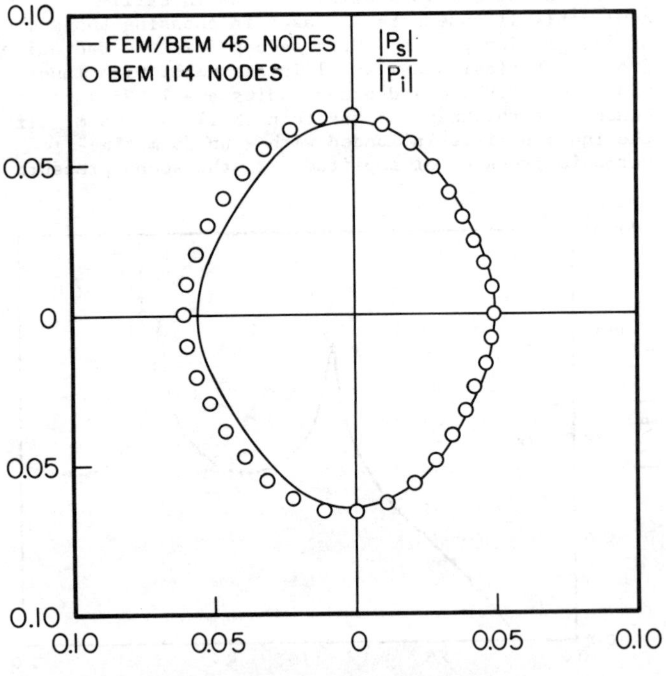

Fig. 8. Polar plot of the ratio of the scattered to the incident pressure at r = 3 m from the center of a spherical steel shell (with radius 1 m and thickness 0.05 m) in sea water when ka = 1.

DISCUSSION AND SUMMARY

We have reformulated the acoustic radiation and scattering problems for elastic bodies in terms of a pair of integral equations, one for the elastic body, the other for the host fluid and both valid at the interface. These integral equations were reduced to a numerical form using quadratic isoparametric elements.

The BEM formulation was tested for several radiation and scattering problems of practical importance. For sound radiation from an internally excited spherical shell, agreement between the BEM data and an exact analytical solution was quite good. Because the only difference between a radiation and a scattering problem is the extra term on the right hand side (see Eq. (15)) due to the incident wave, the validity of the scattering solution data (e.g., Fig. 5) may be insured by verifying the radiation solution. The validity of scattering or radiation data may also be checked by examining the numerical solution at points outside of the acoustic domain, where the right-hand side of Eq. (5) should be zero.

For elastic thin shells, the BEM solution is still quite reliable if more integration points are used to integrate the integral equations for very close elements. A coupled FEM/BEM technique also provides excellent results for thin shell problems.

REFERENCES

1. Faran, J. J., Jr., "Sound Scattering by Solid Cylinders and Spheres," J. Acoust. Soc. Am., Vol. 23, 1951, pp. 405-418.
2. Junger, M. C., "Sound Scattering by Thin Elastic Shells," J. Acoust. Soc. Am., Vol. 24, 1952, pp. 366-373.
3. Hickling, R., "Analysis of Echoes from a solid Elastic Sphere in Water," J. Acoust. Soc. Am., Vol. 34, 1962, p. 1582.
4. Yosioka, K. and Kawasima, Y., "Acoustic Radiation Pressure on a Compressible Sphere," Acustica, Vol. 5, 1955, pp. 167-173.
5. Hasegawa, T.and Yosioka, K., "Acoustic-Radiation Force on a Solid Elastic Sphere," J. Acoust. Soc. Am., Vol. 46, 1969, pp. 1139-1143.
6. Flax, L., Dragonette, L. R., Varadan, V. K. and Varadan, V. V., "Analysis and Computation of the Acoustic Scattering by an Elastic Prolate Spheroid Obtained from the T-matrix Formulation," J. Acoust. Soc. Am., Vol. 71, 1982, pp. 1077-1082.
7. Waterman, P. C., "Survey of T-matrix Methods and Surface Field Representations," in Acoustic, Electromagnetic, and Elastic Wave Scattering, (Ed. Varadan, V. V., and Varadan, V. K.), Pergamon Press, New York, 1980, pp. 61-78.
8. Tobocman, W., "Comparison of the T-matrix and Helmholtz Integral Equation Methods for Wave Scattering Calculations," J. Acoust. Soc. Am., Vol. 77, 1985, pp. 369-374.
9. Hackman, R. H. and Todoroff, D. G., "An Application of the Spheroidal-coordinate-based Transition Matrix: the Acoustic Scattering from High Aspect Ratio Solids," J. Acoust. Soc. Am., Vol. 78, 1985, pp. 1058-1071.
10. Wilton, D. T., "Acoustic Radiation and Scattering from Elastic Structures," Int. J. Num. Meth. Eng., Vol. 13, 1978, pp. 123-138.
11. Schenck, H. A., "Improved Integral Formulation for Acoustic Radiation Problems," J. Acoust. Soc. Am., Vol. 44, 1968, pp. 41-58.
12. Koopman, G. H. and Benner, H., "Method for Computing the Sound Power of Machines Based on the Helmholtz Integral," J. Acoust. Soc. Am., Vol. 71, 1982, pp. 78-89.

13. Seybert, A. F., Soenarko, B., Rizzo, F. J. and Shippy, D. J., "An Advanced Computational Method for Radiation and Scattering of Acoustic Waves in Three Dimensions," J. Acoust. Soc. Am., Vol. 77, 1985, pp. 362-368.
14. Tobocman, W., "Calculation of Acoustic Wave Scattering by Means of the Helmholtz Integral Equation 1.," J. Acoust. Soc. Am., Vol. 76, 1984, pp. 599-607.
15. Seybert, A. F., Soenarko, B., Rizzo, F. J. and Shippy, D. J., "A Special Integral Equation Formulation for Acoustic Radiation and Scattering for Axisymmetric Bodies and Boundary Conditions," J. Acoust. Soc. Am., Vol. 80, 1986, pp. 1241-1247.
16. Seybert, A. F. and Casey, D. K., "An Integral Equation Method for Coupled Fluid/Fluid Scattering in Three Dimensions," J. Acoust. Soc. Am., Vol. 84, 1988, pp. 379-384.
17. Kleinman, R. E. and Roach, G. F., "Boundary Integral Equations for the Three-Dimensional Helmholtz Equation," SIAM Rew., Vol. 16, 1974, pp. 214-236.
18. Seybert, A. F. and Rengarajan, T. K., "The Use of CHIEF to Obtain Unique Solutions for Acoustic Radiation Using Boundary Integral Equations," J. Acoust. Soc. Am., Vol. 81, 1987, pp. 1299-1306.
19. Seybert, A. F. and Rengarajan, T. K., "The High-Frequency Radiation of Sound from Bodies of Arbitrary Shape," ASME Trans. J. Vib. Acoust. Reliabil. Des., Vol. 109, 1987, pp. 381-387.
20. Ewing, W. M., Jardetzky, W. S. and Press, F., Elastic Waves in Layered Media, McGraw-Hill, New York, 1957.
21. Rizzo, F. J. and Shippy, D. J., "A BIE Method for Radiation and Scattering of Elastic Waves in Three Dimensions," Int. J. Num. Meth. Eng., Vol. 21, 1985, pp. 115-129.
22. Seybert, A. F., Wu, T. W. and Wu, X. F., "Radiation and Scattering of Acoustic Waves from Elastic Solids and Shells using the Boundary Element Method," J. Acoust. Soc. Am., Vol. 84, 1988, pp. 1906-1912.

SUBSTRUCTURING TECHNIQUES FOR ACOUSTIC RADIATION FROM COMPLEX SUBMERGED STRUCTURES

R. Vasudevan and J. M. McCormick
Weidlinger Associates
New York, New York

This paper presents a combination of powerful and computationally efficient sub-structuring techniques (combining finite-difference, boundary element, finite element, and analytical approaches) to predict the complete vibratory response of submerged structures and the sound pressure fields radiated from them. These techniques have been thoroughly validated by comparisons with analytical solutions and with experimental results from many tests on both large-scale and small-scale vehicles.

The suggested approach permits the hulls and any attached structures to be of general shape and construction (including composite materials, frames, bulkheads, and insert plates) and incorporates the geometric nonlinearities associated with hydrostatic pressure. The effects of a free surface can be included in the formulation.

For problems where the radiating surface and the major part of the structure are axisymmetric, this approach is very efficient because the solution is based on the modes of the axisymmetric parts of the submerged structure. In practical applications, this results in structures with many thousands of degrees of freedom being represented by only a handful of modes whose underwater frequencies fall into the range of interest. The behavior of the submerged structure is governed by the generalized coordinates associated with those modes, and equations of motion of the structure can be written in terms of these generalized coordinates. An impedance matrix is constructed from the modes of this primary structure. Each of the other major components is considered as a generalized substructure whose impedances are to be combined with the impedance matrix of the primary structure.

1. Introduction

There is great interest in the development of efficient and accurate methods for determining the acoustic radiation from submerged structures. Analytical solutions have been obtained for simple geometries such as spheres and cylinders [1,2,3] and many numerical schemes to solve radiation problems involving more general geometries have been presented (see for example [4,5,6,7]). But none of these methods has proved to be satisfactory for many practical applications. In this paper, we will describe a computationally efficient general approach for calculating the complete vibratory response of submerged structures, and predicting the acoustic fields radiated from such structures. This hybrid approach will combine the strongest features of four different modeling techniques: finite differences, boundary elements, finite elements, and analytical representations of structural components. In particular, we will consider submerged structural systems which have a largely axisymmetric wet surface, as is the case for many practical applications, but may contain arbitrarily attached internal structures, appendages, and regions of entrained fluids. We shall label such configurations as "shell-substructure" systems, where "shell" refers to the axisymmetric parts of the structure, bounded by the wet surface, and "substructure" refers to all other parts of the system, including the infinite fluid region external to the shell and any internal structures or appendages attached to the shell.

To describe such a system mathematically, equations of motion representing the shell and each of the substructures must be constructed. The solution to this system can be achieved by solving the entire set of dynamic equations, but this approach requires the solution of huge sets of equations in practical cases. In this paper we will use an alternate approach, called substructuring, in which impedances are found separately for the shell and each of the substructures, and these separate solutions are then combined to yield the solution for the entire system.

Substructuring techniques have been used extensively in the aerospace industry and have been demonstrated to be a powerful and efficient approach for solving transient structure-fluid interaction problems [8]. In general these substructuring techniques preserve the dynamic characteristics of the structural system, but they allow significant savings in both the formulation of the problem (because of the modularity of the approach) and in the resources required for the solution of the system (sharply reduced computer costs and memory requirements).

The particular power of the substructuring approach presented here is that different analytical methods are used to represent the shell and each of the substructures. The shell is represented by a finite-difference model of a general shell of revolution [9] and the infinite fluid by a Helmholtz boundary element model. Finite elements are used to represent the internal structures and appendages, and analytical models are used, whenever possible, for such simple subsystems as attached masses and springs. Thus each substructure is modeled with the method that best characterizes the behavior of that module, and these individual substructure models can be modified and enhanced at will, without remodeling the shell or any other substructure.

This method is very efficient because it takes advantage of both the axisymmetry of the shell and the modularity of the substructures. The generators of the shell of revolution can be of general shape (including spherical, cylindrical, conical, ellipsoidal, and other special geometries) and the shell can be of general construction (including composite materials) integrated with any bulkheads, frames or other axisymmetric elements.

The infinite fluid can be analyzed using various methods. Except for some special geometries, for which analytical solutions can be achieved by the standard technique of separation of variables, the great majority of investigations have been carried out by using integral equation methods. Although some solutions have been obtained by applying the finite element methods these methods do not compare favorably with the simplicity and economy resulting from the use of integral methods, especially so, because the problems of interest are generally those for which the fluid field is of infinite domain. Using the Helmholtz integral equation method the acoustic field is expressed in terms of surface pressures and velocities of a radiating body.

The fluid-structure interaction due to applied forces on the structures is considered first to obtain the normal velocities of the structure. Upon invoking the compatibility relationship for the normal velocities at the structure-fluid interface, one obtains an integral equation for the unknown surface pressures. Once the surface pressures are determined, the pressures and velocities throughout the fluid may be computed.

The formulation outlined above has been implemented in the HULRAD code [10,11], a group of linked computer programs which employs a sequence of computationally efficient techniques which have been thoroughly validated by comparisons with analytical solutions and with experimental results from many tests on both large-scale and small-scale vehicles.

In section 2 a theoretical formulation to analyze submerged shells of revolution with internal structures only is given. The dynamic equations of motion of the shell, the substructure and the infinite external fluid are developed and assembled together. In section 3, a brief description of the implementation of the theory in the HULRAD system of computer codes is presented. Finally comparisons between numerical calculations using the computer program and analytical or experimental results are presented, in section 4.

2. THEORY

Consider an axisymmetric shell with unsymmetric internal structures submerged in an acoustic medium. The equations of motion of this system are developed in this section. First the infinite external fluid equations are solved using the Helmholtz integral formulation. Then the equations of motion of the axisymmetric shell, expressed in modal form, are altered to include the effects of the external infinite fluid and the internal structures, using displacement compatibility at the interfaces between shell and fluid or internal structures.

(a) Fluid Field

The Helmholtz equation for an inviscid fluid undergoing simple harmonic motion at a frequency ω is,

$$(\nabla^2 + k^2)\phi = 0 \qquad (1)$$

where ϕ is the fluid velocity potential, $k = \omega/c$, in which c is the acoustic velocity.

The particle velocity is expressed as the gradient of ϕ

$$v = -\nabla\phi \qquad (2)$$

and the pressure

$$p = \rho\frac{\partial\phi}{\partial t} \qquad (3)$$

where ρ is the density of the fluid

The solution of eq. (1) corresponding to a unit source, i.e, the free space Green's function representing a divergent spherical wave, can be written as

$$G = \frac{e^{-ikd}}{4\pi d} \qquad (4)$$

where d is the distance from source point Q on the shell surface σ, to field point P. (fig. 1)

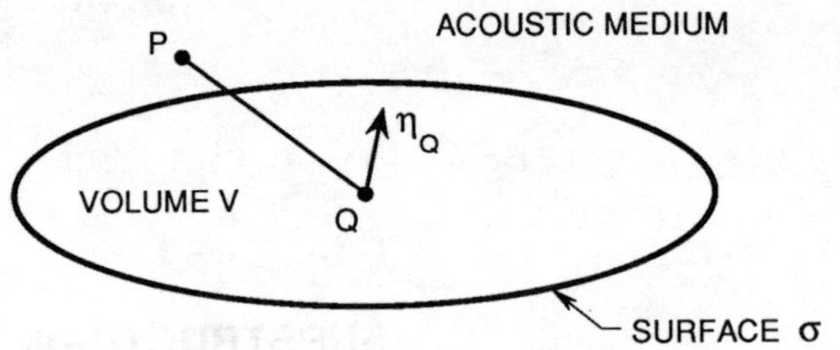

Fig. 1 Source and field points

For convenience, the time factor $e^{i\omega t}$ is supressed here and is omitted in what follows, unless clarity demands its inclusion.

By invoking the Helmholtz integral

$$\epsilon p(P) = \left[-\iint_\sigma i\omega\rho G v \, d\sigma + \iint_\sigma p\frac{\partial G}{\partial n}\, d\sigma\right] \qquad (5)$$

where $\epsilon = 0$ for P not in fluid volume V (see fig. 2), $\epsilon = 1/2$ for P on boundary σ, and $\epsilon = 1$ for P in V, we can express the value of ϕ at a point P in the region V in terms of pressure p and normal velocity v at the fluid bounding surface σ. The derivative of the Green's function in eq. (5) is along the surface normal at source point Q.

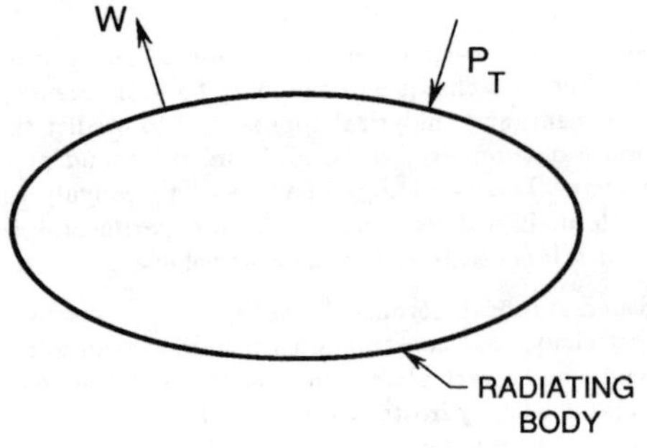

Fig. 2 Geometry and notation

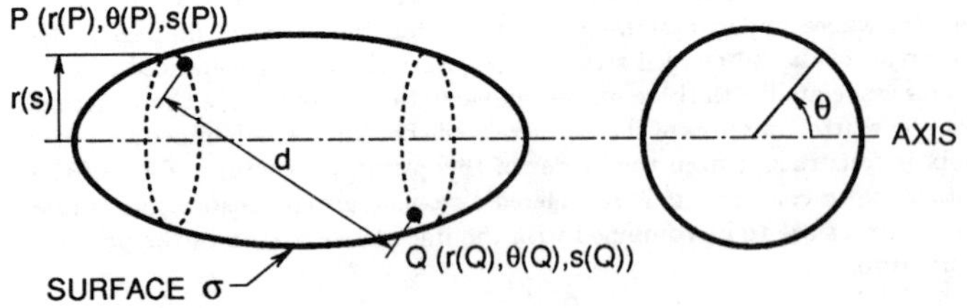

Fig. 3 Geometry of a surface of revolution

The solution to the Helmholtz integral formulation is not unique at frequencies that correspond to the internal resonances of a cavity whose boundary corresponds to the shell radiating surface. A unique solution at all frequencies can be obtained by using either an over determination scheme [4,12] or through a combined Helmholtz integral equation formulation [13], where a linear combination of Helmholtz integral and its differential is used, instead of eq. (5).

When considering surfaces of revolution, (see fig. 3), it is expedient to use a circumferential harmonic expansion in cosine functions,

$$v = \sum_n v_n \cos n\theta \tag{6}$$

$$p = \sum_n p_n \cos n\theta \tag{7}$$

Substituting eq. (6) and eq. (7) in eq. (5), the Helmholtz integral equation for a circumferential wave number n can be written as,

$$\frac{1}{2} p_n(P) = -\iint_\sigma i\omega\rho G(P,Q) v_n(Q) \cos n\theta \, d\sigma + \iint_\sigma \frac{\partial G(P,Q)}{\partial n_Q} p_n(Q) \cos n\theta \, d\sigma \tag{8}$$

Upon subdividing the radiating surface into N bands along the generator and assuming that the pressure and the velocity are constant along the generator within each band, we obtain,

$$\frac{1}{2} p_n(P) = -\sum_{Q=1}^{N} v_n(Q) \left\{ i\omega\rho \int_{S_{Q-1}}^{S_Q} \int_0^{2\pi} r(Q) G(P,Q) \cos n\theta \, d\theta \, ds \right\} + \sum_{Q=1}^{N} p_n(Q) \left\{ \int_{S_{Q-1}}^{S_Q} \int_0^{2\pi} r(Q) \frac{\partial G(P,Q)}{\partial n_Q} \cos n\theta \, d\theta \, ds \right\} \tag{9}$$

where, $r(Q)$ = radius at source band Q and S_{Q-1} and S_Q are respectively the beginning and end coordinates of band Q along the meridian.

Defining influence coefficients as,

$$D_{PQ} = -2i\omega\rho \int_{S_{Q-1}}^{S_Q} \int_0^{2\pi} r(Q) G(P,Q) \cos n\theta \, d\theta \, ds \tag{10}$$

and

$$D_{PQ,n_Q} = 2 \int_{S_{Q-1}}^{S_Q} r(Q) \frac{\partial G(P,Q)}{\partial n_Q} \cos n\theta \, d\theta \, ds \tag{11}$$

and using the definitions of eqs. (10,11) in eq. (9) we obtain,

$$p_n(P) = \sum_{Q=1}^{N} v_n(Q) D_{PQ} + \sum_{Q=1}^{N} p_n(Q) D_{PQ,n_Q} \tag{12}$$

(b) Surface Expansion Functions - Impedance Functions

Since we are concerned with surfaces of revolution, a circumferential modal expansion is used. The solution for each circumferential wave number is obtained and the final result obtained by superposition.

For each circumferential mode, an orthogonal set of J fluid surface expansion functions ψ_j is obtained using the Gram-Schmidt orthogonalization procedure, for use as an expansion basis for the normal component of the surface displacement W,

$$W = \sum_{j=1}^{J} \hat{q}_j \, \psi_j \, e^{i\omega t} \tag{13}$$

It is to be understood that all of the above ψ_j's have the same harmonic circumferential variation, and that they have been normalized with respect to the surface area A, so that

$$\iint_\sigma \psi_j \, \psi_l \, d\sigma = A \, \delta_{jl} \tag{14}$$

The function Z^j is defined as the surface pressure field p^j, due to the surface velocity $i\omega\psi_j$, and is determined by equating the normal components of the fluid velocity at the surface to that associated with the function ψ_j

$$v = i\omega\psi_j \tag{15}$$

Substitution of eq. (15) in eq. (12) results in the corresponding function in matrix form,

$$\{Z_P^j\} = [D_{PQ}][i\omega\psi_j^P] + [D_{PQ,n_Q}][Z_Q^j] \tag{16}$$

The solution of eq. (16) then yields the vectors Z^j which are due to the surface velocity distribution $i\omega\psi_j$.

In turn, we represent Z^j in terms of the surface expansion functions, so that

$$Z^j = \sum_{l=1}^{J} \frac{\gamma_{lj}}{A} \psi_l \tag{17}$$

where

$$\gamma_{lj} = \iint_\sigma Z^j \psi_l \, d\sigma = \pi(1+\delta_{on}) \int_0^s r(s) Z^j \psi_l ds \tag{18}$$

$$\delta_{on} = \begin{cases} 1, n = 0 \\ 0, n \neq 0 \end{cases} \tag{19}$$

Observe that γ_{jl} is the generalized force associated with the ψ_l - displacement and the pressure field arising from ψ_j.

(c) Fluid-Structure Interaction

The normal component of the displacement of the shell can be expressed as

$$W = \sum_{m=1}^{M} q_m W_m \tag{20}$$

where W_m is the normal component of the mth in vacuo shell mode, and q_m, the mth generalized coordinate. Again it should be emphasized that all the above M modes are associated with one particular circumferential wave number.

For harmonic steady-state motion, the shell equations of motion become

$$M_m(\omega_m^2 - \omega^2 + C_m)q_m = Q_m^E + Q_m \tag{21}$$

where ω_m is the in-vacuo frequency of the mth mode, C_m, the damping for the mth node, $C_m = 2i\omega\zeta\omega_m$, where ζ is the ratio of included damping to critical damping or $C_m = i\eta\omega_m^2$, where η is the loss factor. The mth component of the generalized mass and force respectively are

$$M_m = \iint_\sigma \rho h(U_m^2 + V_m^2 + W_m^2) d\sigma \tag{22}$$

$$Q_m = -\iint_\sigma p_T W_m d\sigma \tag{23}$$

and Q_m^E = generalized force associated with the applied loads. U_m, V_m, W_m are the shell displacements along the generator and in the circumferential and normal directions, respectively. Q_m is the generalized force due to surface pressure p_T due to fluid - structure interaction.

Expressing the surface pressure in the form of series expansions in both the surface expansion functions and the in vacuo structural modes, we obtain the following

$$p_T = -\sum_{m=1}^{M} \frac{Q_m}{\mu_m} W_m = -\sum_{j=1}^{J} \frac{\hat{Q}_j}{A} \psi_j \tag{24}$$

where,

$$\hat{Q}_j = -\iint_\sigma p_T \psi_j d\sigma \tag{25}$$

$$\mu_m = \iint_\sigma W_m W_m \, d\sigma \tag{26}$$

At the shell-fluid interface the normal shell velocity must be equal to the normal fluid velocity. Hence,

$$i\omega W = i\omega \sum_{m=1}^{M} q_m W_m = i\omega \sum_{j=1}^{J} \hat{q}_j \psi_j \tag{27}$$

Applying the orthogonality condition in eq. (14) to eq. (27), we have

$$\hat{q}_j = \sum_{m=1}^{M} \lambda_{jm}\, q_m \tag{28}$$

$$\text{where, } \lambda_{jm} = \frac{1}{A} \iint_\sigma W_m \psi_j \, d\sigma \tag{29}$$

The generalized forces defined in eq. (24) can be related as follows, using the orthogonality relationship of eq. (14):

$$Q_m = \sum_{j=1}^{J} \frac{\hat{Q}_j}{A} \iint_\sigma W_m \psi_j \, d\sigma = \sum_{j=1}^{J} \lambda_{mj} \hat{Q}_j \tag{30}$$

From eq. (16) Z^j is the surface pressure due to the surface displacement ψ_j. For the actual surface velocity of eq. (27), the related surface pressure p_T is,

$$p_T = \sum_{j=1}^{J} \hat{q}_j \, Z^j \tag{31}$$

Upon substituting eq. (31) into eq. (25), we obtain

$$\hat{Q}_l = -\sum_{j=1}^{K} \hat{q}_j \iint_\sigma Z^j \psi_l \, d\sigma \tag{32}$$

Using the definitions in eqs. (18) and (32), eq. (32) can also be written as,

$$\hat{Q}_l = -\sum_{j=1}^{J} \gamma_{lj} \hat{q}_j = -\sum_{j=1}^{J} \gamma_{lj} \sum_{m=1}^{M} \lambda_{jm} q_m \tag{33}$$

Substituting for $\hat{Q}_l$ from eq. (33) into eq. (30),

$$Q_h = -\sum_{l=1}^{J} \sum_{j=1}^{J} \sum_{m=1}^{M} \lambda_{hj} \lambda_{lm} \gamma_{jl} q_m \tag{34}$$

The equations of motion of the shell can be rewritten by using eq. (34) in eq. (21). Upon rearranging, the resulting equation in matrix notation is,

$$[K - \omega^2 M + L][q] = [Q^E] \tag{35}$$

where

$$[K] = [\omega_m^2 M_m + C_m] \tag{36}$$

$$[\omega^2 M] = [\omega^2 M_m] \tag{37}$$

$$[L] = [\lambda_{lm}]^T [\gamma_{lj}][\lambda_{jm}] \tag{38}$$

The solution of eq. (35) yields the generalized coordinate vector q. The shell displacements are obtained by substituting the solution vector q into eq. (27).

By substituting q into Eq. (28) and then into eq. (31), we obtain an expansion for the total surface pressures at the i bands (i = 1→N)

$$[p_{T_i}] = [Z_{ij}][\hat{q}_j] \tag{39}$$

where $Z_{ij} = Z_i^j$ = the value of the impedance function at the ith node associated with the ψ_j surface displacement.

The far-field pressure is obtained by substituting the surface velocities and pressures into the discretized form of eq. (5). Thus the far-field pressure at point P,

$$p(P) = \sum_{Q=1}^{N} D_{PQ} v_n(Q) + \sum_{Q=1}^{N} D_{PQ,nQ} p_n(Q) \tag{40}$$

where p_n and v_n are surface pressures and velocities and D_{PQ} and $D_{PQ,nQ}$ are defined as in eq. (10) and eq. (11).

(d) Shell - Internal Structure Interaction

So far we have dealt with a submerged shell possessing axisymmetry which leads to uncoupling of circumferential modes. However, if an arbitrary internal structure is attached to the shell all the modes of the shell will be coupled due to the non-axisymmetric nature of the internal structure. In this section we will derive the equations of motion of a shell-internal structure- external fluid system.

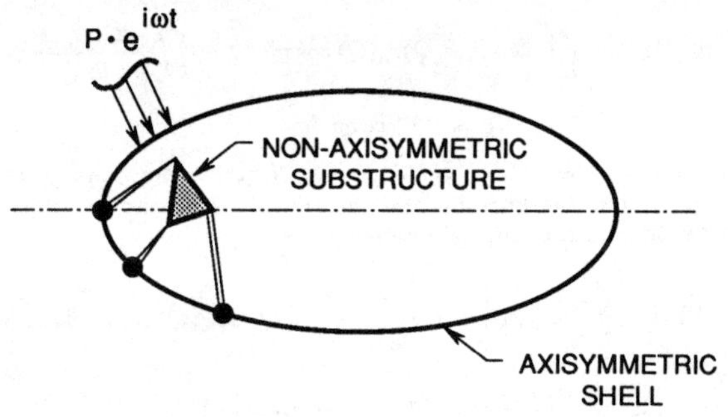

Fig. 4 Shell - substructure system

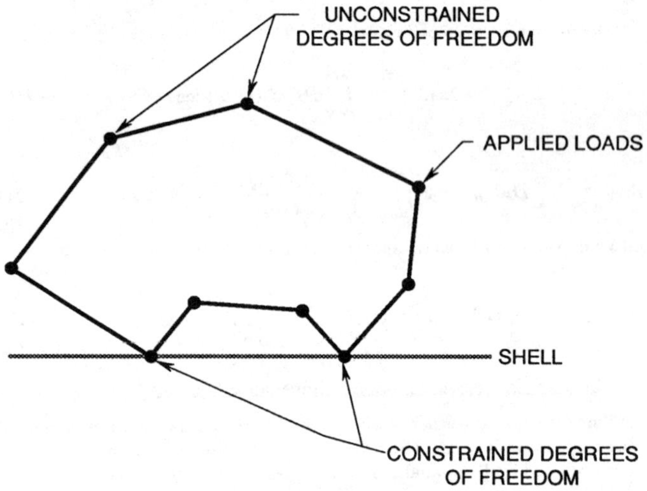

Fig. 5 Typical substructure

Consider a substructure internal to the shell, see fig. (4) and fig. (5), consisting of N degrees of freedom. If N_2 degrees of freedom of the substructure are constrained to move with the shell and the remaining N_1 degrees of freedom are unconstrained, then

$$N = N_1 + N_2 \tag{41}$$

The equations of motion of the substructure not connected to the shell (free-free) can be written in matrix form as,

$$\begin{bmatrix} M_{11} & M_{12} \\ M_{21} & M_{22} \end{bmatrix} \begin{bmatrix} \ddot{X}_1 \\ \ddot{X}_2 \end{bmatrix} + \begin{bmatrix} C_{11} & C_{12} \\ C_{21} & C_{22} \end{bmatrix} \begin{bmatrix} \dot{X}_1 \\ \dot{X}_2 \end{bmatrix} + \begin{bmatrix} K_{11} & K_{12} \\ K_{21} & K_{22} \end{bmatrix} \begin{bmatrix} X_1 \\ X_2 \end{bmatrix} = \begin{bmatrix} F_1 \\ F_2 \end{bmatrix} \tag{42}$$

where,

X_1 = Unconstrained degrees of freedom
X_2 = Constrained degrees of freedom
F_1 = Forces at unconstrained degrees of freedom
F_2 = Forces at constrained degrees of freedom
$(\cdot)$ over variable X denotes time differentiation

M, C and K are mass, damping and stiffness matrices, respectively, in partitioned form, with subscripts (1) corresponding to unconstrained degrees of freedom and (2) to constrained degrees of freedom. The damping matrix C in eq. (42) can be expressed assuming Raleigh damping as

$$[C] = \alpha[M] + \beta[K] \tag{43}$$

where α, β are constants to be chosen.

At an excitation frequency ω, after dropping $e^{i\omega t}$ terms for convenience, eq. (42) is rewritten as,

$$\begin{bmatrix} Z_{11} & Z_{12} \\ Z_{21} & Z_{22} \end{bmatrix} \begin{bmatrix} X_1 \\ X_2 \end{bmatrix} = \begin{bmatrix} F_1 \\ F_2 \end{bmatrix} \tag{44}$$

where,

$$Z_{ij} = (K_{ij} + i\omega C_{ij} - \omega^2 M_{ij}) \qquad i = 1,2, \; j = 1,2 \tag{45}$$

and Z_{ij} is a $(N_i \, X \, N_j)$ matrix.

For the substructure, the forces are applied at the unconstrained degrees of freedom while the motion is prescribed at the constrained degrees of freeedom. Thus in eq. (44) F_1 and X_2 are prescribed. In this section in what follows we will drop the brackets around the matrices for convenience.

Eliminating unconstrained degrees of freedom X_1 from eq. (44) produces,

$$F_2 = (Z_{22} - Z_{21}Z_{11}^{-1}Z_{12})X_2 + Z_{21}Z_{11}^{-1}F_1 \tag{46}$$

in which, superscript -1 indicates the inverse of a matrix, F_2 is the set of forces at the constrained degrees of freedom, and the forces to be applied on the shell are reactions, $-F_2$.

The right hand side of eq. (46) is the sum of two reactions. The first term on the right is due to base motion and the second term is due to the applied forces at the unconstrained degrees of freedom. The force contained in the last term is amplified at the resonance of the substructure with its constrained degrees of freedom fixed.

In eq. (46) we have solved for the interaction forces at the shell-substructure interface. These forces are applied on the shell to produce the equations of motion of shell-internal structure-external fluid system.

Using eq. (35), the equation of motion of the shell for the jth mode is,

$$M_j\ddot{q}_j + (c_j/i\omega)\dot{q}_j + M_j\omega_j^2 q_j = Q^j_{\text{fluid}} + Q^j_{\text{applied}} + Q^j_{\text{substructure}} \tag{47}$$

where, the loading terms on the right hand side are,

Q^j_{fluid} = generalized forces due to fluid-structure interaction which can be obtained using eq. (34)

Q^j_{applied} = generalized forces due to any applied forces

$Q^j_{\text{substructure}}$ = generalized forces due to substructure base forces

$Q^j_{\text{substructure}}$ = generalized forces due to substructure base forces

Let $Q_{\text{substructure}} = -\phi^T_{\text{sub}} F_2$ (48)

where ϕ_{sub} = matrix containing mode shapes of the shell at the shell-substructure connection points, in the substructure coordinate system. These mode shapes are for all the circumferential wave numbers. The superscript T denotes a transpose.

Substituting for F_2 from eq. (46) into eq. (48) we get,

$$Q^j_{\text{substructure}} = -\phi^T_{\text{sub}}(Z_{22} - Z_{21}Z_{11}^{-1}Z_{12})X_2 - \phi^T_{\text{sub}}Z_{21}Z_{11}^{-1}F_1 \tag{49}$$

The first term on the right-hand side of eq. (49) is unknown, since X_2 is not known. However, if we enforce continuity between the shell and the internal structure,

$$X_2 = d_{\text{sub}} \tag{50}$$

where d_{sub} = shell displacements at the points of connection between the shell and the substructure.

Now, using the definition of ϕ_{sub} in eq. (48) we can write d_{sub} in terms of generalized coordinates of the shell as,

$$d_{\text{sub}} = \phi_{\text{sub}} \, q \tag{51}$$

Substituting eqs. (50) and (51) in eq. (49) we get,

$$Q_{\text{substructure}} = -\phi^T_{\text{sub}}(Z_{22} - Z_{21}Z_{11}^{-1}Z_{12})\phi_{\text{sub}} \, q - \phi^T_{\text{sub}} \, Z_{21}Z_{11}^{-1}F_1 \tag{52}$$

The second term on the right hand side of eq. (52) is known, since F_1 is prescribed and its calculation is straightforward. The bracketed term on the right hand side of eq. (52) is the impedance of the substructure to be added to the impedance of the shell to get the total impedance of shell-substructure-external fluid system.

It should be emphasized here that the substructure impedance defined in eq. (52) couples all the shell modes due to the non-axisymmetric nature of the internal structure.

(e) Applied Forces

The forces that induce harmonic motion of the structure to cause sound radiation, are applied directly on the shell or a substructure.

It is assumed that the loads applied on the shell are of two types, namely (1) concentrated forces and (2) circumferential line loads. The loads can be in axial or radial or circumferential directions. The generalized forces for line loads are calculated as

$$Q^E_m = \int_{\theta_1}^{\theta_2} F_E \phi_{Em} r \, d\theta \tag{53}$$

in which

Q^E_m are as defined in eq. (21)

F_E are applied forces

ϕ_{Em} are mode shapes of the mth mode at points of force application

r is the radius of the shell at point of load application

θ is the circumferential coordinate

For concentrated forces, the integral of eq. (53) is evaluated and then the limit θ_2 tending to θ_1 is taken to obtain generalized forces.

The loads on a substructure are specified in the local coordinates of that substructure. These forces correspond to the load vector F_1 in eq. (42).

3. Implementation

The theoretical formulation presented in section 2 to compute the near field and the far-field responses of submerged structures, has been implemented in a system of computer programs called HULRAD [10,11]. This implementation is computationally very efficient because the solution is based on the modes of the axisymmetric parts of the submerged structure. In practical applications, this results in structures with many thousands of degrees of freedom being represented by only a handful of modes whose underwater frequencies fall into the range of interest. The equations of motion expressed in terms of the generalized coordinates associated with these modes, govern the behavior of the submerged structure. An impedance matrix is constructed from the modes of this primary structure. Each of the other major components, external fluid boundary elements and analytical or finite element representations of attached structures and appendages, are considered as generalized substructures whose impedances are to be combined with the impedance matrix of the primary structure.

The major modules of this implementation are:

- a heavily modified version of the BOSOR4 [9] computer program developed by David Bushnell of Lockheed, which uses a finite difference approach to obtain the structural modes of the primary axisymmetric structure (the hydrodynamic and pressure hulls, including all bulkheads and frames). All HULRAD computational models must include a BOSOR4 model since the entire solution is based on the modes of this model and the geometry of the cavity in the external fluid is derived from this model.
- RADISH is a computer program developed by the authors to calculate the acoustic impedances associated with a surface of revolution. The program utilizes a Helmholtz integral boundary element formulation to calculate the impedances due to prescribed velocity distributions (surface expansion functions) on the wet surface of the primary structure. The surface expansion functions (which form an orthogonal set on the wet surface) are chosen so as to facilitate the calculation of these acoustic impedances. All HULRAD computational models must include a RADISH model.
- SAPIV is a heavily modified version of the SAPIV [14] computer program developed by K.J. Bathe and E.L. Wilson of University of California, Berkeley. SAPIV is used to model the internal substructures as assemblages of finite elements. The SAPIV finite element library includes truss elements, beam elements, shell elements, solid elements etc. HULRAD computational models require one or more SAPIV models only if the primary structure contains attached structures, machinery, or appendages.
- SUBSTRU is a computer program developed by the authors to calculate the sound radiated by an excited submerged shell of revolution with non-axisymmetric substructures internally attached to the shell. SUBSTRU is the driver program used to set up and solve the equations of motion obtained using the BOSOR4, SAPIV and RADISH programs.

4. Numerical Results

To validate the theoretical formulation of section 2 and the implementation described in section 3, calculations were made and compared against analytical solutions and experimental results. A description of some of these problems is as follows.

(a) Spherical shell with point excitation

Consider a spherical shell submerged in water and excited by a concentrated force F at the pole at $\theta = 0$, fig. 6a. The exact solution of this problem is calculated using a series solution [15]. Since this problem is axisymmetric, only the circumferential wave number n=0 need be used. The following parameters were used to obtain the responses of the shell: F = 4.45 N, E = 6.894 x 10^6 KPa, $\nu = 0.3$, $\rho_{shell} = 2681.9 Kg/m^3$, a = 2.54 m, t=0.127 m, $\rho_{water} = 1000.0 Kg/m^3$ and velocity of sound in water = 1524 m/sec. The comparisons between the exact solutions and the calculations are shown in figs. 7,8,9,10 for two different loss factors, $\eta = 0.0$ and $\eta = 0.01$. The exact solution was obtained in the non-dimensional frequency range ka = $(\frac{\omega a}{c})$ = 0.0 to ka = 3.0 at ka = 0.0052 intervals and the HULRAD calculations were performed in the same frequency range at 0.052 intervals. In figs. 7,9 the calculated structural velocities at the point of application of the force are compared against the exact solution. At resonances the responses show very sharp peaks and even at these resonances the comparisons are excellent. In figs. 8,10 the far-field pressures at $\theta = 0^o$, normalized to R, R= distance to the far-field, are shown for both the exact solution and the calculations.

(b) Spherical shell with an internal structure

To illustrate the effects of a substructure, the system of a simple oscillator and the shell described in section 4a was analyzed by applying a concentrated force on the oscillator fig. 6b. Both the shell and the oscillator have zero damping and in all four oscillators were used in the calculations to obtain the responses.

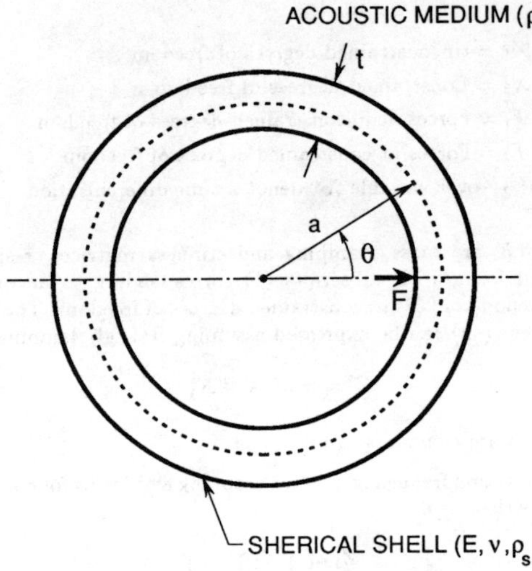

(a) Sperical shell with a concentrated force

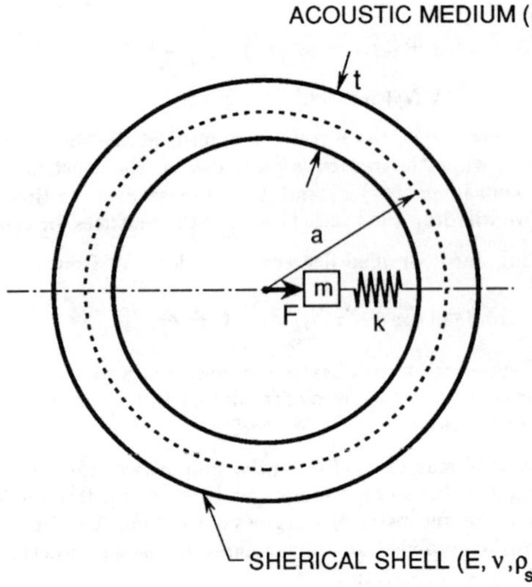

(b) Spherical shell with a substructure (oscillator)

Fig. 6 Spherical shell geometry

Substructure	Natural Frequency of the oscillator	Stiffness of the oscillator
A	4747.5 Hz	1.751 x 10^5 N/m
B	4747.5 Hz	1.751 x 10^{11} N/m
C	150.1 Hz	1.751 x 10^8 N/m
D	47.54 Hz	1.751 x 10^7 N/m

In figs. 11,12, the velocity of the shell and the far-field pressures at $\theta = 0^o$ are shown. These results are plotted against the responses of the point driven shell of section 4a for a damping value of zero. The system consisting of the shell and the substructure A which has a very high natural frequency and a small impedance (eq. (47)), behaves identically to the point driven shell. Although substructure B, with identical resonant frequency of 4747.5 Hz but a large impedance, behaves similarly, the coupling of the shell nodes is much stronger so that the resonant peaks in figs. 11,12 are shifted.

In figs. 13,14 the velocities and the far-field pressures at $\theta = 0^o$ of the point driven shell are plotted against the responses obtained using substructures B, C and D. The responses using substructure B does not exhibit any substructure resonances but has shifted shell resonances. The responses using substructure C exhibit shifted shell resonances as well as amplification at substructure resonance at a frequency of 147.5 Hz. The substructure D responses have an indication of substructure resonance at 47.5 Hz. Compared to all other substructures, substructure C, having resonance of 147.5 Hz has a very strong effect on the responses, since the substructure resonant frequencies are very close to shell resonance frequencies.

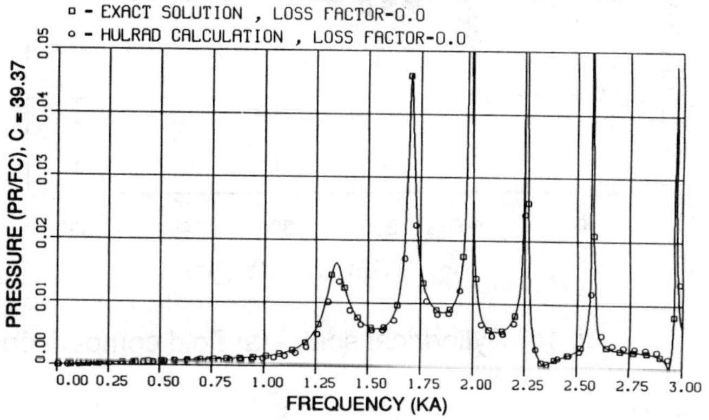

Fig. 7 Spherical shell - far field comparisons

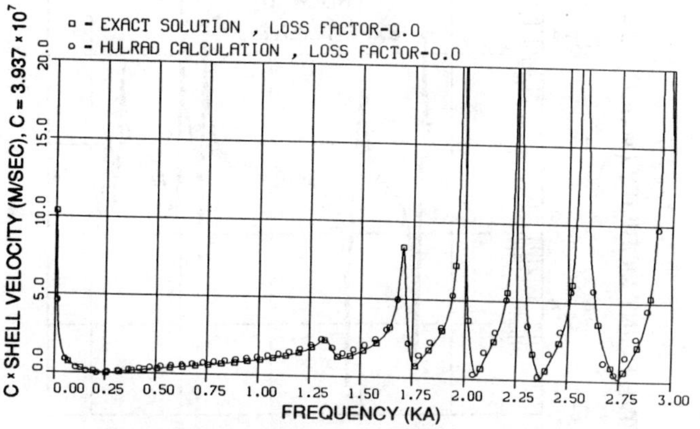

Fig. 8 Spherical shell - velocity comparisons

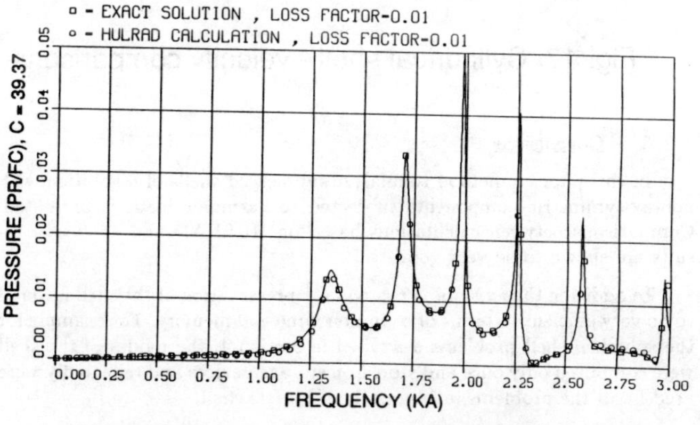

Fig. 9 Spherical shell - far field comparisons

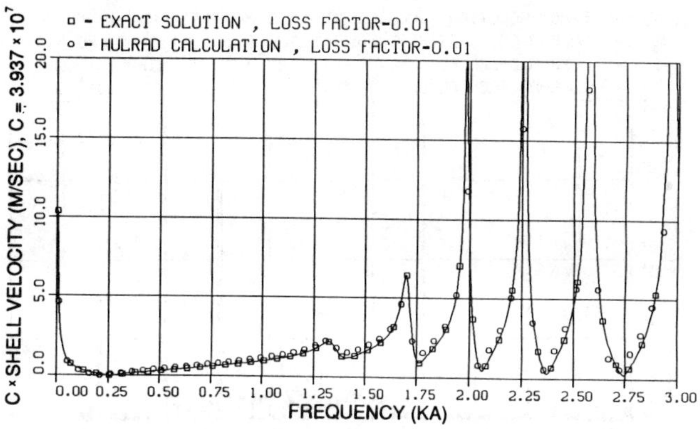

Fig. 10 Spherical shell - velocity comparisons

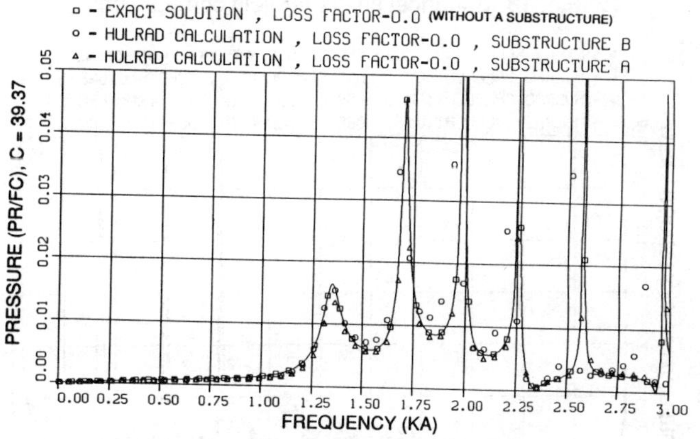

Fig. 11 Spherical shell - far field comparisons

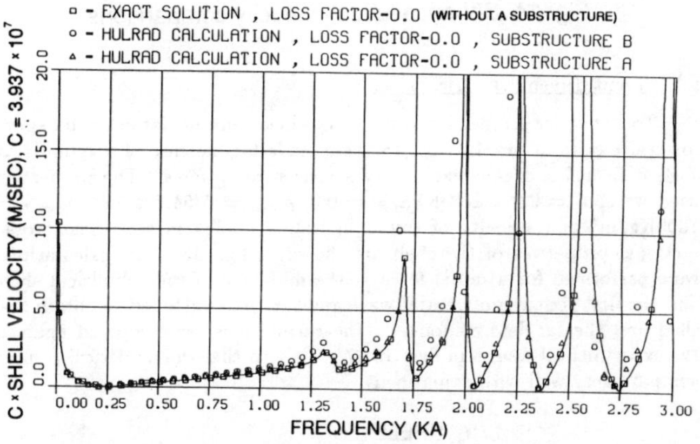

Fig. 12 Spherical shell - velocity comparisons

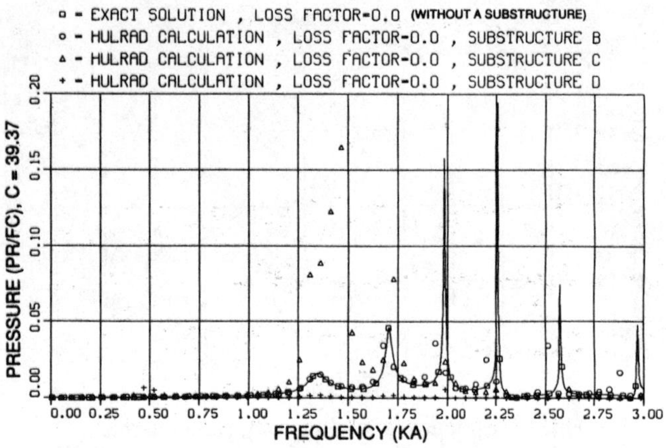

Fig. 13 Spherical shell - far field comparisons

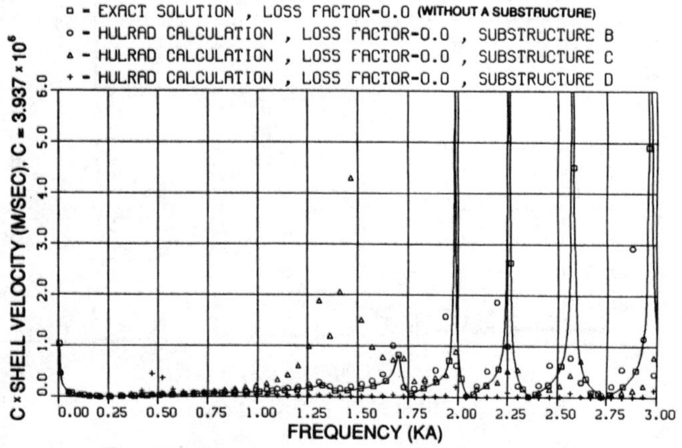

Fig. 14 Spherical shell - velocity comparisons

(c) Point driven cylindrical shell

To test other geometries, comparisons between calculations and experimental results of problem 5a [16] were made for the case of a cylindrical shell with flat end caps excited by a concentrated force. The properties used were, E = 2.0×10^8 KPa, $\nu = 0.3$, $\rho_{shell} = 7754.8 kg/m^3$, $\rho_{water} = 1000 Kg/m^3$ and velocity of sound in water = 1524 m/sec. The cross-sectional properties of the shell are shown in fig. 15. The calculations were performed for a radial force at the mid-span of the cylindrical shell and the first six circumferential wave numbers were included to obtain the shell and the far-field responses. The calculations are compared against the experimental results in figs. 16,17. It is seen that the numerical results compare very well with experiment.

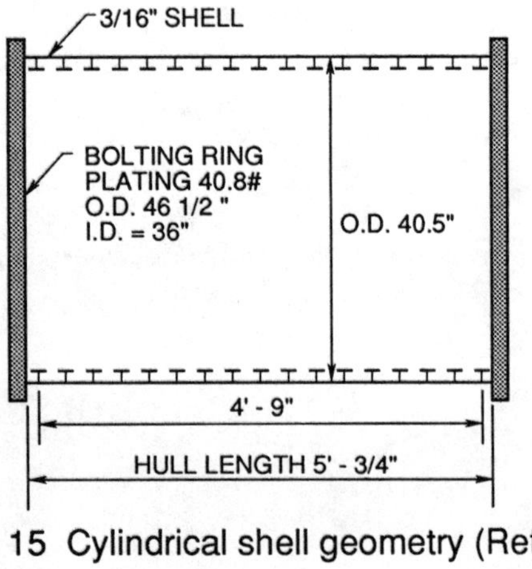

FIG. 15 Cylindrical shell geometry (Ref. 16)

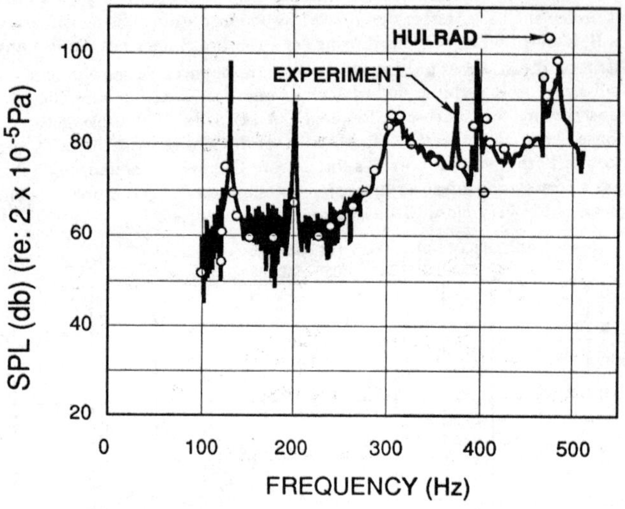

Fig. 16 Cylindrical shell - far field comparisons

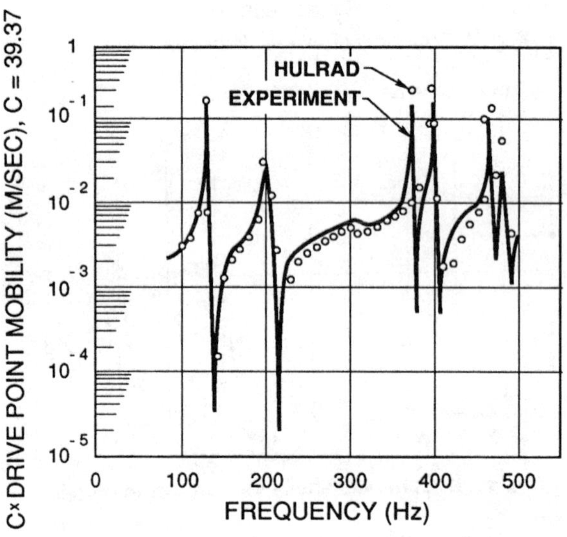

Fig. 17 Cylindrical shell - velocity comparisons

5. Conclusions

In this paper a method to analyze submerged shells of revolution with non-axisymmetric components subjected to harmonic loads is presented. Comparisons between calculations based on HULRAD and analytical results are shown to be very good.

In addition the axisymmetric modal representation of the shell is found to be very efficient in terms of computer time and memory. For example for the spherical shell problems described in section 4, the modes of the shell were computed only once and stored on storage devices and repeatedly were used in all the problems involving the spherical shell.

The method presented is modular in construction so that it can easily be extended to include the effects of internal regions such as liquid filled tanks. These regions of fluid can be modelled as assemblages of finite elements. Also, using imaging methods the presence of a free surface can be accounted for while calculating impedances of the exterior fluid field.

REFERENCES

1. Lamb, H., Hydrodynamics, Dover, New York, 1945.

2. Morse, P.M., Vibration and Sound, McGraw-Hill, New York, 1936.

3. Junger, M.C., "Radiation Loading of Spherical and Cylindrical Surfaces," J. Acoust. Soc. Amer., Vol. 24, No. 3, pp. 288-289, 1952.

4. Schenck, Harry A., "Improved Integral Formulation for Acoustic Radiation Problems", J. Acoust. Soc. Am. 44, pp. 41-58, 1968.

5. Baron, M.L. and McCormick, J.M., "Sound Radiation From Submerged Cylindrical Shells of Finite Length," Paul Weidlinger, Consulting Engineer, Office of Naval Research, Technical Report No. 2, June 1964.

6. Bell, W.A., Meyer, W.L., and Zinn, B.T., "Predicting the Acoustics of Arbitrarily Sharped Bodies Using an Integral Approach," AIAA J., Vol 15, pp. 813-820, 1977.

7. Wilton, D.T., "Acoustic Radiation and Scattering from Elastic Structures", International Journal for Numerical Methods in Engineering, 13, pp. 123-138, 1978.

8. Ranlet, D., DiMaggio, F.L., Bleich, H.H. and Baron, M.L., "Elastic Response of Submerged Shells with Internally Attached Structures to Shock Loading," Computers and Structures, Vol. 7, pp. 355-364, 1977.

9. Bushnell, D. "Stress, Stability and Vibration of Complex Branched Shells of Revolution: Analysis and User's Manual for BOSOR4", LM SC-D243605, Lockheed Missiles and Space Co., Inc., Sunnyvale, California, March 1972.

10. Vasudevan, R. and McCormick, J.M., "HULRAD User's Manual," Technical Report No. WARAD-88/001 Weidlinger Associates, New York, New York, March 1988.

11. Vasudevan, R. and McCormick, J.M., "HULRAD Theoretical Manual", Technical Report No. WARAD-88/002 Weidlinger Associates, New York, May 1988.

12. Piaszczyk, C., and Klosner, J., "Acoustic Radiation from Vibrating Surfaces at Characteristic Frequencies", J. Acoust. Soc. Amer., 75, pp. 363-375, 1984.

13. Burton, A., and Miller, G., "The Application of Integral Equation Methods to the Numerical Solution of Some Boundary Value Problems", Proc. R. Soc., London, Ser. A323, pp. 201-210, 1971.

14. Bathe, K.J. , Wilson, E.L. and Peterson, F.E., "SAPIV - A Structural Analysis Program for Static and Dynamic Response of Linear Systems", Earthquake Engineering Research Center, Report No. EERC 73-11, College of Engineering, University of California, Berkeley, California, April 1974.

15. Junger, M.C., and D. Feit, Sound, Structures, and Their Interaction, second edition, The MIT Press, Cambridge, Massachusetts, 1986.

16. Chen, L.H., "Acoustic Emission from Submerged Structures", Developments in Boundary Element Methods 2, edited by Banerjee, P.K. and Shaw, R.P., Applied Science Publishers Ltd., England, 1982.

THE ANALYSIS OF THE RADIATED AND SCATTERED ACOUSTIC FIELDS FROM SUBMERGED SHELL STRUCTURES USING A MODAL FINITE ELEMENT/BOUNDARY ELEMENT FORMULATION

R. D. Miller
NKF Engineering Incorporated
Arlington, Virginia

H. Huang
Naval Surface Warfare Center
Silver Spring, Maryland

E. T. Moyer, Jr.
Department of Civil and Mechanical Engineering
The George Washington University
Washington, D.C.

H. Uberall
Physics Deparment
Catholic University of America
Washington, D.C.

ABSTRACT

The analysis of submerged elastic shell structures using a combined finite element/boundary element approach is presented. The finite element program NASTRAN (NASA Structural Analysis) program is used to formulate the structural matrices. The SIERRAS (Surface Integral Equation Radiation and Scattering) program is then used to solve the coupled fluid-structure interaction problem. A superparametric boundary element with nine nodes is employed. Two modal formulations are developed. In the first approach, the structural velocities are expanded in terms of the mode shapes in vacuo. In the second approach, the fluid unknowns on the surface are expanded in terms of the eigenvectors of a homogeneous structure. The CHIEF (Combined Helmholtz Integral Equation Formulation) is employed to insure a unique solution for all frequencies. The methods are demonstrated for the scattering of plane waves from a submerged elastic spherical shell and from a cylinder with hemispherical endcaps.

NOMENCLATURE

a = radius of body
"A" = antisymmetric fluid-borne plate wave
$\underset{\sim}{A}$, $\underset{\sim}{B}$ = fluid matrices
c = velocity of sound in the fluid
c_l = velocity of longitudinal plate wave
$\underset{\sim}{C}$ = structural damping matrix
$\underset{\sim}{D}$ = structural influence matrix
DOF = degrees of freedom
f = pressure loading function
$\underset{\sim}{F}$ = kernel of an integral equation
$\underset{\sim}{F}_A$ = force applied directly to structure
$\underset{\sim}{F}_I$ = fluid interaction force on structure
g = normal vector component
G = Green's function
h = shell thickness
h_i = interpolation function
i = $(-1)^{1/2}$
k = ω/c = wavenumber
$\underset{\sim}{K}$ = structural stiffness matrix
L = matrix relating fluid pressure to structural load
L' = matrix relating structural displacements to normal fluid displacements
m = structural mode number
$\underset{\sim}{M}$ = structural mass matrix
n = number of Gauss integration points
n_ξ = normal to S at point ξ
N^S = number of surface elements
$|N|$ = the Jacobian
p = acoustic pressure
p^i = incident pressure
p^s = scattered pressure
r = distance between two points
R_i = region interior to the surface S
R_o = region interior to the surface S
$\underset{\sim}{S}$ = surface of the body
S_o = symmetric plate wave
S = harmonic structural matrix
t = time
$\underset{\sim}{U}$ = displacement vector
$\underset{\sim}{V}$ = velocity vector
w = Gauss weighting function
x,y,z = Cartesian coordinates
δ_{ij} = Kronecker delta symbol
ξ = a general point on the surface S of the body
η = a general point in three-dimensional space
$\underset{\sim}{\phi}$ = structural eigenvector matrix
$\underset{\sim}{\psi}$ = fluid eigenvector matrix
ρ = fluid density
ω = radian frequency

Superscripts

T = transpose
$*$ = complex conjugate
$-$ = a generalized vector or matrix

INTRODUCTION

One of the most complex problems in computational mechanics is analytically modeling the interaction between an elastic body and surrounding fluid. The finite element method (FEM) is available for the structure but is not effective for modeling the three-dimensional fluid surrounding the structure. The boundary element method (BEM), when applied to the Helmholtz integral equation, is the most recent development for solving general acoustic fluid problems.

This Helmholtz integral equation may be developed by a treatment of Green's Second Theorem as in Baker and Copson (1939). The derivation of the Helmholtz integral equation is based on Green's second theorem which transforms the three-dimensional problem into a two-dimensional formulation on the boundary. As with many integral formulations, the Helmholtz integral equation was not fully appreciated until the computer became available to perform the integrals required. There are several approaches which have been employed to solve the Helmholtz equation for selected classes of problems. One of the first approximate solutions was obtained by Williams et al. (1964) using an expansion in terms of spherical Hankel functions. This method works best for bodies which are nearly spherical. Another method of solution, introduced by Kupradze (1965), is based upon a distribution of monopole sources on the boundary of the surface. By requiring that the sources agree with the boundary conditions on the surface, the velocities and pressures are known at any point. This method is known as an "indirect" BEM since the formulation is written in terms of source strengths rather than the primary physical unknowns.

The Helmholtz integral equation was applied to the radiation of acoustic waves from structures with a prescribed velocity on the surface by Chertock (1971). Relationships were developed relating the surface pressures and velocities through the Helmholtz integral equation. This method is a "direct" BEM since the matrices developed relate the physical unknowns of pressure and velocity. The advantage of the direct formulation is that it allows a physical interpretation of the fluid matrix coefficients.

Given the Helmholtz integral equation, there are several ways to obtain a matrix form similar to that of the FEM. In fact, many of the two-dimensional shape functions in three-dimensional space for the FEM may be adapted for the BEM (Brebbia and Walker, 1980). An isoparametric quadratic boundary element for radiation and scattering from rigid bodies was introduced by Seybert et al. (1985). This element was used to obtain the solution for the scattered field from a rigid sphere and from two rigid spheres (Seybert et al., 1985).

A difficulty with the Helmholtz integral formulation for exterior fluid problems is that it fails to provide a unique solution at certain "internal" frequencies associated with the resonances of the interior fluid region. These resonances are non-physical when solving an exterior fluid problem and yield undesired peaks in the radiated noise spectrum or scattering form function. One method for eliminating the "internal resonances" proposed by Chertock (1971) uses a combination of the normal derivative of the Helmholtz equation and the Helmholtz equation. However, it was not until Burton and Miller (1971) reduced the higher order singularities in the normal equation to integrals which could be solved in the Cauchy value sense that this method became effective. Another method for regularizing the normal equation was developed by Meyer et al. (1978). Another combined Helmholtz integral equation formulation (CHIEF) was proposed by Schenck (1968). This method is based on combining the Helmholtz equation on the surface with the interior equation to insure a unique solution. This combined approach involves adding several equations for selected interior points to the surface integral equations. The overdetermined set of equations is then solved by the least squares approach. The main difficulty with CHIEF has been in selecting the interior point locations. However, it has recently been suggested that a unique solution is obtained by the CHIEF provided that there is at least one interior point which does not correspond to a node point associated with the interior mode (Seybert and Rengarajan, 1987). The use of CHIEF has been found to be very effective in providing unique solutions for the problems presented in this paper. Piaszcyk and Klosner (1984) developed a combined formulation of the surface Helmholtz equation with the exterior equation. This approach employs the use of a high frequency fluid impedance on the surface to obtain an overdetermined set of equations which may be solved in a least-squares sense.

The acoustic radiation from forced elastic bodies using the FEM for the structure and the BEM for the fluid was discussed by Engblom and Nelson (1975). The method involves the use of a "consistent" formulation involving the same shape function for both the structure and the fluid. The finite element/boundary element (FE/BE) method was further outlined in a general form for radiation and scattering of elastic structures by Wilton (1978).

The first FE/BE approach for the solution of a fluid-structure radiated noise problem using the Burton and Miller combined Helmholtz integral formulation for the elimination of internal resonances was demonstrated by Mathews (1980). An isoparametric quadratic element was implemented for both the fluid and the structure and the FE/BE coupled equations of Wilton (1978) were employed. An advanced integration scheme of Lachat and Watson (1976) was used along with a symmetric formulation by Zeinkiewitz et al. (1977). The radiated noise from an elastic sphere was presented by Mathews (1984). The modal formulation for the structure has been presented in terms of the closed form analytical eigensolution by Chen (1982). The matrix formulation of the coupled radiated noise problem in terms of the structural modes has been presented without implementation by Mathews (1979).

The problem of scattering due to a fluid target of arbitrary shape has been considered by Tobocman (1984). The first paper indicates that as of 1983, the only applications of the Helmholtz formula to scattering were due to Brundrit (1965). The problems solved by Tobocman include the scattering due to rigid spheres, oblate spheroids and prolate spheroids. Two Helmholtz integral equations were solved simultaneously, one for the exterior fluid and one for the interior fluid. The CHIEF methodology was used to remove the internal resonances from the exterior solution. Problems solved include the acoustic scattering from spherical air bubbles, and prolate spheroidal CCl^4 inclusions. The normal equation of Meyer et al. (1978) was also used to solve scattering problems. It was shown that the normal equation approach is superior to that of the surface Helmholtz equation formulation for problems of increasing nonsphericity.

The first scattering application which employs the FE/BE approach was given by Huang (1984). In this case, the monostatic and bistatic form functions are found for the scattering of an incoming plane wave by an elastic sphere. Also, the solution of the scattering from a submerged spherical shell using the BEM has been considered by Everstine et al. (1987) and Seybert et al. (1988). The computations presented for the scattering from elastic shells have been limited to the low wavenumbers (ka<5).

In this paper, the FE/BE method is applied to problems involving the scattering of plane waves from elastic shells. The structural matrices are generated by the NASTRAN program (Butler and Michel, 1971). The formulation of boundary element matrices representing the fluid and the simultaneous solution of the fluid-structure equations is provided by the SIERRAS (Surface Integral Equation Radiation and Scattering) program written by R.D. Miller. The SIERRAS program solves the fluid-loaded structure problem for either radiated noise or scattering of incident plane waves for arbitrarily shaped bodies.

In the following paragraphs, the theoretical development of the SIERRAS program is presented. The Helmholtz integral equation is given and a four- to nine-noded superparametric discontinuous boundary element is derived. The direct form of the fluid variable methodology (FVM) for the simultaneous solution of the fluid-structure equations is reviewed followed by the modal formulations for the structure and fluid. The CHIEF method is used to insure a unique solution for all formulations. Resonant peaks are found by using a peak search routine.

The methodologies are demonstrated for two examples. In the first example, the computed form function for the scattering of plane waves from an elastic spherical shell is compared with an "exact" solution using the direct FVM. In the second example, the NASTRAN/SIERRAS results for the backscattered form function of a cylindrical shell with hemispherical endcaps are compared with experimental results. Complete FE/BE form function computations are presented for the scattering solutions. Results for the direct and modal formulations are presented for selected peaks. The modes associated with peaks in the scattering form function are identified from the structural velocity phase plots across the surface of the shell at the resonances.

ACOUSTIC RADIATION AND SCATTERING FORMULATION

The analysis of radiation and scattering problems using the FE/BE approach involves the development of the boundary element for the fluid and the simultaneous solutions of the structural FEM and fluid BEM equations. A variable-noded, superparametric discontinuous direct boundary element is developed below with FVM coupling and CHIEF for the direct and modal approaches.

The Helmholtz Integral Equation

For acoustics problems the BEM is applied to the Helmholtz integral equation. It is possible to solve the fluid and structure equations simultaneously to obtain the pressures in the fluid and the velocities of the structure. Following this, far-field radiated or scattered pressures may be obtained from the exterior Helmholtz integral equation by simple quadrature. The advantage of using this method is that it is exact, approaching the solution of the acoustic wave equation in the limit as the discretization approaches the continuum. A derivation of the Helmholtz integral equation is given in Baker and Copson (1939).

The Helmholtz integral equation for the pressure p and normal velocity v in a three-dimensional fluid under the assumptions of a compressible inviscid elastic fluid with small disturbances is given by:

$$\iint_{S_\xi} \left[p(\xi) \frac{\partial G(\xi,\eta)}{\partial n_\xi} - G(\xi,\eta) \frac{\partial p(\xi)}{\partial n_\xi} \right] dS_\xi = \begin{cases} 4\pi p^s(\eta), & \eta \in R_o \\ (2\pi p(\eta) - 4\pi p^i(\eta)), & \eta \in S \\ 4\pi p^i(\eta), & \eta \in R_i \end{cases} \quad \ldots (1)$$

where the interior region R_i and the exterior region R_o are related to the surface S of the body as shown in Figure 1, n is the outward normal to the surface of the body. The Green's function G is given by:

$$G = \frac{e^{ikr}}{r} e^{-i\omega t} \quad \ldots (2)$$

where r is the magnitude of the distance between two points:

$$r = |r_\eta - r_\xi| \quad \ldots (3)$$

and the geometry of the points is depicted in Figure 1.

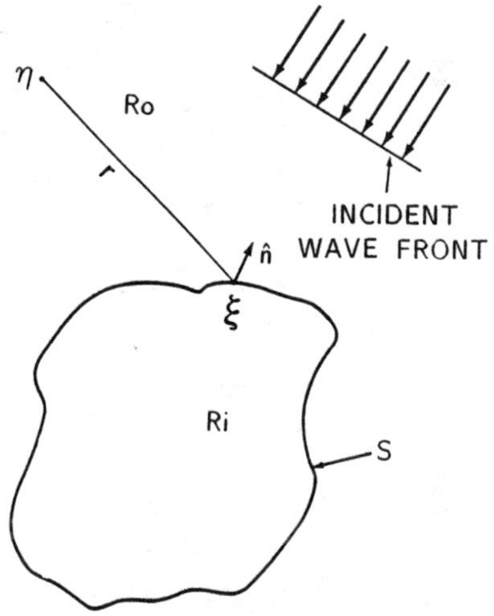

Figure 1. Geometry of the Fluid-Structure Regions

The Superparametric Discontinuous Element

The SIERRAS program currently employs a four- to nine-noded superparametric discontinuous element as shown in Figure 2. This element may be used for any arbitrary shape of structure using from four to nine nodes per element as required to fit any surface. A superparametric element is one in which the order of the assumed geometry exceeds the order of the unknowns on the boundary. The surface is therefore fit with quadratics and the surface unknowns are held constant over the element. A discontinuous element is one in which the point at which unknowns are specified are well within the boundaries of the element. The fact that this type of element is allowed is a direct result of the integral equation formulation (Brebbia and Walker, 1980). The discontinuous constant element has been demonstrated to have accuracy superior to that of a continuous linear element and approaching that of a continuous quadratic boundary element (Brebbia et al., 1984).

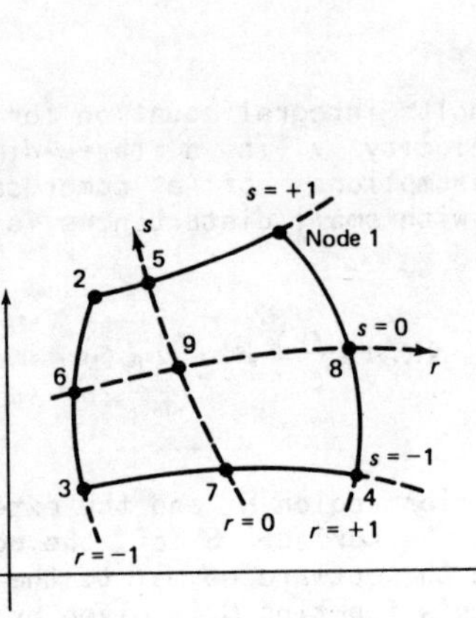

Figure 2. A Four- to Nine-Noded Two-Dimensional Element in Three-Dimensional Space (Bathe, 1984)

Writing equation (1) as a sum of integrals over which the pressures and velocities are taken to be constant, an equation for each such point η_j of area dS is given by:

$$\sum_{i=1}^{N} \iint_{S_{\xi_i}} \left[\frac{\partial G(\xi_i, \eta_j)}{\partial n_{\xi_i}} - 2\pi \delta_{ij} \right] dS_{\xi_i} \; p(\xi_i)$$

$$= \sum_{i=1}^{N} \iint_{S_{\xi_i}} i\rho\omega G(\xi_i, \eta_j) dS_{\xi_i} \; v_n(\xi_i) + f(\eta_j) \quad j=1,N \quad \ldots (4)$$

where

$$f(\eta_j) \equiv \begin{cases} 4\pi p^s(\eta_j) & \eta_j \in R_o \\ -4\pi p^i(\eta_j) & \eta_j \in S \\ 4\pi p^i(\eta_j) & \eta_j \in R_i \end{cases}$$

and $2\pi\delta_{ij}$ applies only for η_j on s where δ_{ij} is the Kronecker delta.

Equation (4) may also be written in matrix form:

$$\underset{\sim}{A}\, \underset{\sim}{p} = \underset{\sim}{B}\, \underset{\sim}{v}_n + \underset{\sim}{f} \quad \ldots (5)$$

The general form of the integrals of equation (4) is given by:

$$\iint_{S_{\xi_i}} F\left[x_{\xi_i}, y_{\xi_i}, z_{\xi_i}\right] dS_i \quad \ldots (6)$$

For the axisymmetric problems considered in this paper, it is sufficient to handle off-diagonal (non-singular) terms by taking the function F in equation (6) outside the integral. In this case the points at the center of each element are used to evaluate F. Then the function F is simply multiplied by the elemental area.

Equation (6) may be written for Gaussian integration as follows:

$$\iint_{S_{\xi_i}} F\left[x_{\xi_i}(r,s), y_{\xi_i}(r,s), z_{\xi_i}(r,s)\right] |N|\, dr\, ds \quad \ldots (7)$$

For a general l-point Gaussian integration equation (7) becomes:

$$\sum_{j=1}^{n} \sum_{k=1}^{n} F\left[x_{\xi_i}(r_j, s_k), y_{\xi_i}(r_j, s_k), z_{\xi_i}(r_j, s_k)\right] |N|\, w_j\, w_k \ldots (8)$$

where r_j, s_k are n integration points,

w_j, w_k are n weighting functions,

and

$|N|$ is the Jacobian.

For the four- to nine-noded element of Bathe (1982), the location of an arbitrary point in the element in terms of the nodes of the element are given by:

$$x(r,s) = \sum_{i=1}^{9} h_i(r,s)\, x_i$$

$$y(r,s) = \sum_{i=1}^{9} h_i(r,s)\, y_i \quad \ldots (9)$$

$$z(r,s) = \sum_{i=1}^{9} h_i(r,s)\, z_i$$

where the interpolation functions h (r,s) are given by Bathe (1982).

Given the location of any point in the element through equation (9), the tangents to the element surface are given by:

$$\frac{\partial \underset{\sim}{r}}{\partial r} = \left(\frac{\partial x}{\partial r}, \frac{\partial y}{\partial r}, \frac{\partial z}{\partial r}\right)$$

$$\frac{\partial \underset{\sim}{r}}{\partial s} = \left(\frac{\partial x}{\partial s}, \frac{\partial y}{\partial s}, \frac{\partial z}{\partial s}\right) \quad \ldots (11)$$

and the normal vector is given by:

$$\underset{\sim}{n} = \frac{\partial \underset{\sim}{r}}{\partial r} \times \frac{\partial \underset{\sim}{r}}{\partial s} = (g_1, g_2, g_3) \quad \ldots (12)$$

where X denotes the vector cross-product and the magnitude of the normal vector is the Jacobian given as:

$$|N| = \sqrt{g_1^2 + g_2^2 + g_3^2} \quad \ldots (13)$$

Expression (4) may be written in matrix form for points on the surface S as follows:

$$\underset{\sim}{A}\, \underset{\sim}{p} = \underset{\sim}{B}\, \underset{\sim}{v} - 4\pi \underset{\sim}{p}^i \quad \ldots (14)$$

__Fluid Variable Methodology__

Given the FEM equations for the structure and the BEM equations for the fluid there are several ways to solve the simultaneous equations. Of the various formulations were considered (Mathews, 1984), the fluid variable methodology (FVM) was selected due to the efficiency of the method.

The fundamental equations for acoustic scattering from a submerged elastic structure are obtained from the matrix structural equation for harmonic motion of a linear structure and the Helmholtz integral equation. The matrix form of the Helmholtz integral equation is given in (14) and the governing equation for the structural finite element model is:

$$[-\omega^2 \underset{\sim}{M} - i\omega \underset{\sim}{C} + \underset{\sim}{K}]\underset{\sim}{u} = \underset{\sim}{S}\underset{\sim}{u} = \underset{\sim}{F}_I + \underset{\sim}{F}_A \qquad \ldots (15)$$

where $\underset{\sim}{M}$, $\underset{\sim}{K}$, $\underset{\sim}{C}$ constitute the structural mass, stiffness and damping matrices respectively, $\underset{\sim}{F}_A$ is the force on the structure, $\underset{\sim}{F}_I$ is the force of the fluid on the structure and $\underset{\sim}{u}$ is the displacement.

The structural matrices of equation (15) are provided by the NASTRAN program. The coupling of SIERRAS with NASTRAN does not limit the modeling capability of NASTRAN. All of the structural elements in NASTRAN are available for modeling and coupling with the SIERRAS program. It is possible, for example, to analyze a fluid-loaded structure with damped internal structure and/or exterior coating. The SIERRAS program allows frequency-dependent structural damping to be included in the analysis. In this way, the true resonances and damping factors associated with the structure may be considered. The use of SIERRAS is not limited to NASTRAN; the only requirement is that the structural mass and stiffness and/or eigensolution matrices of the finite element program be made available to SIERRAS. Guyan reduction may be used to reduce the number of structural degrees of freedom (DOF) for the structure prior to running SIERRAS for the coupled problem.

Let the matrix $\underset{\sim}{L}$ relate the fluid pressures in the fluid to the forces on the structure:

$$\underset{\sim}{F}_I = -\underset{\sim}{L}\,\underset{\sim}{p} \qquad \ldots (16)$$

The matrix $\underset{\sim}{L}$ consists of areas associated with the normal direction to the surface.

The continuity equation at the fluid structure interface requires that the normal velocity of the structure and fluid follow equation (17).

$$\underset{\sim}{v}_n = -i\omega \underset{\sim}{L}'\,\underset{\sim}{u} \qquad \ldots (17)$$

The matrix $\underset{\sim}{L}'$ transforms structural vectors to normal vectors at the surface.

Equations (14) and (15) together with continuity and compatibility conditions of equations (16) and (17) may be solved simultaneously for the pressure p on the surface of the structure using expression (18).

$$[\underset{\sim}{A} + \underset{\sim}{D}\underset{\sim}{L}]\,\underset{\sim}{p} = \underset{\sim}{D}\,\underset{\sim}{F}_A - 4\pi \underset{\sim}{p}^i \qquad \ldots (18)$$

where

$$\underset{\sim}{D} = -i\omega \underset{\sim}{B}\,\underset{\sim}{L}'\,\underset{\sim}{S}^{-1}$$

In the FVM the fluid variable (pressure) is solved for prior to solving for the structural response. The FVM appears to be the most efficient formulation for general fluid-structure interaction problems. In most practical applications the DOF required to model the structure exceed the fluid DOF. Two inversions are required for FVM as observed in equation (18), one for the structural matrix and another for the fluid-loaded structure. The advantage in using the FVM is that the matrix for the structure is often real (for the case of no damping), symmetric, and well-banded (provided that Guyan reduction is not used). The final inversion requires the inverse of a non-banded, non-symmetric, complex matrix. The size this matrix of inversion for FVM is limited to the number of fluid DOF which is generally less than the number of structural DOF.

Given the pressure on the surface, equation (14) may be used to obtain the velocities on the surface. The same surface elements used in developing equation (14) on the surface are then used to obtain the far-field solution:

$$\underset{\sim}{A}\,\underset{\sim}{p} = \underset{\sim}{B}\,\underset{\sim}{v} + 4\pi \underset{\sim}{p}^s \qquad \ldots (19)$$

An effort has been made to improve the computational efficiency of the SIERRAS program since it was first written in 1983. The improvements include a vectorized inversion routine, discontinuous super-parametric elements, sparse matrix multiplication and enhanced program efficiency. Several types of symmetry have been implemented to reduce computation time including symmetry in up to three planes, as well as axisymmetry. Guyan reduction has been used in NASTRAN to reduce the number of structural degrees of freedom prior to running SIERRAS.

The Problem of Uniqueness of Solution for Certain Frequencies

One of the major difficulties with the Helmholtz integral formulation has been non-uniqueness at certain frequencies associated with the frequencies of the internal region bounded by the structure. CHIEF has been used previously to obtain the solutions of elementary problems in radiation and scattering of rigid bodies (Schenck, 1968, and Tobocman, 1984). In this paper, CHIEF is demonstrated for solving problems of uniqueness related to the scattering of acoustic waves by elastic shells.

The CHIEF method of Schenck (1968), and the combined formulation using the normal equation of Meyer et al. (1978) were considered in the development of SIERRAS. The CHIEF was selected over the normal equation method. The primary reason for selecting CHIEF is that the internal and surface integral equations are of the same order. The normal equation used by Meyer is of a higher order due to the fact that it is obtained by taking a partial derivative with respect to the normal of the entire Helmholtz integral equation. Use of the normal equation would therefore require a finer discretization in order to ensure an accurate solution.

The CHIEF method involves the use of equation (1) for selected points in the interior region R_i of the body in addition to equation (1) at the surface S. The problem for the fluid then becomes overdetermined, having more equations than unknowns, and must be solved in a least squared sense. The least squares form of equation (18) for scattering is written:

$$\underset{\sim}{A}^{T*}[\underset{\sim}{A} + \underset{\sim}{D}\,\underset{\sim}{L}]\,\underset{\sim}{p} = \underset{\sim}{A}^{T*}[\underset{\sim}{D}\,\underset{\sim}{F}_A - 4\pi \underset{\sim}{p}^i] \qquad \ldots (20)$$

where the structural force $\underset{\sim}{F}_A$ applies for the radiated noise case, * denotes the complex conjugate of a matrix, and the incoming wave pressure function $\underset{\sim}{p}^i$ applies for the scattering case.

The CHIEF method has been found to be quite adequate for insuring uniqueness at all interior resonances as long as there is at least one interior point selected which is not on a node of the interior mode (Seybert and Rengarajan, 1987). For the axisymmetric problems solved in this paper, it was found that the solution was not particularly sensitive to the placement of the interior points, and that only ten points were required to insure uniqueness to ka = 30.

The Modal Formulations

The modal formulation for the structure has been used in the past when the closed form analytical eigensolution is known (Chen, 1982). The matrix formulation in terms of the structural modes has been presented previously without implementation (Mathews, 1979).

In the SIERRAS modal formulation, the structural eigenvalue analysis is performed by NASTRAN prior to running the SIERRAS program. The structural displacements may be represented as:

$$u = \phi \bar{u} \qquad \ldots (21)$$

where ϕ is the eigenvector solution for the structure, and $\bar{u}$ is the generalized displacement vector.

The introduction of equation (21) into any of the formulations developed to this point results in the replacement of the structural inverse by:

$$S^{-1} = \phi \bar{S}^{-1} \phi^T \qquad \ldots (22)$$

where

$$\bar{S} = \bar{K} - \omega^2 \bar{M}$$

and $\bar{K}$ and $\bar{M}$ are the in-vacuo generalized stiffness and mass matrices.

The formulation is quite interesting from a physical viewpoint since the orthogonal in-vacuo modes of the structure are coupled together and loaded by the integral equations of the fluid (see equation 18). The modal formulation is much more efficient than the direct method of equation (18) in that the inversion of the structural matrix is replaced by a single matrix multiplication.

The modal analysis method allows the user to increase the number of modes kept in the analysis as frequency increases. This leads to an analysis which is adaptive and minimizes computation time.

In the second modal formulation, the pressures on the surface are also expressed in terms of structural modes.

Suppose that the fluid-loaded normal displacements are related to the entire structure as follows:

$$\psi = L' \phi \qquad \ldots (23)$$

Then express the pressure at the surface in terms of the modal pressures through equation (21).

$$p = \psi \bar{p} \qquad \ldots (24)$$

Substitution into the general fluid-structure coupled equation yields:

$$\psi^T (A + D L) \psi \bar{p} = \psi^T (D E_A + 4\pi p^i) \qquad \ldots (25)$$

and for the least squares problem this results in:

$$\psi^T \left\{A^{T*}(A + D L)\right\} \psi \bar{p} = \psi^T \left\{A^{T*}(D E + 4\pi p^i)\right\} \ldots (26)$$

The problem size for the final inversion may be reduced without regeneration of a mesh by selecting a few modes at the lower frequency and increasing the number of modes kept as frequency increases. The modal approach then becomes adaptive in the sense that the size of the structural and fluid equations may be reduced to the number of modes required to represent the response of each.

RESONANCE CLASSIFICATION

The analysis of a general submerged shell structure involves the search for peaks in the scattered form function occurring at shell mode frequencies. The program SIERRAS uses an advanced peak response searching routine called GOLDEN (Press, 1986) to locate these resonances during the frequency sweep. It is this capability which allows SIERRAS to determine the frequency spectrum completely in a single computer run.

The modes associated with peaks in the radiated noise spectrum or scattering form function are identified from surface structural velocity response or phase plots at the resonances. This technique is also useful in determining whether a peak is due to the presence of an internal resonance.

RESULTS

Several examples have been considered to demonstrate the accuracy of the NASTRAN/SIERRAS approach in solving fluid-structure interaction problems. The results presented here include the backscattered echo of plane waves from a spherical shell and from a cylindrical shell with hemispherical endcaps. The direct FVM of equation (18) was used for the spherical shell analysis to verify the NASTRAN/SIERRAS approach. The modal formulations are demonstrated for the case of the scattering from a cylindrical shell with hemispherical endcaps by comparison with the direct formulation. Experimental results are also provided.

The Scattering of Plane Waves by an Elastic Spherical Shell

To verify the NASTRAN/SIERRAS approach for solving fluid-structure interaction problems of scattering, the solution for the backscattering from a spherical shell was computed and compared with an "exact" solution. A spherical shell of brass with a h/a (thickness-to-radius ratio) of 6.6 percent was selected. The properties of brass for this analysis are given in Table 1. For this example, 90 fluid DOF and 180 structural DOF were used in an axisymmetric formulation representing 8100 fluid DOF. The azimuthal increment for the axisymmetric elements is 4 degrees. A nine by nine integration scheme was used for evaluation of the singular terms of the fluid matrices. It was found that no integration was required for obtaining the off-diagonal (non-singular) terms. Rather, the kernals were found using the center points of each element, taken outside the integral in equation (6), and multiplied by the elemental area.

Table 1. Properties of Brass

Dilational Wave Speed	= 4.70 x 10^5 cm/s
Shear Wave Speed	= 2.11 x 10^5 cm/s
Density	= 8.64 g/cm^3

The "exact" form function obtained by a closed-form expansion is given together with the NASTRAN/SIERRAS results in Figure 3. The wavenumber k= ω/c where c is the speed of sound in the fluid. Due to the narrowness of the peaks between ka=3 and ka=15, resonances have been omitted from either the "exact" or the numerical solutions, or both. Although it is possible to extract all of the peaks with SIERRAS (Talmant, 1989), it is an academic exercise due to the fact that the radiation efficiencies of the missing modes are so low that they would not be found in an actual experiment (cf. Figure 10).

A comparison of resonance frequencies is given in Table 2. Agreement is considered excellent with a maximum error of two percent for the majority of resonances. Comparison of antisymmetric resonant frequencies below ka = 12 is not provided due to the lack of resolution of the "exact" solution.

Table 2. Comparison of SIERRAS and Exact Resonance Values for a Brass Spherical Shell with h/a = 6.6%

Branch 1 (= Lamb "A" antisymmetric bending waves for ka>3)

Mode Number (m)	"Exact" (ka)	NASTRAN/SIERRAS (ka)	Error (%)
18	13.53	13.24	2.1
19	14.82	N/A	N/A
20	15.91	15.71	1.3
21	17.12	16.98	0.8
22	18.49	19.51	1.7
23	19.84	19.51	1.7
24	21.10	20.82	1.3
25	22.32	22.09	1.0
26	23.59	23.37	0.9
27	24.95	24.64	1.2
28	26.96	26.73	0.9
29	28.24	28.04	0.7
30	29.54	29.15	1.3

Branch 2 (= So Lamb symmetric longitudinal waves for ka>3)

Mode Number (m)	"Exact" (ka)	NASTRAN/SIERRAS (ka)	Error (%)
2	3.47	3.47	0.0
3	7.01	7.04	0.4
4	9.63	9.61	0.2
5	12.34	12.31	0.2

The resonances of the spherical shell have been described in a previous paper (Reference 31). The structural mode number m, is determined from the structural velocity or phase plot over the surface as shown in Figure 4 for the m=18 "A" wave. The modes of a submerged spherical shell have been classified previously (Junger, 1986)) as branch 1 and branch 2 modes. The branch 1 modes are primarily bending modes which involve the entire shell at low frequency. The frequencies associated with these resonances are highly sensitive to changes in the h/a ratio. These modes may be described by "A" fluid-borne waves and Ao antisymmetric Lamb waves as curvature effects become negligible (ka>3). The branch 2 modes are primarily longitudinal waves along the shell. These waves are sensitive to the material properties but nearly independent of the thickness. For the submerged case, the longitudinal waves are acoustically fast waves (c_l>c) and hence highly damped in the radial direction so that the longitudinal wave rapidly dominates the response. The branch 2 waves therefore degenerate to So symmetric Lamb waves as ka increases and curvature effects become negligible (ka>3).

The Scattering by a Cylindrical Shell with Hemispherical Endcaps

The analysis of the cylindrical shell with hemispherical endcaps was performed using the NASTRAN/SIERRAS approach. The dimensions of the shell are given in Figure 5. The radius of the cylinder and hemispherical endcaps is exactly that of the spherical shell just considered. Originally, the properties of brass were taken as those of Table 1.

In this analysis 125 fluid DOF and 250 structural DOF were used in an axisymmetric formulation representing 11250 fluid DOF. The FVM formulation of equation (18) was used to obtain the results for the direct formulation. Nine integration points were used for singular integrands. Two by two Gauss integration was used in determining the areas of the elements. No integration points were used for determining the off-diagonal terms. Rather, the kernals were found using the center points of each element, taken outside the integral in equation (6), and multiplied by the appropriate elemental area. It has been found that for the problems presented here, the solution is not sensitive to the number of integration points used for off-diagonal integration. It is suspected that this simplification may only be true for axisymmetric bodies. However, more research is required to determine the exact limitations of this approximation. Two by two Gauss integration was used in evaluating the pressure in the far-field with equation (1) in R_o.

The NASTRAN/SIERRAS analysis was performed using solid CHEXA elements as shown in Figure 6. The form function obtained by the direct method of equation (18) for the end-on scattering of plane harmonic waves is shown in Figure 7.

The form functions obtained by the direct method of equation (18) and the structural modal formulation (in which equation (22) replaces the structural inverse in equation (18)) are compared in Figure 8 for "A" waves with m = 29, 30 and 31. The structural modal analysis was performed using 120 structural modes out of a total of 250.

The results of the full modal formulation (using the structural modes to represent the structure and the fluid) for 60 and 120 modes is plotted against the direct result in Figure 9.

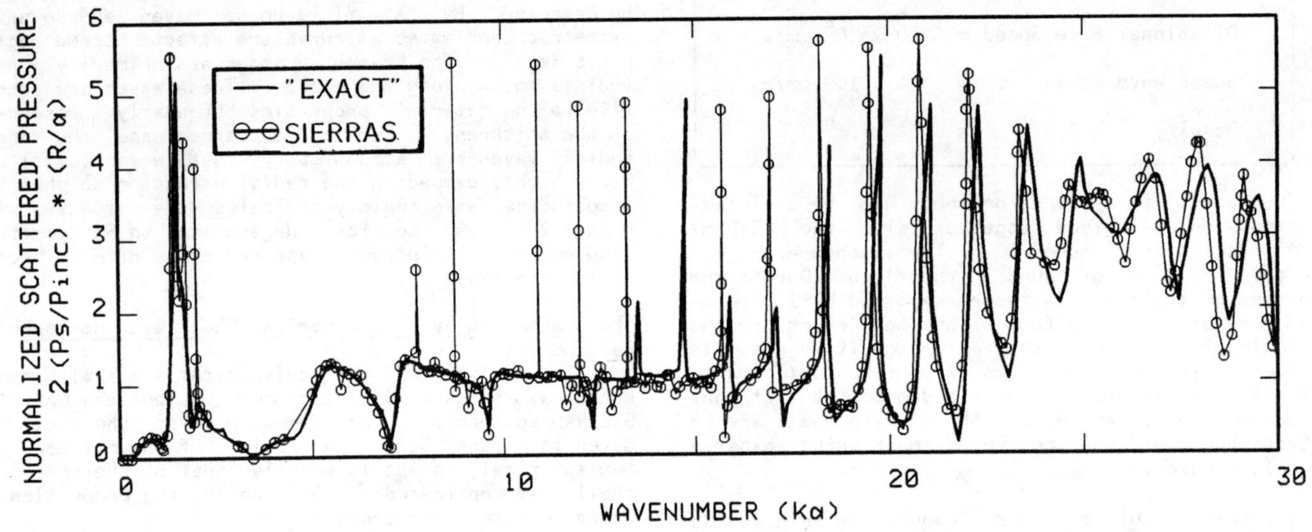

Figure 3. Backscattered Form Function from a Brass Spherical Shell

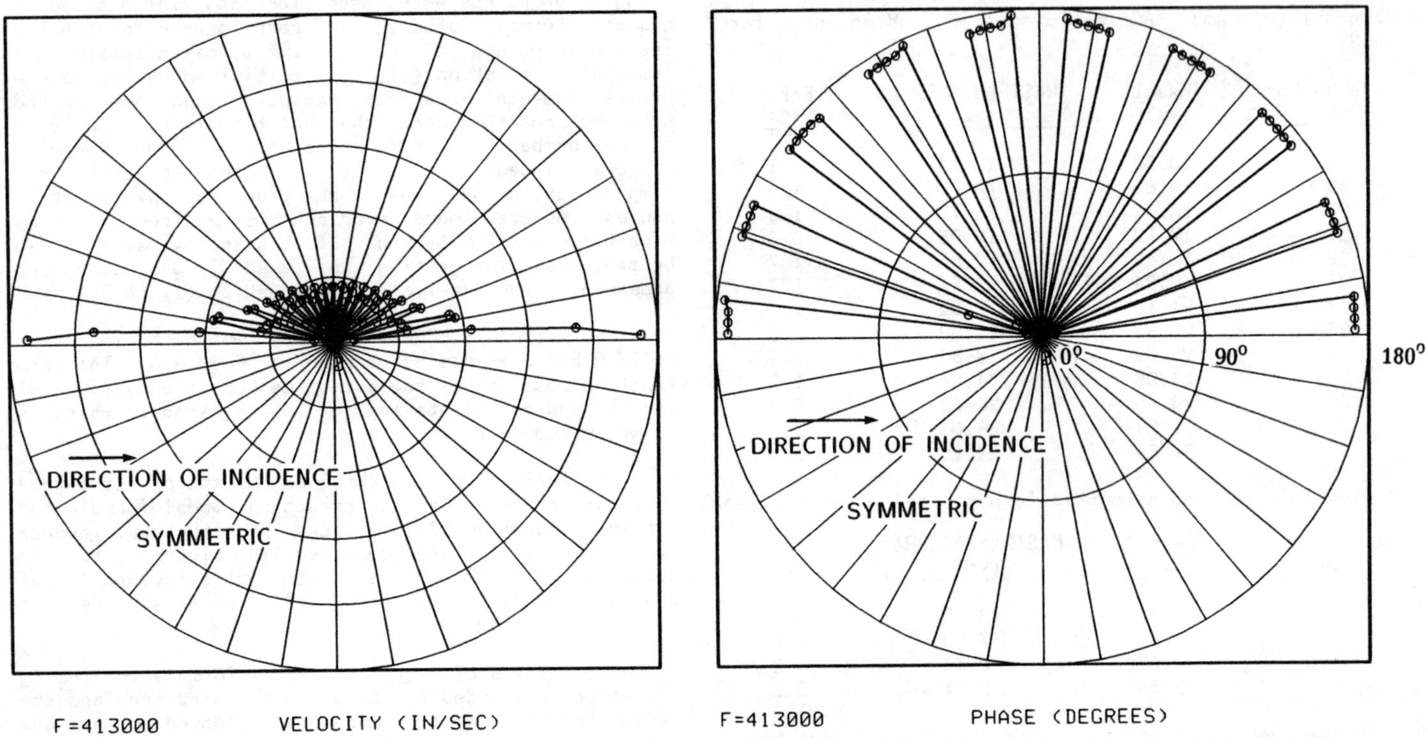

Figure 4. NASTRAN/SIERRAS (a) Surface Velocity and (b) Phase Plot for the m=18 "A" Wave of the ka=13.24 Resonance

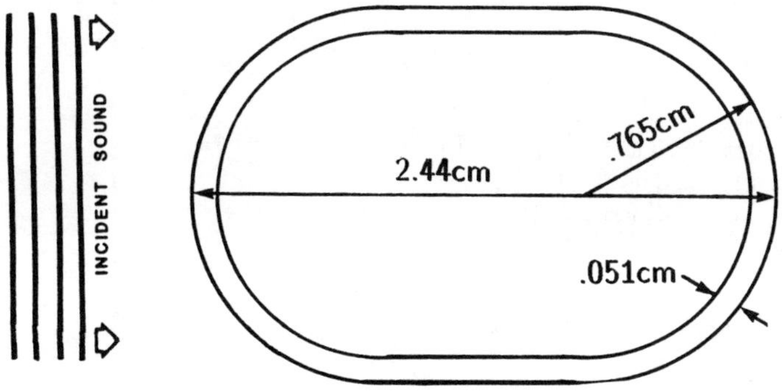

Figure 5. Dimensions of a Hollow Cylindrical Brass Shell with Hemispherical Endcaps

Figure 6. Axisymmetric Finite Element Model of a Cylindrical Shell with Hemispherical Endcaps Using Solid Elements and an Azimuthal Increment of 4 Degrees.

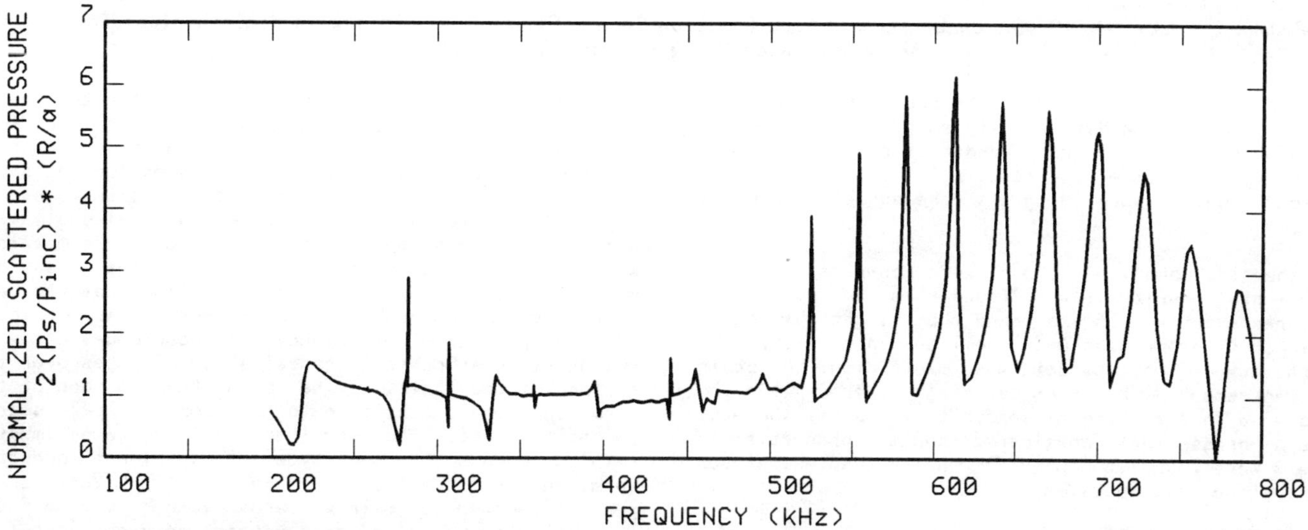

Figure 7. NASTRAN/SIERRAS Backscattered Form Function from a Cylindrical Shell with Hemispherical Endcaps for End-On Incidence Using the Direct Method

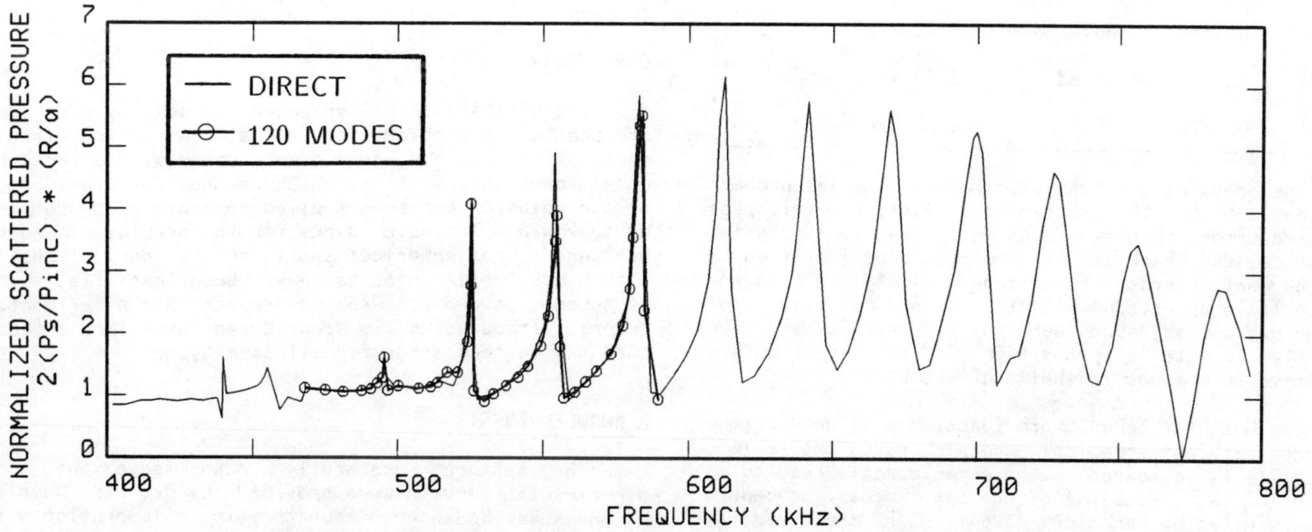

Figure 8. NASTRAN/SIERRAS Backscattered Form Function from a Cylindrical Shell with Hemispherical Endcaps for End-On Incidence Using the Structural Modal Formulation

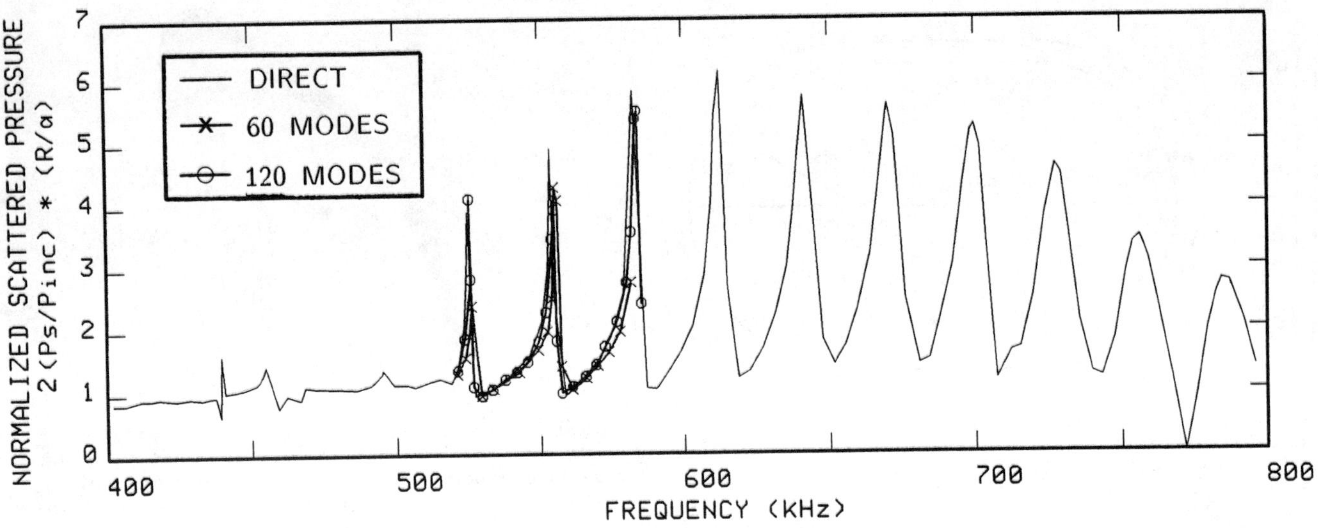

Figure 9. NASTRAN/SIERRAS Backscattered Form Function from a Cylindrical Shell with Hemispherical Endcaps for End-On Incidence Using the Full Modal Formulation

The maximum difference in the resonant frequencies for the modal versus direct analyses of Figures 8 and 9 is 0.3 percent. It is concluded that the modal formulation is effective in reducing the size of the matrices while providing a reasonably accurate solution.

When the analyses above were compared with the experimental results it was found that the results did not agree exactly. It was found that slight changes in the material properties and shell thickness within the uncertainty of the target were sufficient to obtain exact agreement. The exact material properties for the brass used in the target were not available. Several types of brass were considered and the properties in Table 3 were found to provide the best agreement for the first several So waves.

Table 3. Revised Properties of Brass

Dilational Wave Speed	= 4.28×10^5 cm/s
Shear Wave Speed	= 2.03×10^5 cm/s
Density	= 8.56 g/cm^3

The So waves are sensitive to the material properties but not to the thickness. Also, the thickness measured after machining was not readily available. Therefore, the thickness of the model was increased by five percent in order to match the "A" wave frequencies. It is observed that the So and "A" wave resonances may be adjusted separately; the So waves are sensitive to material properties and the "A" waves are sensitive to changes in shell thickness.

The NASTRAN/SIERRAS form function obtained for the cylinder with hemispherical endcaps using the revised properties is compared with experimental results in Figure 10. A discussion of the experimental procedure used was given by Numrich (1984). The model used is identical to that used to obtain Figure 6 except that the CQUAD4 and CTRIA3 elements of MSC NASTRAN were used to model the shell in order to demonstrate that the solution is not overly sensitive to the finite element used in obtaining the structural matrices. The form function of Figure 10 is for the case of backscattering for end-on incidence. The flexural modes for the shell were identified by phase plots similar to the plots of Figure 4. Resonances of the cylindrical shell with hemispherical endcaps were identified by comparison with those of the spherical shell. The S_0 and "A" waves are identified in Figure 10. Note that the very narrow resonances of Figure 3 do not appear in Figure 10. The radiation efficiency of these modes are very low due to near perfect cancellation of lobes over the surface of the shell. The absolute magnitude of the form function is sensitive to the number of peak search points. Also, the magnitude of the experimental results depends on the type of experiment conducted (e.g. whether transient or harmonic). Very narrow peaks extending to fairly large values may be found theoretically which do not appear in experiment (cf. Figures 3 and 10). For these reasons, the difference in the magnitude of the form function between modal and direct methods shown in Figures 8 and 9 are not a major concern.

CONCLUSION

The NASTRAN/SIERRAS approach to solving radiation and scattering problems for acoustic waves and elastic shell structures has been demonstrated with direct and modal formulations. The CHIEF method for providing a unique solution has been applied to scattering problems up to ka=30. The wave types for the problem of end-on scattering from spherical shells and cylindrical shells with hemispherical endcaps have been identified. The agreement between SIERRAS and "exact" and experimental resonant frequencies is considered excellent at less than two percent error for all wave types.

ACKNOWLEDGEMENTS

The authors gratefully acknowledge that the experimental results were provided by Dr. S.K. Numrich at the Naval Research Laboratory in collaboration with Dr. G. Maze at the University of Le Havre.

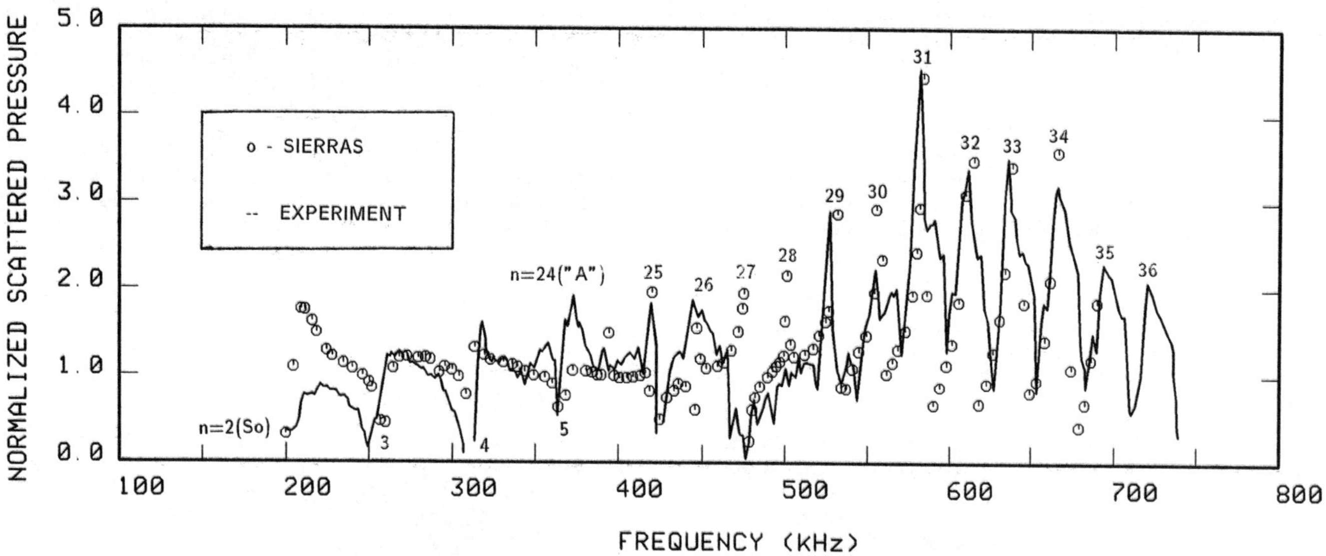

Figure 10. Backscattered Form Function from a Cylindrical Shell with Hemispherical Endcaps for End-On Incidence Experimental Versus the Direct Formulation

REFERENCES

Baker, B.B., and Copson, E.T., 1939, The Mathematical Theory of Huygen's Principle, Part I, McGraw-Hill, New York, pp. 23-32.

Bathe, K., 1982, Finite Element Procedures in Engineering Analysis, Prentice-Hall, Inc., New Jersey, pp. 200-201.

Brebbia, C.A., and Walker S., 1980, Boundary Element Techniques in Engineering, Newnes-Butterworths, Boston.

Brebbia, C.A., Telles, J.C.F., and Wrobel, L.C., 1984, Boundary Element Techniques, Springer-Verlag, New York.

Brundrit, G.B., 1965, "A Solution to the Problem of Scalar Scattering from a Smooth, Bounded Obstacle Using Integral Equations," The Quarterly Journal of Mechanics and Applied Mathematics, Vol. XVIII, Part 4, pp. 473-489.

Burton, A.J., and Miller, G.F., 1971, "The Application of Integral Equation Methods to the Numerical Solution of Some Exterior Boundary Value Problems," The Proceedings of the Royal Society of London Ser. A, Vol. 323, pp. 201-210.

Butler, T.G., and Michel, D., 1971, "NASTRAN: A Summary of Functions and Capabilities of the NASA Structural Analysis Computer System," NASA SP-260, Washington, DC.

Chen, L.H., 1982, "Acoustic Emission from Submerged Structures," Development in Boundary Element Methods 2, Chap. 9, edited by P.K. Banerjee, and R.P. Shaw, Applied Science Publishers Ltd., Ripple Road, Barking, Essex, England.

Chertock, G., 1971, "Integral Equation Methods in Sound Radiation and Scattering from Arbitrary Structures," NSRDC Technical Report 3538, Washington, DC.

Engblom, J.J., and Nelson, R.B., 1975, "Consistent Formulation of Sound Radiation from Arbitrary Structures," The Journal of Applied Mechanics, Vol. 42, Ser E, No. 2, pp. 295-300.

Everstine, G.C., Henderson, F.M., and Schuetz, L.S., 1987, NASTRAN/Boundary Element Formulation for Acoustic Scattering," Fifteenth NASTRAN User's Colloquium, NASA CP-2481, National Aeronautics and Space Administration, Washington, DC, pp.250-265.

Huang., H., 1984, "Helmholtz Integral Equations for Fluid-Structure Interaction," in Advances in Fluid-Structure Interaction, ASME, AMD, Vol. 64, pp. 19-38.

Junger, M.C., and Feit, D, 1986, Sound, Structures, and Their Interaction, The MIT Press, Cambridge.

Kupradze, W.D., 1965, Potential Methods in the Theory of Elasticity, Daniel Davey, New York.

Lachat, J.C., and Watson, J.O., 1976, "Effective Numerical Treatment of Boundary Integral Equations," The International Journal of Numerical Methods in Engineering, Vol. 10, pp. 991-1005.

Mathews, I.C., 1979, "Acoustic Radiation from Vibrating Elastic Structures of Arbitrary Shape," Ph.D. Thesis, Imperial College, London.

Mathews, I.C., 1980, "Acoustic Radiation for Three Dimensional Elastic Structures," in Innovative Numerical Analysis in Applied Engineering Science, Montreal.

Mathews, I.C., 1984, "A Symmetric Integral - Finite Element Approach for 3-D Fluid Structure Interaction," in Advance in Fluid-Structure Interaction, ASME, AMD, Vol. 64, pp. 39-48.

Meyer, W.L., Bell, W.A., Zinn, B.T., and Stallybrass, M.P., 1978, "Boundary Integral Solutions of Three-Dimensional Acoustic Radiation Problems," The Journal of Sound and Vibration, Vol. 59, pp. 245-262.

Numrich, S.K., Dale, N.H., and Dragonette, L.R., 1984, "Generation and Exploitation of Plate Waves in Submerged, Air-Filled Shells," in Advances in Fluid-Structure Interaction, ASME, AMD, Vol. 64, pp. 59-74.

Piaszczyk, C.M., and Klosner, J.M., 1984, "Acoustic Radiation from Vibrating Surfaces at Characteristic Frequencies," The Journal of the Acoustical Society of America, Vol. 75, pp. 363-375.

Press, W.H., Flannery, B.P., Teukolsky, S.A, and Vetterling, W.T., 1986, Numerical Recipes, Cambridge University Press, pp. 27- 282.

Schenck, H.A., 1968, "Improved Integral Formulation for Acoustic Radiation Problems," The Journal of the Acoustical Society of America, Vol. 44, pp. 41-58.

Seybert, A.F., Soenarko, B., Rizzo, F.J., and Shippy, D.J., 1985, "An advanced Computational Method for Radiation and Scattering of Acoustic Waves in Three Dimensions," The Journal of the Acoustical Society of America, Soc. Am., Vol. 77, pp. 362-368.

Seybert, A.F., Soenarko, B., Rizzo, F.J., and Shippy, D.J., 1986, "A Special Integral Equation Formulation for Acoustic Radiation and Scattering for Axisymmetric Bodies and Boundary Conditions," The Journal of the Acoustical Society of America, Vol. 80, pp. 1241-1247.

Seybert, A.F., and Rengarajan, T.K., 1987, "The use of CHIEF to obtain Unique Solutions for Acoustic Radiation using Boundary Integral Equations," The Journal of the Acoustical Society of America, Vol. 81, pp. 1299-1306.

Seybert, A.F., Wu, T.W., and Wu, X.F., 1988, "Radiation and Scattering of Acoustic Waves from Elastic Solids and Shells Using the Boundary Element Method," The Journal of the Acoustical Society of America, Vol. 84, pp. 1906-1912.

Talmant, M., Uberall, H., Miller, R.D., Werby, M.F., and Dickey, J.W., (to be published), "Lamb Waves and Fluid-Borne Waves on Water-Loaded, Air-Filled Thin Spherical Shells," The Journal of the Acoustical Society of America.

Tobocman, W., 1984, "Calculation of Acoustic Wave Scattering by Means of the Helmholtz Integral Equation. I," The Journal of the Acoustical Society of America, Vol. 76, pp. 599-607.

Tobocman, W., 1984, "Calculation of Acoustic Wave Scattering by Means of the Helmholtz Integral Equation. II," The Journal of the Acoustical Society of America, Vol. 76, pp. 1549-1554.

Williams, W.W., Parke, N.G., Moran, D.A., and Sherman, C.H., 1964, "Acoustic Radiation from a Finite Cylinder," The Journal of the Acoustical Society of America, Vol. 36, pp. 2316-2322.

Wilton, D.T., 1978, "Acoustic Radiation and Scattering from Elastic Structures," The International Journal of Numerical Methods in Engineering , Vol. 13, pp. 123-128.

Zienkiewicz, O.C., Kelly, D., and Bettess, P., 1977, "The Coupling of Finite Element and Boundary Solution Procedures," The International Journal of Numerical Methods in Engineering, Vol. 11, pp. 355-375.

PROBABILISTIC ANALYSIS OF A FLUID-SHELL SYSTEM WITH RANDOM IMPERFECTIONS

W. K. Liu, J. S. Chen and Y. Y. Lu
Department of Mechanical Engineering
Northwestern University
Evanston, Illinois

1. Introduction

The consideration for the uncertainties inherent in the submerged system can be made by understanding the influences of random parameters on the energy transfer between the structural system and the acoustic system. These random parameters can be the various random multiscales, the random mass distribution, the random material properties, the random geometric boundary conditions, the random imperfections, the random external loading and the random locations of machineries. Since the acoustic interaction with the structure is coupled with the transmission of the vibration energy from the structural components to the acoustic fluid and vice versa, the effect of the secondary structural system on the scattering pressure should also be understood.

This paper focuses on the formulation of the probabilistic and reliability analysis of an acoustic fluid-imperfection cylindrical shell system. The general probabilistic formulation is given using the finite element method. Other uncertainties described above can also be studied using a similar procedure developed in this paper.

Section 2 describes the general fluid-structure interaction problem. The probabilistic finite element method (PFEM) is given in Section 3. In Section 4, the linear transformation procedure from the correlated random variables to uncorrelated random variables and the Kuhn-Tucker problem for reliability analysis using PFEM are constructed. In Section 5, the out-of-roundness imperfection is described via a random function. Numerical examples, using both the perturbation method and PFEM, are given in Section 6, along with discussions and conclusions.

2. General Fluid-Structure Interaction Problem

In the fluid-structure interaction problem, two sets of governing equations defining the structure and fluid separately, in each domain, are coupled at the fluid-structure interface. In the following statement, it is assumed that the structure is submerged in the acoustic fluid, and that the fluid is inviscid, compressible and barotropic.

In the structure, the momentum balance is given, in the frequency domain, by

$$\mathrm{div}\,\underset{\sim}{\sigma} + \underset{\sim}{f} = -\omega^2 \rho_S \underset{\sim}{u} \quad \text{in } \bar{\Omega}_S \tag{1a}$$

$$\underset{\sim}{\sigma} = \underset{\sim}{\sigma}^T \quad \text{in } \bar{\Omega}_S \tag{1b}$$

$$\underset{\sim}{\sigma} \cdot \underset{\sim}{n} = \underset{\sim}{h} \quad \text{on } \Gamma_{\underset{\sim}{h}} \tag{1c}$$

$$\underset{\sim}{\sigma} \cdot \underset{\sim}{n} = -p\underset{\sim}{n} \quad \text{on } \Gamma_{FS} \tag{1d}$$

with the kinematics given by

$$\underset{\sim}{\varepsilon} = \nabla^s \underset{\sim}{u} \quad \text{in } \bar{\Omega}_S \tag{1e}$$

$$\underset{\sim}{u} = \underset{\sim}{g} \quad \text{on } \Gamma_{\underset{\sim}{g}} \tag{1f}$$

and the constitutive relation given by

$$\underset{\sim}{\sigma} = \underset{\sim}{D}\,\underset{\sim}{\varepsilon} \quad \text{in } \bar{\Omega}_S \tag{2a}$$

where

$$\bar{\Omega}_S = \Omega_S \cup \Gamma_S \tag{2b}$$

$$\Gamma_S = \Gamma_{\underset{\sim}{h}} \cup \Gamma_{FS} \cup \Gamma_{\underset{\sim}{g}} \tag{2c}$$

In the above equations, Ω, Γ, $\underset{\sim}{\sigma}$, $\underset{\sim}{u}$, $\underset{\sim}{f}$, ρ_S, ω, $\underset{\sim}{h}$, $\underset{\sim}{g}$ and $\underset{\sim}{D}$ are the domain, boundary, Cauchy stress, displacement, body force, density, harmonic frequency, traction, prescribed displacement, and material response matrix, respectively. The ∇^s denotes the symmetric part of gradient operator; Γ_{FS} denotes the interface boundary; and $\underset{\sim}{n}$ denotes the unit normal whose positive is outward from the structure region. Subscripts S and FS denote structure and fluid-structure interface; and all the functions are Fourier transformed. For example

$$\underset{\sim}{\sigma}(\underset{\sim}{x},\omega) = \int_{-\infty}^{\infty} \underset{\sim}{\sigma}(\underset{\sim}{x},t)e^{-j\omega t}\,dt \tag{3}$$

where $\underset{\sim}{x}$ is the spatial coordinate; t is the time; and j represents $\sqrt{-1}$. The pressure p is assumed to be decomposed into the incident part and the scattered part

$$p = p^i + p^s \tag{4}$$

In the fluid region, the motion of the fluid is governed by

$$-\omega^2\rho_F p^s - B\nabla^2 p^s = 0 \quad \text{in } \bar{\Omega}_F \tag{5a}$$

$$\nabla p^s \cdot \underset{\sim}{n} = -\omega^2\rho_F(\underset{\sim}{u}^i - \underset{\sim}{u}) \cdot \underset{\sim}{n} \quad \text{on } \Gamma_{FS} \tag{5b}$$

$$\nabla p^s \cdot \underset{\sim}{m} = -j\omega p^s/c \quad \text{on } \Gamma_{RB} \tag{5c}$$

$$\nabla p^i \cdot \underset{\sim}{n} = \omega^2\rho_F \underset{\sim}{u}^i \cdot \underset{\sim}{n} \quad \text{on } \Gamma_{FS} \tag{5d}$$

where

$$\bar{\Omega}_F = \Omega_F \cup \Gamma_F \tag{5e}$$

$$\Gamma_F = \Gamma_{FS} \cup \Gamma_{RB} \tag{5f}$$

In the above equations, ρ_F, B and c denote the fluid density, the bulk modulus and the wave speed, respectively; $\underset{\sim}{u}^i$ represents the incident displacement at the interface, and Γ_{RB} is the radiation boundary condition which is given by Eq.(5c). The outward unit normal from the fluid region is denoted by $\underset{\sim}{m}$, which is equal to $-\underset{\sim}{n}$. A dot designates the inner product.

The weak form of the above equations is

$$G(\underset{\sim}{u}, \underset{\sim}{\eta}, p, \beta) = -\int_{\Omega_S} (\omega^2 \rho_S \underset{\sim}{u} + \underset{\sim}{f}) \cdot \underset{\sim}{\eta}\, d\Omega + \int_{\Omega_S} \underset{\sim}{\sigma} : \nabla^s \underset{\sim}{\eta}\, d\Omega$$
$$- \int_{\Gamma_h} \underset{\sim}{h} \cdot \underset{\sim}{\eta}\, d\Gamma + \int_{\Gamma_{FS}} (p^i + j\omega q)\underset{\sim}{n} \cdot \underset{\sim}{\eta}\, d\Gamma - \int_{\Omega_F} j\omega^3 \rho_F q\beta\, d\Omega$$
$$+ \int_{\Omega_F} j\omega B \nabla q \cdot \nabla\beta\, d\Omega - \int_{\Gamma_{FS}} B\underset{\sim}{n} \cdot (\nabla p^i - \omega^2 \rho_F \underset{\sim}{u})\, \beta\, d\Gamma$$
$$- \int_{\Gamma_{RB}} \omega^2 \frac{B}{c} q\beta\, d\Gamma = 0 \tag{6}$$

for any admissible variations $\underset{\sim}{\eta} \in H^1(\bar{\Omega}_S)$ such that $\underset{\sim}{\eta} = \underset{\sim}{0}$ on $\underset{\sim}{\Gamma}_g$ and $\beta \in H^1(\bar{\Omega}_F)$. A function $\beta \in H^1(\Omega)$ represents that β^2 and $(\nabla\beta)^2$ are square integrable in Ω. It is noted that the scattering pressure p^s has been replaced by $j\omega q$ in equation (6) to obtain a symmetric formulation [Everstine, 1981].

3. Probabilistic Finite Element Method

The PFEM can be developed by applying the perturbation technique [Liu, et al., 1986] to the weak form (Eq.(6)). The application of second order perturbation techniques involves the expansion of all random functions about a "design point" of the random field, $\bar{b}(\underset{\sim}{x})$ (not necessary the mean value), where $b(\underset{\sim}{x})$ is the random field and retaining up to second order terms. For example, the random functions $\underset{\sim}{u}$ and p^s are expanded about $\bar{b}(\underset{\sim}{x})$ as follows:

$$\underset{\sim}{u} = \underset{\sim}{u}^o + \gamma\underset{\sim}{u}' + \gamma^2\underset{\sim}{u}'' \tag{7a}$$

$$q = q^o + \gamma q' + \gamma^2 q'' \tag{7b}$$

where γ is a small parameter; and the zeroth, prime and double primes, represent the zeroth, first and second order variations, respectively, due to variations in $\underset{\sim}{b}$. Using the finite element techniques, the surface and volume integrations are evaluated via parameteric representations:

$$d\Gamma = J_A dA \tag{7c}$$

$$d\Omega = J_V dV \tag{7d}$$

where J_A and J_V represent the determinants of the surface and volume Jacobian matrices, respectively. The random functions for $\underset{\sim}{\varepsilon}$, p^i (hence,$\underset{\sim}{u}^i$), $\underset{\sim}{h}$, $\underset{\sim}{n}$, J_A, J_V, $\nabla^s\underset{\sim}{u}$, ∇p^s, $\nabla^s\underset{\sim}{\eta}$ and $\nabla\beta$ are expanded in a similar manner to Eqs.(7a) and (7b). For example,

$$\nabla^s\underset{\sim}{u} = (\nabla^s)^o\underset{\sim}{u}^o + \gamma\left[(\nabla^s)'\underset{\sim}{u}^o + (\nabla^s)^o\underset{\sim}{u}'\right]$$
$$+ \gamma^2\left[(\nabla^s)''\underset{\sim}{u}^o + (\nabla^s)'\underset{\sim}{u}' + (\nabla^s)^o\underset{\sim}{u}''\right] \tag{8}$$

Note that $\underset{\sim}{\eta}$ and β are not perturbed since they are arbitrary quantities. Therefore, the expansion of the random gradient operators $\nabla^s\underset{\sim}{\eta}$ and $\nabla\beta$ are:

$$\nabla^s\underset{\sim}{\eta} = (\nabla^s)^o\underset{\sim}{\eta} + \gamma(\nabla^s)'\underset{\sim}{\eta} + \gamma^2(\nabla^s)''\underset{\sim}{\eta} \tag{9a}$$

$$\nabla\beta = \nabla^o\beta + \gamma\nabla'\beta + \gamma^2\nabla''\beta \tag{9b}$$

Such expansions can be interpreted as an expansion of the weak form about $\bar{b}$ via a second order perturbation at a given $\underset{\sim}{x}$:

$$G = G^o + \gamma G' + \gamma^2 G'' \tag{10}$$

Expanding the random functions and collecting terms of zeroth and first orders of γ yield the following weak forms

$$G^0 = -\int_{V_S} (\omega^2 \rho_S \underset{\sim}{u}^0 + \underset{\sim}{f}) \cdot \underset{\sim}{\eta}\, J_V^0\, dV + \int_{V_S} \underset{\sim}{\underset{\sim}{D}}\underset{\sim}{\varepsilon}^0 : (\nabla^s)^0\underset{\sim}{\eta}\, J_V^0\, dV$$
$$- \int_{A_h} \underset{\sim}{h}^o \cdot \underset{\sim}{\eta} J_A^0 dA + \int_{A_{FS}} (p^i + j\omega q^o)\underset{\sim}{n}^o \cdot \underset{\sim}{\eta} J_A^0 dA$$

$$-\int_{V_F} j\omega^3\rho_F q^o\beta J_V^o dV + \int_{V_F} jB\omega(\nabla^o q^o \cdot \nabla^o\beta)J_V^o dV$$

$$-\int_{A_{FS}} B\underset{\sim}{n}^o \cdot (\nabla p^i - \omega^2\rho_F\underset{\sim}{u}^o)\beta J_A^o dA - \int_{A_{RB}} \omega^2 \frac{B}{c} q^o\beta J_A^o dA \tag{11a}$$

$$G' = -\int_{V_s} \omega^2 \rho_s (\underset{\sim}{u}' \cdot \underset{\sim}{\eta}\, J_V^0 + \underset{\sim}{u}^0 \cdot \underset{\sim}{\eta} J_V') \, dV - \int_{V_s} \underset{\sim}{f} \cdot \underset{\sim}{\eta} J_V' \, dV$$

$$+\int_{V_S} \left[\underset{\sim}{D}\underset{\sim}{\varepsilon}': (\nabla^s)^o\underset{\sim}{\eta} J_V^o + \underset{\sim}{D}\underset{\sim}{\varepsilon}^o: (\nabla^s)'\underset{\sim}{\eta} J_V^o + \underset{\sim}{D}\underset{\sim}{\varepsilon}^o: (\nabla^s)^o\underset{\sim}{\eta} J_V'\right] dV$$

$$-\int_{A_h} (\underset{\sim}{h}' \cdot \underset{\sim}{\eta} J_A^o + \underset{\sim}{h}^o \cdot \underset{\sim}{\eta} J_A') \, dA + \int_{A_{FS}} j\omega \, (q'\underset{\sim}{n}^o \cdot \underset{\sim}{\eta} J_A^o + q^o\underset{\sim}{n}' \cdot \underset{\sim}{\eta} J_A^o$$

$$+ q^o\underset{\sim}{n}^o \cdot \underset{\sim}{\eta} J_A') dA + \int_{A_{FS}} p^i(\underset{\sim}{n}' \cdot \underset{\sim}{\eta} J_A^o + \underset{\sim}{n}^o \cdot \underset{\sim}{\eta} J_A') dA$$

$$-\int_{V_F} j\omega^3\rho_F(q^o\beta J_V' + q'\beta J_V^o) dV + \int_{V_F} j\omega B[(\nabla' q^o + \nabla^o q')$$

$$\cdot \nabla^o\beta J_V^o + \nabla^o q^o \cdot \nabla'\beta J_V^o + \nabla^o q^o \cdot \nabla^o\beta J_V'] dV$$

$$+\int_{A_{FS}} \omega^2\rho_F B(\underset{\sim}{n}' \cdot \underset{\sim}{u}^o\beta J_A^o + \underset{\sim}{n}^o \cdot \underset{\sim}{u}'\beta J_A^o + \underset{\sim}{n}^o \cdot \underset{\sim}{u}^o\beta J_A') dA$$

$$-\int_{A_{FS}} B(\underset{\sim}{n}' \cdot \nabla p^i\beta J_A^o + \underset{\sim}{n}^o \cdot \nabla p^i\beta J_A') dA$$

$$-\int_{A_{RB}} \omega^2 \frac{B}{c} (q'\beta J_A^o + q^o\beta J_A') dA \tag{11b}$$

As in the finite element approximation, the random functions which are functions of $b(\underset{\sim}{x})$ at $\underset{\sim}{x}_i$, are discretized with q interpolation functions where q is the number of random parameters [Liu, et al., 1986]. For example, the determinant of the volume Jacobian matrix for the random domain can be approximated as [Liu,et al., 1987]

$$J_V \simeq \sum_{i=1}^{q} \phi_i(\underset{\sim}{x})(J_{Vi}^o + \gamma J_{Vi}' + \gamma^2 J_{Vi}'') \tag{12a}$$

with

$$J_{Vi}' \simeq \sum_{k=1}^{q} J_{Vi,b_k}\Delta b_k \tag{12b}$$

$$J_{Vi}'' = \frac{1}{2} \sum_{k,n=1}^{q} J_{Vi,b_k b_n}\Delta b_k \Delta b_n \ , \tag{12c}$$

where J_{Vi,b_k} denotes the partial derivative of J_{Vi} with respect to b_k evaluated at $\bar{b}$ and $J_{Vi,b_k b_n}$ denotes the mixed partial derivatives of J_{Vi} with respect to b_k and b_n evaluated at $\bar{b}$. The b_i's are the values of $b(\underset{\sim}{x})$ evaluated at $\underset{\sim}{x}_i$, i = 1, ..., q.

Applying the Galerkin finite element approximation, the finite element matrix equations can be obtained:

(Zeroth order equation)

$$\underset{\sim}{H}\underset{\sim}{d} = \underset{\sim}{F} \tag{13a}$$

(First order equation):

$$\underset{\sim}{H}\underset{\sim}{d}' = \underset{\sim}{F}' \tag{13b}$$

In the above equations, $\underset{\sim}{H}$, $\underset{\sim}{d}$, $\underset{\sim}{F}$ are defined by

$$\underset{\sim}{H} = -\omega^2\begin{bmatrix} \underset{\sim}{M} & \underset{\sim}{0} \\ \underset{\sim}{0} & -\underset{\sim}{Q} \end{bmatrix} + j\omega\begin{bmatrix} \underset{\sim}{C}_S & \underset{\sim}{M}_{SF} \\ \underset{\sim}{M}_{SF}^T & -\underset{\sim}{C}_B \end{bmatrix} + \begin{bmatrix} \underset{\sim}{K} & \underset{\sim}{0} \\ \underset{\sim}{0} & -\underset{\sim}{G} \end{bmatrix} \tag{14a}$$

$$\underset{\sim}{d} = \begin{Bmatrix} \underset{\sim}{u}^0 \\ \underset{\sim}{q}^0 \end{Bmatrix}; \qquad \underset{\sim}{F} = \begin{Bmatrix} \underset{\sim}{f}^{(S)} \\ \frac{-1}{j\omega}\underset{\sim}{f}^{(F)} \end{Bmatrix} \tag{14b,c}$$

and $\underset{\sim}{d}'$, $\underset{\sim}{F}'$ are defined by

$$\underset{\sim}{d}' = \begin{Bmatrix} \underset{\sim}{u}' \\ \underset{\sim}{q}' \end{Bmatrix} \tag{15a}$$

$$\underset{\sim}{F}' = \begin{Bmatrix} \underset{\sim}{f}'^{(S)} + \omega^2\underset{\sim}{f}^M - j\omega\underset{\sim}{f}^{DS} - \underset{\sim}{f}^K \\ \frac{-1}{j\omega}\underset{\sim}{f}'^{(F)} - \omega^2\underset{\sim}{f}^Q - j\omega(\underset{\sim}{f}^{DF} - \underset{\sim}{f}^C) + \underset{\sim}{f}^G \end{Bmatrix} \tag{15b}$$

where the structural damping, denoted by $\underset{\sim}{C}_S$, has been added to Eq.(15a). The coefficient matrices and vectors at an element level are given by

$$\underset{\sim}{M}_{IJ} = \int_{V_F} \rho_s N_I N_J \, J_V^o \, dV \, \underset{\sim}{I} \ ; \ (\underset{\sim}{M}_{SF})_{IJ} = \int_{A_{FS}} N_I N_J \, \underset{\sim}{n}^o \, J_A^o \, dA \tag{16a,b}$$

$$K_{IJ} = \int_{V_S} B_I^{oT} D\, B_J^o\, J_V^o\, dV \; ; \; Q_{IJ} = \int_{V_F} \frac{1}{B} L_I L_J J_V^o\, dV \qquad (16c,d)$$

$$(C_B)_{IJ} = \int_{A_{RB}} \frac{1}{\rho_F c} L_I L_J J_A^o\, dA \; ; \; G_{IJ} = \int_{V_F} \frac{1}{\rho_F} A_I^T A_J J_V^o\, dV;$$

$$A_I^T = [L_{I,1}, L_{I,2}, L_{I,3}] \qquad (16e,f,g)$$

$$f_I^{(S)} = \int_{V_s} N_I\, f\, J_V^0\, dV + \int_{A_h} N_I\, h^0 J_A^0\, dA$$

$$- \int_{A_{FS}} N_I\, p^i\, n^0 J_A^0\, dA \qquad (16h)$$

$$f_I^{(F)} = \int_{A_{FS}} \frac{1}{\rho_F} L_I\, n^{oT} \bar{p}^i J_A^o\, dA \; ; \; (\bar{p}^i)^T = [p^i_{,1}, p^i_{,2}, p^i_{,3}] \qquad (16i)$$

where N and L are the shape functions for the displacement and scattering pressure field, respectively; and B and A are the gradients of N and L respectively.

$$f_I^{'(S)} = \int_{V_s} N_I\, f\, J_V'\, dV + \int_{A_h} N_I\, h' J_A^0\, dA + \int_{A_h} N_I\, h\, J_A'\, dA$$

$$- \int_{A_{FS}} N_I\, p^i\, n'\, J_A^o\, dA - \int_{A_{FS}} N_I\, p^i\, n^o\, J_A'\, dA \qquad (16j)$$

$$f_I^M = \int_{V_F} \rho_S\, N_I\, u^o\, J_V^o\, dV \qquad (16k)$$

$$f_I^{DS} = \int_{A_{FS}} N_I\, q^o\, n^o\, J_A^o\, dA \qquad (16l)$$

$$f_I^K = \int_{V_S} B_I^{oT} \sigma^o\, J_V^o\, dV \qquad (16m)$$

$$f_I^{'(F)} = \int_{A_{FS}} \frac{1}{\rho_F} L_I\, n'^T \bar{p}^i J_A^o\, dA + \int_{A_{FS}} \frac{1}{\rho_F} L_I\, n^{oT} \bar{p}^i J_A'\, dA \qquad (16n)$$

$$f_I^Q = \int_{V_F} \frac{1}{B} L_I\, q^o\, J_V^o\, dV \qquad (16o)$$

$$f_I^{DF} = \int_{A_{FS}} N_I\, u^{oT} n^o\, J_A^o\, dA \qquad (16p)$$

$$f_I^c = \int_{A_{RB}} \frac{1}{\rho_F c} L_I\, q^o\, J_A^o\, dA \qquad (16q)$$

$$f_I^G = \int_{V_F} \frac{1}{\rho_F} A_I^T \bar{q}^o\, J_V^o\, dV \; ; \; A_I^T = [L_{I,1}, L_{I,2}, L_{I,3}] \,,$$

$$\bar{q}^{oT} = [q^o_{,1}, q^o_{,2}, q^o_{,3}] \qquad (16r)$$

4. Reliability Analysis using PFEM

To improve the computational efficiency in solving Eqs.(13a) and (13b), the linear transformation from the correlated random variables into uncorrelated random variables is employed. If the random variables are denoted by

$$b^T = (b_1, b_2, \cdots, b_q) \qquad (17a)$$

then the covariance matrix G is diagonalized

$$\Lambda = \psi^T G\, \psi \qquad (17b)$$

where

$$g_{nk} = \mathrm{Cov}(b_n, b_k) \qquad 1 \le n,k \le q \qquad (17c)$$

$$\Lambda_{nk} = \mathrm{Var}(c_k)\, \delta_{nk} \qquad 1 \le n,k \le q \qquad (17d)$$

In the above equations, c_n are the uncorrelated random variables . The relation between the uncorrelated random variable c and the correlated random variables b is given by

$$c = \psi\, b \qquad (17e)$$

Using Eq.(17b), the partial derivatives with respect to b_n in Eq.(12) are replaced by those with respect to c_n, and the covariance $\mathrm{Cov}(b_n, b_k)$ by the variance, $\mathrm{Var}(c_k)\, \delta_{nk}$.

The derivatives of the coefficient matrices and vectors with respect to the uncorrelated random variables can be evaluated at element e by , for example,

$$H^e_{,c_n} = \sum_{k=1}^{q} H^e_{,c_k} \psi_{kn} \qquad (18a)$$

$$H^e_{,c_n c_k} \delta_{nk} = \sum_{s,t=1}^{q} H^e_{,b_s b_t} \psi_{sn} \psi_{tn} \qquad (18b)$$

The reliability analysis using PFEM in this paper is completed by constructing the Kuhn-Tucker problem. We consider the problem of minimizing an objective function,

$$\left(\sum_{i=1}^{q} y_i^2 \right)^{1/2} \equiv \beta(c) \qquad (19a)$$

subject to Eq.(87a) and the inequality constraint:

$$g(c, d^R) \le 0 \qquad (19b)$$

Here $\tilde{d}^R$ is the real part of $\tilde{d}\, e^{-j\omega t}$; q is the number of uncorrelated random variables ($\tilde{c}$); and y_i are the standard normal random variables denoted by

$$y_i = (c_i - \bar{c}_i) / \sigma_i \qquad i=1,\cdots,q \qquad \text{no sum on } i \quad (19c)$$

where $\bar{c}_i$ are the mean values of c_i and σ_i are the standard deviations of c_i.

To form the Kuhn-Tucker problem, Eqs.(13a),(19a), and (19b) are combined to yield the following Lagrangian:

$$L(\tilde{c}, \tilde{d}^R, \tilde{\mu}, \lambda, \alpha) = \beta + \tilde{\mu} \cdot \{\tilde{d}^R - \hat{\tilde{d}}\} + \lambda (g - \alpha^2) \qquad (20)$$

where $\hat{\tilde{d}} = \mathrm{Re}(\tilde{H}^{-1}\tilde{F}e^{-j\omega t})$; $\tilde{\mu}$ is a vector of Lagrange multipliers; α is the slack variable; and $\lambda > 0$. The Kuhn-Tucker necessary conditions for the Lagrangian then lead to

$$L_{,\tilde{c}} = \beta_{,\tilde{c}} - \tilde{\mu} \cdot \hat{\tilde{d}}_{,\tilde{c}} + \lambda\, g_{,\tilde{c}} = \tilde{0} \qquad (21a)$$

$$L_{,\tilde{d}^R} = \tilde{\mu} + \lambda\, g_{,\tilde{d}^R} = \tilde{0} \qquad (22b)$$

$$L_{,\tilde{\mu}} = \tilde{d}^R - \hat{\tilde{d}} = \tilde{0} \qquad (23c)$$

$$L_{,\lambda} = g - \alpha^2 = 0 \qquad (24d)$$

$$L_{,\alpha} = -2\lambda\alpha = 0 \qquad (24e)$$

Equations (21a) and (21b) can be combined to yield

$$\beta_{,\tilde{c}} + \lambda [\, g_{,\tilde{c}} + g_{,\tilde{d}^R} \cdot \hat{\tilde{d}}_{,\tilde{c}}] = \tilde{0} \qquad (25)$$

where $\hat{\tilde{d}}_{,\tilde{c}}$ can be determined from the first-order probabilistic finite element Eqs.(13a) and (13b), it is given by

$$\hat{\tilde{d}}_{,\tilde{c}} = \mathrm{Re}\{-\tilde{H}^{-1}[\,\tilde{H}_{,\tilde{c}}\,\tilde{d} - \tilde{F}_{,\tilde{c}}\,]\, e^{-j\omega t}\} \qquad (26)$$

The Kuhn-Tucker problem is to find $\tilde{c}^* = (c_1^*, c_2^*, \cdots, c_q^*)$, λ and α^2 that satisfy Eqs.(24c)-(24e), and (26). It is noted that when β is evaluated at $\tilde{c}^*$, $\beta(\tilde{c}^*)$ is called the reliability index.

In interpreting the reliability index, it is noted that the higher the reliability index, the better is the performance of the system. This method requires a nonhomogeneous linear mapping of the primitive random variables into a set of uncorrelated standard normal variables. For example, the basic variables b_n can be transformed into the standard normal variables y_n by

$$\tilde{b} = T(\tilde{y}) \qquad (27a)$$

Therefore, the failure surface in the b-space is also mapped onto the corresponding failure surface in the y-space by

$$g(\tilde{b}) = g(T(\tilde{y})) \equiv G(\tilde{y}) = 0 \qquad (27b)$$

The reliability index β_{HL} can then be defined

$$\beta_{HL} = \min_{G(\tilde{y})=0} (\tilde{y}^T \tilde{y})^{1/2} \qquad (27c)$$

Many methods for solving the above optimization problem are available. The tangent matching method will be used here to determine the solution of the above equation. The iterative algorithm in the y-space is [Ang and Tang, 1984]

$$\tilde{y}^{i+1} = \{ [(\tilde{y}^i)^T \tilde{\alpha}^i] + \frac{G(\tilde{y}^i)}{|\nabla G(\tilde{y}^i)|} \} \tilde{\alpha}^i \qquad (27d)$$

$$\tilde{\alpha}^i = - \frac{\nabla G(\tilde{y}^i)}{|\nabla G(\tilde{y}^i)|} \qquad (27e)$$

where ∇G is the gradient of the performance function, $|\nabla G|$ is its magnitude, and the superscript i denotes the iteration number. A superscript " T " designates the transpose.

An important index for structural acoustics is the scattering cross section which indicates the degree of interaction between the incident wave and the structure. The reliability of this imperfect-shell-acoustic fluid system can be measrued by the probability that the scattering cross section is less than a given desired threshold. This means that the smaller the scattering cross section, the better is the system design. The performance function is defined by

$$g(\tilde{b}) = (E_s(\tilde{b}) - E_p) / E_0 \qquad (28a)$$

where $E_s(\tilde{b})$ represents the scattering cross section which is a function of the random variables b_n, that is

$$E_s(\tilde{b}) = p^t / I_0 \qquad (28b)$$

$$\tilde{b}^T = (b_1, b_2, \cdots, b_q) \qquad (28c)$$

where $I_0 = |p_0|^2/\rho c$ and $E_0 = E_s(\tilde{\lambda} = \tilde{0})$ are the incident density and the scattering cross section of the perfect shell, respectively. P_0 is the amplitude of the incident pressure, and E_p is the threshold of interest. In this application, E_p takes the form

$$E_p = (1+\delta) E_0 \qquad \delta \ge 0 \qquad (28d)$$

p^t is the average of the total scattered power over one period scattered by the imperfect shell per unit length defined by

$$p^t = \frac{1}{2\rho c} \int_s \mathrm{Re}(p^s)\mathrm{Re}((-j\omega u)^* \cdot \tilde{n})\, ds \qquad (29)$$

5. Description of the Out-of-Roundness Imperfection via a Random Function

The out-of-roundness imperfection is described by

$$R(\phi) = \bar{R}\,[1+\varepsilon\, \lambda_R(\phi)] \tag{30a}$$

where $\lambda_R(\phi)$ assumes the form of a cosine series

$$\lambda_R(\phi) = \sum_n^q b_n \cos n\phi \tag{30b}$$

The coeffieints b_n are given by

$$b_n = \frac{\mu_n}{2\pi}\int_{-\pi}^{\pi} \lambda_R(\phi) \cos n\phi \, d\phi \tag{31a}$$

and

$$\mu_n = 1 \quad \text{for } n=0; \qquad \mu_n = 2 \quad \text{for } n>0 \tag{31b}$$

The radius $R(\phi)$ is assumed to be a random function with mean $\bar{R}$. The imperfection $\lambda_R(\phi)$, therefore, is a stochastic function with zero mean , i.e.,

$$E[\, R(\phi)\,] = \bar{R} \tag{32a}$$

$$E[\, \lambda_R(\phi)\,] = 0 \tag{32b}$$

where E [•] represents the expected value operator. For simplicity, $\lambda_R(\phi)$ is assumed to be an even function so that Eq.(30b) is valid and b_n are random variables.

In order to obtain the coefficients b_n (i.e., Eq.(30b)), we need the explicit function of imperfection $\lambda_R(\phi)$. The variance and autocorrelation between the imperfection for any two points, i.e., $\rho_{\phi_1\phi_2}$, are considered.

From Eqs.(30) and (32b), the mean value and covariance of the coefficients b_n are given by

$$E[b_n\,] = 0 \tag{33a}$$

$$\mathrm{Cov}[b_n\, b_k] = \frac{\mu_n\mu_k}{4\pi^2\varepsilon^2}\int_{-\pi}^{\pi}\int_{-\pi}^{\pi} \mathrm{Cov}[R(\phi_1)\, R(\phi_2)] \cos n\phi_1 \cos k\phi_2 \, d\phi_1\, d\phi_2 \tag{33b}$$

where Cov[•] represents the covariance operator.

The autocorrelation function of $R(\phi)$ is assumed to be of the form

$$\rho_{\phi_1\phi_2} = F(\Delta\phi)\, e^{-|\Delta\phi|\bar{R}/2\pi L} \tag{34a}$$

where

$$F(\Delta\phi) = \Big(\sum_i a_i \cos i\,\Delta\phi\Big) \Big/ \sum_i a_i \tag{34b}$$

a_i are selected coefficients. For example, a_i=0, for $i \neq 0$, is an exponential correlation, and a_0=0 is an exponential cosine correlation. The difference between any two angles ϕ_1 and ϕ_2, is given by

$$\Delta\phi = \min\{\, |\phi_1 - \phi_2|\, ,\ 2\pi - |\phi_1 - \phi_2|\ \} \tag{35}$$

ϕ is measured from $-\pi$ to $+\pi$, and L is the correlation length.

By the definition of the correlation coefficient, the covariance of $R(\phi_1)$ and $R(\phi_2)$ can be expressed as

$$\mathrm{Cov}[R(\phi_1)\, R(\phi_2)] = \sigma_R(\phi_1)\sigma_R(\phi_2)\ \rho_{\phi_1\phi_2} \tag{36a}$$

where

$$\sigma_R(\phi) = \{\, E[R(\phi)R(\phi)] - (\, E[R(\phi)]\,)^2\, \}^{1/2}$$

$$= (\, E[\lambda_R(\phi)\lambda_R(\phi)] - \bar{R}^2)^{1/2} \tag{36b}$$

is the standard deviation of $R(\phi)$. It should be noted that

$$\sigma_R(\phi) = \varepsilon\, \sigma_\lambda(\phi) \tag{36c}$$

Substituting the above equation into (33b) and assuming $\sigma_\lambda(\phi)$ is a constant, the covariance of b_n and b_k becomes

$$\mathrm{Cov}[b_n\, b_k] = \frac{\mu_n\mu_k\sigma_R^2}{4\pi^2\varepsilon^2}\int_{-\pi}^{\pi}\int_{-\pi}^{\pi} F(\Delta\phi)\, e^{-|\Delta\phi|\bar{R}/2\pi L} \cos n\phi_1 \cos k\phi_2 \, d\phi_1\, d\phi_2 \tag{37}$$

6. Numerical Results

The formulation derived in the preceding sections is tested by numerical examples. The problem statement for an infinitely long imperfect shell submerged in an acoustic medium is depicted in Fig.1.

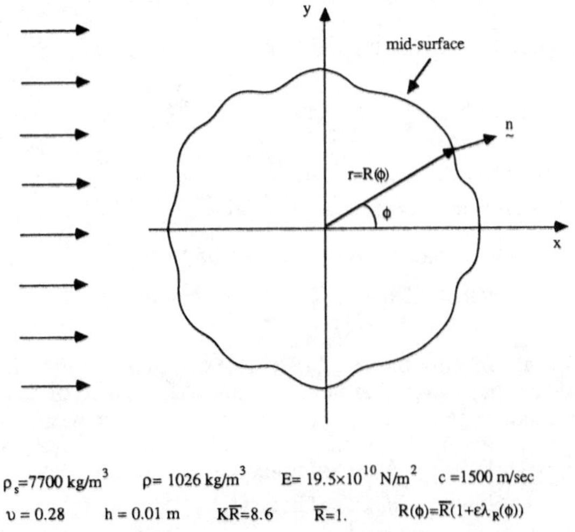

Figure 1. Problem statement.

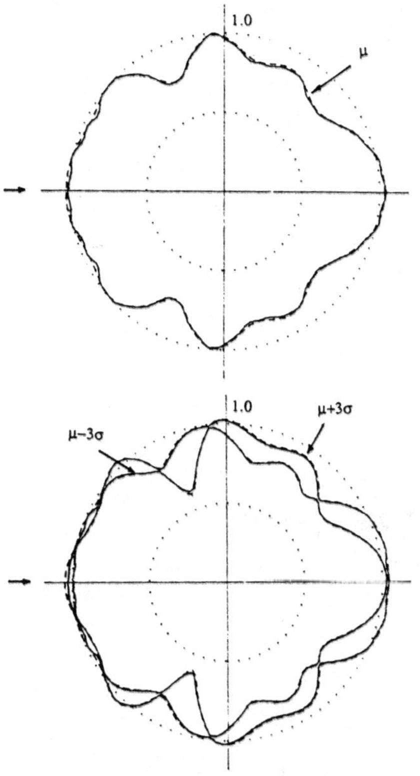

Figure 2. μ_p and $\mu_p \pm 3\,\sigma_p$ of the scattered pressure amplitude with the correlation model $a_3 = 1$, and $a_i = 0$, $i \neq 3$, perturbation method (dashed line), finite element method (solid line).

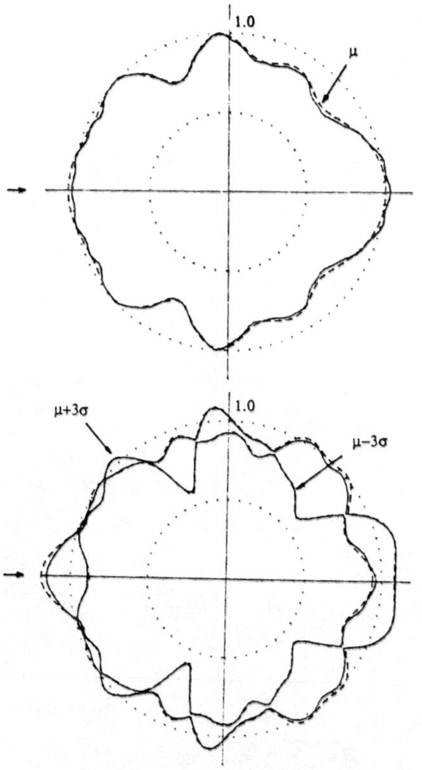

Figure 3. μ_p and $\mu_p \pm 3\,\sigma_p$ of the scattered pressure amplitude with the correlation model $a_5 = 1$, and $a_i = 0$, $i \neq 5$, perturbation method (dashed line), finite element method (solid line).

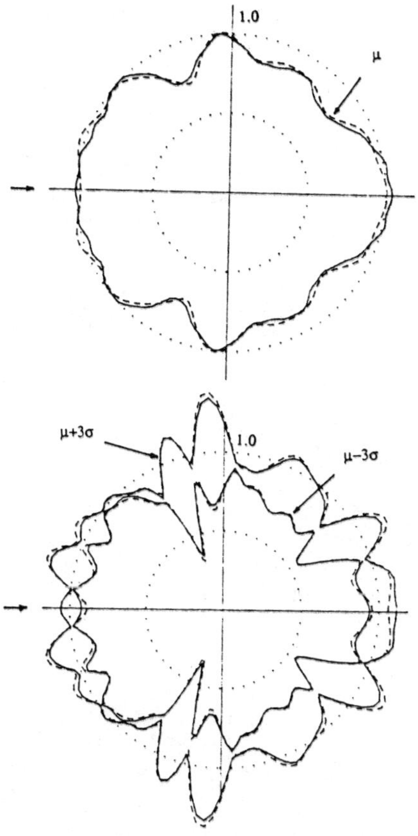

Figure 4. μ_p and $\mu_p \pm 3\,\sigma_p$ of the scattered pressure amplitude with the correlation model $a_7 = 1$, and $a_i = 0$, $i \neq 7$, perturbation method (dashed line), finite element method (solid line).

The exponential cosine correlation defined in Eq.(34) with $\varepsilon = 0.02$ for various a_i is considered to described the random out-of-roundness imperfection. $\lambda_R(\phi)$ has zero mean and unit variance. Figures 2 through 5 present the mean μ_p and $\mu_p \pm 3\,\sigma_p$ of the scattered pressure amplitude on the shell surface using both the perturbation method (dashed line) [Liu, et al.,1989] and PFEM (solid line) for different correlation coefficients (i.e., a_i). σ_p is the standard deviation of the scattered pressure amplitude. Figure 6 shows the influence of the correlation length on the coefficients of variation (σ_p/μ_p) of the scattered pressure amplitude using L=1 and L=1000. It is observed that correlation length does not has much influence on the distribution of the scattered pressure.

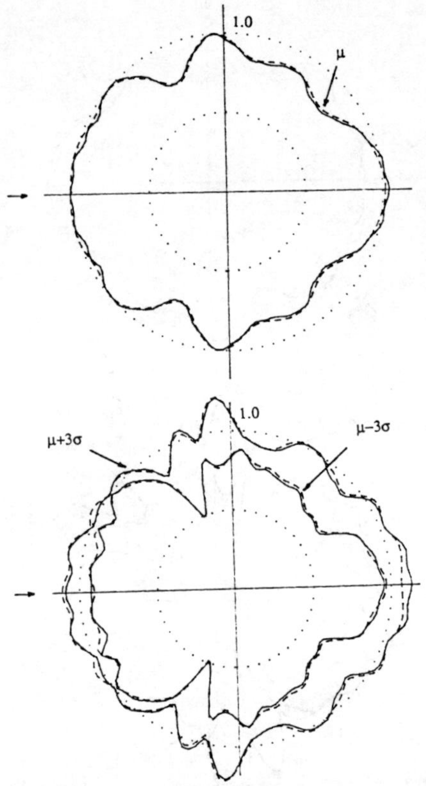

Figure 5. μ_p and $\mu_p \pm 3\,\sigma_p$ of the scattered pressure amplitude with the correlation model $a_3 = a_5 = a_7 = 1$ and $a_i = 0$, $i \neq 3,5,7$; perturbation method (dashed line), finite element method (solid line).

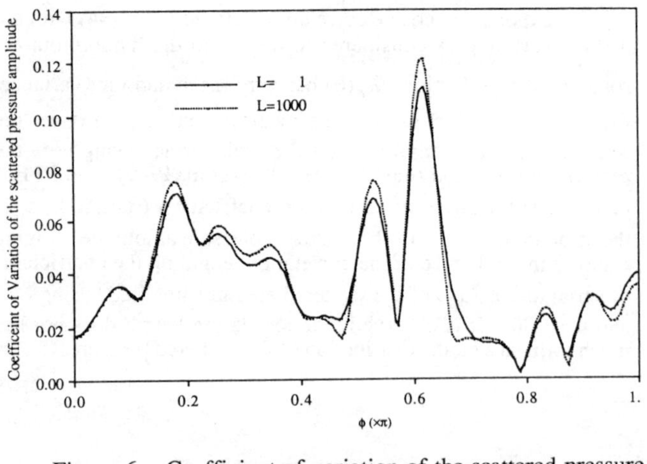

Figure 6. Coefficient of variation of the scattered pressure amplitude for the correlation length L=1 and L=1000; $a_3 = a_5 = a_7 = 1$ and $a_i = 0$, $i \neq 3,5,7$.

ACKNOWLEDGMENT

The support of this research by an ONR grant N00014-87-K-0756 and the encouragement of Dr. Rembert F. Jones, Jr., are gratefully acknowledged.

Reference

1. Ang,A. H-S. and Tang,W. H., Probabilistic Concepts in Engineering Planning and Design: Volume II— Decision, Risk, and Reliability, John Wiley & Sons, Inc., New York, 1984.

2. Everstine, G.C., "A symmetric potential formulation for fluid-structure interaction", Journal Sound & Vibration, **79**(1),1981, pp.157-160.

3. Liu,W.K., Lu,Y.Y. and Chen,J.S., "Probabilistic and Reliability Analysis of a Structural-Acoustic System", (accepted by Journal Sound & Vibration) 1989.

4. Liu,W.K., Belytschko,T. and Mani,A., "Random field finite elements", Int. Journal Numerical Methods in Engineering, **23**, 1986,pp.1831-1845.

5. Liu,W.K., Mani,A. and Belytschko,T., "Finite element methods in probabilistic mechanics", Probabilistic Engineering Mechanics, **2**, 1987,pp.201-213.

BOUNDARY ELEMENT METHODS IN PROBABILISTIC ACOUSTIC RADIATION PROBLEMS

R. Daddazio and M. Ettouney
Applied Science Division
Weidlinger Associates
New York, New York

ABSTRACT

Boundary Element Methods (BEM) are suited to a wide range of engineering problems, especially those of a semi–infinite nature. Examples of such problems can be found in the fluid–structure interactions of acoustic radiation and scattering problems and in the soil–structure interactions of earthquake and machine foundation problems. The required input parameters, dynamic loads, and system properties for such problems are not in general well defined and can be considered random variables. Probabilistic structural analysis through the use of the BEM has been introduced by Ettouney et al. (1989a, 1989b) for the solution of plane elastic systems with uncertain material properties. The method was applied to solution of problems in the fields of structural and geotechnical engineering. In this work, we extend the use of this probabilistic approach to area of fluid–structure interaction by applying this technique to the problem of acoustic radiation from structures. The advantages of the probabilistic viewpoint are discussed with regard to current engineering practices. The importance of confidence estimates for criteria of non–exceedance response is emphasized.

BOUNDARY ELEMENT APPROACH

The basic idea of Boundary Element Methods is to reduce a problem in n dimensions to one in $n-1$ dimensions by using integral equation techniques to express the solution over a domain in terms of its surface values. This approach has been applied to problems in acoustic radiation and scattering by among others Schenck (1968), Burton and Miller (1971), Meyer et al. (1978), Rizzo et al. (1985), Everstine et al. (1986), Vasudevan and McCormick (1988) and Seybert et al. (1988). Application to problems in such fields as elastodynamics by Ahmad and Banerjee (1988) and magnetoelasticity by Sladek and Sladek (1988) has been undertaken. Additional details can be found in Banerjee and Butterfield (1981), and Brebbia, Telles, and Wrobel (1984). We briefly describe its application to the exterior fluid problem.

The propagation of acoustic waves through a homogeneous fluid of density ρ with velocity c is described by the linear wave equation

$$\nabla^2 P(\underline{x},t) = \frac{1}{c^2}\frac{\partial^2}{\partial t^2}P(\underline{x},t) \tag{1}$$

where $P(\underline{x},t)$ is the sound pressure at a point in the fluid $\underline{x}$ at time t. The pressure $P(\underline{x},t)$ is related to the fluid particle velocity $\underline{V}(\underline{x},t)$ through

$$\underline{\nabla} P(\underline{x},t) = -\rho\frac{\partial \underline{V}}{\partial t}(\underline{x},t). \tag{2}$$

For steady state harmonic motion at angular frequency ω, the total fluid pressure may be expressed as

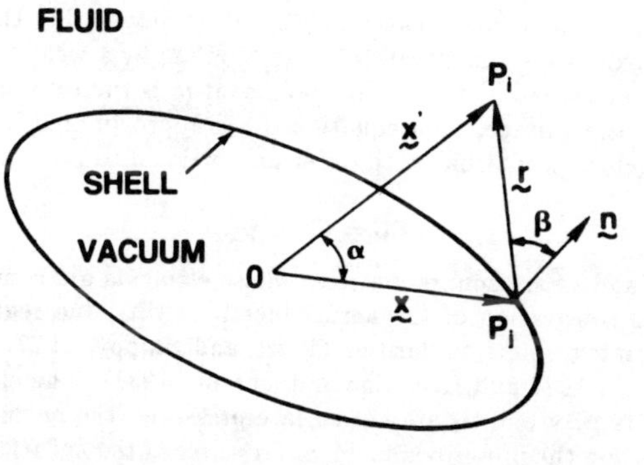

Figure 1. Notation for Helmholtz integral equation

$$P(\underline{x},t) = p(\underline{x})e^{i\omega t} \tag{3}$$

Substituting equation (3) into equation (1) yields the Helmholtz equation

$$\nabla^2 p(\underline{x}) + k^2 p(\underline{x}) = 0 \tag{4}$$

where $k = \omega/c$ is the acoustic wave number.

Applying Green's Theorem together with the appropriate boundary conditions, the pressure and velocity can be related through the Helmholtz integral equation (Baker and Copson, 1987). Following the notation of Everstine and Quezon (1988), this is written as

$$\int_S p(\underline{x})\frac{\partial D(\underline{r})}{\partial \underline{n}}\,dS - \int_S q(\underline{x})D(\underline{r})\,dS$$

$$= \begin{cases} p(\underline{x}')/2 - p_I(\underline{x}'), & (\underline{x}') \text{ on } S \\ p(\underline{x}') - p_I(\underline{x}'), & (\underline{x}') \text{ in } E \end{cases} \quad (5)$$

where S and E denote the surface and exterior fluid points, respectively, p_I is the pressure of an acoustic wave incident upon the closed surface S and $\underline{r}$ is the vector defined by $\underline{x} - \underline{x}'$ (Figure 1) in which $\underline{x}$ is the location of a point on S and $\underline{x}'$ is either on S or in E. The expression

$$D(\underline{r}) = \frac{e^{-ik\underline{r}}}{4\pi\underline{r}} \quad (6)$$

is the three dimensional free space Green's function for equation (4), and $\partial/\partial\underline{n}$ denotes differentiation in the outward normal direction ($\underline{n}$) on S at $\underline{x}$. At any point $\underline{x}$ on S, equation (2) gives

$$q(\underline{x}) = \frac{\partial p(\underline{x})}{\partial \underline{n}} = -i\omega\rho v_n(\underline{x}) \quad (7)$$

where $v_n(\underline{x})$ is the outward normal surface velocity.

Substituting equations (6) and (7) into equation (5) leads to the following integral equation relating the total pressure, $p(\underline{x})$ and normal velocity $v_n(\underline{x})$ on S .

$$\frac{1}{2}p(\underline{x}') - \int_S p(\underline{x})\left(\frac{e^{-ik\underline{r}}}{4\pi\underline{r}}\right)\left(ik + \frac{1}{r}\right)\cos\beta\,dS$$

$$= i\omega\rho \int_S v_n(\underline{x})\left(\frac{e^{-ik\underline{r}}}{4\pi\underline{r}}\right)dS + p_I(\underline{x}') \quad (8)$$

Once surface values are properly defined, the problem specification is complete. We assume that surface pressures are specified over certain portions of the surface and that normal velocities are prescribed over the remaining portions of S. Equation (8) is used in the probabilistic the analysis.

A numerical solution can be obtained by discretizing the surface and approximating the integrals in equation (8) by a weighted sum of functions evaluated at the surface points. If m is the total number of points on the surface, then equation (8) is approximated by a set of linear algebraic equations with order proportional to m:

$$\mathbf{Ep} = \mathbf{Cv} + \mathbf{p_I}. \quad (9)$$

where $\mathbf{E}$ and $\mathbf{C}$ are square matrices whose elements are computed by numerical integration of the kernel functions with due regard being given to integrable singularities (Rizzo and Shippy, 1977, Benthien and Barach, 1987 and Everstine and Quezon, 1988). The elements of the vectors $\mathbf{p}$, $\mathbf{v}$ and $\mathbf{p_I}$ are the surface pressures, the normal surface velocities and the prescribed incident pressures at the m surface points, respectively. In the case of acoustic radiation only, the total surface pressure $\mathbf{p}$ consists of solely radiated pressure and the term $\mathbf{p_I}$ is not present in equation (9).

Since $\mathbf{p}$ and $\mathbf{v}$ contain both unknown and prescribed surface values, equation (9) must be rewritten as follows

$$\mathbf{My} = \mathbf{Nz} + \mathbf{p_I}. \quad (10)$$

in which $\mathbf{y}$ contains the unknown pressures and normal velocities and $\mathbf{z}$ contains the specified surface pressures and velocities.

Solution of these equations yields approximate values for the unknown surface pressures and normal velocities. Once the surface pressures and normal velocities are known on the entire surface, the value of the pressure anywhere in the exterior fluid can be calculated using equation (8).

UNCERTAINTY MODELING

In general, uncertainties can arise in material properties, specified surface values, the geometry of the body, or any combination of these. In extending the formulation to include uncertainties, we distinguish between various classes of problems. The first distinction is between deterministic and random systems. The concept of randomness or uncertainty may be distinguished as follows:

(1) uncertainty in constant system parameters resulting in a *spread* or density of possible parameter values, i.e., random variable model,
(2) uncertainty in the behavior of parameters as time–dependent processes, where the statistics of each process may be either time–dependent or –independent, i.e., random or stochastic function model, and
(3) uncertainty with respect to the behavior of parameter properties as functions of spatial coordinates where, again, the statistics of the functional behavior may be either space–dependent or –independent, i.e., random or stochastic field model.

Except for the case where only surface values are random, these are, in general, all very difficult problems to solve, both mathematically and in application to existing problems. We are initially interested in the random variable model, since this will be of most practical value for the applications in mind. Extensions of the developed ideas will be subsequently possible.

A case of considerable interest in acoustic radiation from structures is one where the statistics of the material parameters c, ρ and ω are available so that their mean values are known and deviations relative to the mean are not too large. We consider this case and limit the problem to a body of one material with given surface values and geometry. Then the unknown pressures and normal velocities on the surface can be expanded in a Taylor series about the mean values. This technique has been used in probabilistic finite element methods. Benaroya and Rehak, 1988 presented a well documented summary of its applications to structural analysis. Ettouney, Benaroya and Wright, 1989a, 1989b used this approach in probabilistic BEM as applied to structural analysis. We will use the same method for the purpose of this study.

PERTURBATION EXPANSIONS

The dependence on c, ρ and ω in equation (10) for the case of radiation is indicated explicitly by writing

$$\mathbf{M}(c,\rho,\omega)\mathbf{y}(c,\rho,\omega) = \mathbf{N}(c,\rho,\omega)\mathbf{z} + \mathbf{p_I}(c,\rho,\omega) \quad (11)$$

where $\mathbf{z}$ is specified, and therefore has no c,ρ,ω dependence. Expansions in Taylor series of $\mathbf{y}$ about the mean values $\overline{c}$,$\overline{\rho}$ and $\overline{\omega}$, with $\varepsilon_c = (c - \overline{c})$, $\varepsilon_\rho = (\rho - \overline{\rho})$ and $\varepsilon_\omega = (\omega - \overline{\omega})$, yields

$$\mathbf{y}(c,\rho,\omega) = \overline{\mathbf{y}} + \varepsilon_c\overline{\mathbf{y}}_c + \varepsilon_\rho\overline{\mathbf{y}}_\rho + \varepsilon_\omega\overline{\mathbf{y}}_\omega$$

$$+ \frac{1}{2}\varepsilon_c^2\overline{\mathbf{y}}_{cc} + \frac{1}{2}\varepsilon_\rho^2\overline{\mathbf{y}}_{\rho\rho} + \frac{1}{2}\varepsilon_\omega^2\overline{\mathbf{y}}_{\omega\omega} + \varepsilon_c\varepsilon_\rho\overline{\mathbf{y}}_{c\rho} + \cdots \quad (12)$$

where the subscripts on $\overline{\mathbf{y}}$ denote partial differentiation (no summation implied for repeated *Greek* subscripts) and the overline denotes evaluation of the functions at the respective mean values. Similar notation will be used for Taylor series expansions of $\mathbf{M}(c,\rho,\omega)$, $\mathbf{N}(c,\rho,\omega)$ and other quantities below. Substitution of equation (12) into equation (11), and equating terms of the same order in ve_c, ε_ρ and ε_ω, produces the usual set of recursion relations which result from perturbation expansion techniques (Bellman, 1964):

$$\overline{\mathbf{M}}\overline{\mathbf{y}} = \overline{\mathbf{N}}\mathbf{z} \tag{13a}$$
$$\overline{\mathbf{M}}\overline{\mathbf{y}}_c = \overline{\mathbf{N}}_c\mathbf{z} - \overline{\mathbf{M}}_c\overline{\mathbf{y}} \tag{13b}$$
$$\overline{\mathbf{M}}\overline{\mathbf{y}}_\omega = \overline{\mathbf{N}}_\omega\mathbf{z} - \overline{\mathbf{M}}_\omega\overline{\mathbf{y}} \tag{13c}$$
$$\overline{\mathbf{M}}\overline{\mathbf{y}}_\rho = \overline{\mathbf{N}}_\rho\mathbf{z} - \overline{\mathbf{M}}_\rho\overline{\mathbf{y}} \tag{13d}$$
$$\cdots$$

The recursive system provides an efficient procedure for calculating the terms in equation (12) since the coefficient matrix $\overline{\mathbf{M}}$ is the same for all orders. Once the elements of this matrix have been computed and the matrix has been factorized into lower and upper triangular matrices, most of the computational work has been done (order m^3 operations), and the equations can be solved quickly (order m^2 operations). It is relatively easy to assess the rate of convergence of the series and, as expected, the expansion converges rapidly for physically reasonable uncertainties in the values of c,ρ and ω.

Specifically, with reference to equation (9),

$$\mathbf{p}(c,\rho,\omega) = \overline{p} + \varepsilon_c\overline{\mathbf{p}}_c + \varepsilon_\rho\overline{\mathbf{p}}_\rho + \varepsilon_\omega\overline{\mathbf{p}}_\omega + \frac{1}{2}\varepsilon_c^2\overline{\mathbf{p}}_{cc} + \frac{1}{2}\varepsilon_\rho^2\overline{\mathbf{p}}_{\rho\rho} + \frac{1}{2}\varepsilon_\omega^2\overline{\mathbf{p}}_{\omega\omega} + \varepsilon_\rho\varepsilon_\omega\overline{\mathbf{p}}_{\rho\omega} + \cdots \tag{14a}$$

$$\mathbf{v}(c,\rho,\omega) = \overline{\mathbf{v}} + \varepsilon_c\overline{\mathbf{v}}_c + \varepsilon_\rho\overline{\mathbf{v}}_\rho + \varepsilon_\omega\overline{\mathbf{v}}_\omega + \frac{1}{2}\varepsilon_c^2\overline{\mathbf{v}}_{cc} + \frac{1}{2}\varepsilon_\rho^2\overline{\mathbf{v}}_{\rho\rho} + \frac{1}{2}\varepsilon_\omega^2\overline{\mathbf{v}}_{\omega\omega} + \varepsilon_\rho\varepsilon_\omega\overline{\mathbf{v}}_{\rho\omega} + \cdots \tag{14b}$$

$$\mathbf{E}(c,\rho,\omega) = \overline{\mathbf{E}} + \varepsilon_c\overline{\mathbf{E}}_c + \varepsilon_\rho\overline{\mathbf{E}}_\rho + \varepsilon_\omega\overline{\mathbf{E}}_\omega + \frac{1}{2}\varepsilon_c^2\overline{\mathbf{E}}_{cc} + \frac{1}{2}\varepsilon_\rho^2\overline{\mathbf{E}}_{\rho\rho} + \frac{1}{2}\varepsilon_\omega^2\overline{\mathbf{E}}_{\omega\omega} + \varepsilon_\rho\varepsilon_\omega\overline{\mathbf{E}}_{\rho\omega} + \cdots \tag{14c}$$

$$\mathbf{C}(c,\rho,\omega) = \overline{\mathbf{C}} + \varepsilon_c\overline{\mathbf{C}}_c + \varepsilon_\omega\overline{\mathbf{C}}_\omega + \varepsilon_\rho\overline{\mathbf{C}}_\rho + \frac{1}{2}\varepsilon_c^2\overline{\mathbf{C}}_{cc} + \frac{1}{2}\varepsilon_\rho^2\overline{\mathbf{C}}_{\rho\rho} + \frac{1}{2}\varepsilon_\omega^2\overline{\mathbf{C}}_{\omega\omega} + \varepsilon_\rho\varepsilon_\omega\overline{\mathbf{C}}_{\rho\omega} + \cdots \tag{14d}$$

$$\mathbf{p_I}(c,\rho,\omega) = \overline{p}_I + \varepsilon_c\overline{\mathbf{p}}_{\mathbf{I}c} + \varepsilon_\rho\overline{\mathbf{p}}_{\mathbf{I}\rho} + \varepsilon_\omega\overline{\mathbf{p}}_{\mathbf{I}\omega} + \frac{1}{2}\varepsilon_c^2\overline{\mathbf{p}}_{\mathbf{I}cc} + \frac{1}{2}\varepsilon_\rho^2\overline{\mathbf{p}}_{\mathbf{I}\rho\rho} + \frac{1}{2}\varepsilon_\omega^2\overline{\mathbf{p}}_{\mathbf{I}\omega\omega} + \varepsilon_\rho\varepsilon_\omega\overline{\mathbf{p}}_{\mathbf{I}\rho\omega} + \cdots \tag{14e}$$

represent the Taylor series expansions of the functions $\mathbf{p}$, $\mathbf{v}$, $\mathbf{E}$, $\mathbf{C}$, and $\mathbf{p_I}$ in terms of the respective c,ρ and ω *sensitivity functions.* Four terms in the series will be needed to account for first order perturbations, while ten terms will be needed in order to include up to second order perturbations. The resulting ten equations corresponding to (9) which must be solved are:

$$\overline{\mathbf{E}}\overline{\mathbf{p}} = \overline{\mathbf{C}}\overline{\mathbf{v}} + \overline{\mathbf{p}}_\mathbf{I} \tag{15a}$$
$$\overline{\mathbf{E}}\overline{\mathbf{p}}_\rho = \overline{\mathbf{C}}\overline{\mathbf{v}}_\rho - (\overline{\mathbf{E}}_\rho\overline{\mathbf{p}} + \overline{\mathbf{C}}_\rho\overline{\mathbf{v}}) + \overline{\mathbf{p}}_{\mathbf{I}\rho} \tag{15b}$$
$$\overline{\mathbf{E}}\overline{\mathbf{p}}_c = \overline{\mathbf{C}}\overline{\mathbf{v}}_c - (\overline{\mathbf{E}}_c\overline{\mathbf{p}} + \overline{\mathbf{C}}_c\overline{\mathbf{v}}) + \overline{\mathbf{p}}_{\mathbf{I}c} \tag{15c}$$
$$\overline{\mathbf{E}}\overline{\mathbf{p}}_\omega = \overline{\mathbf{C}}\overline{\mathbf{v}}_\omega - (\overline{\mathbf{E}}_\omega\overline{\mathbf{p}} + \overline{\mathbf{C}}_\omega\overline{\mathbf{v}}) + \overline{\mathbf{p}}_{\mathbf{I}\omega} \tag{15d}$$
$$\overline{\mathbf{E}}\overline{\mathbf{p}}_{\rho\rho} = \overline{\mathbf{C}}\overline{\mathbf{v}}_{\rho\rho} - (2\overline{\mathbf{E}}_\rho\overline{\mathbf{p}}_\rho + 2\overline{\mathbf{C}}_\rho\overline{\mathbf{v}}_\rho + \overline{\mathbf{E}}_{\rho\rho}\overline{\mathbf{p}} + \overline{\mathbf{C}}_{\rho\rho}\overline{\mathbf{v}}) + \overline{\mathbf{p}}_{\mathbf{I}\rho\rho} \tag{15e}$$
$$\overline{\mathbf{E}}\overline{\mathbf{p}}_{cc} = \overline{\mathbf{C}}\overline{\mathbf{v}}_{cc} - (2\overline{\mathbf{E}}_c\overline{\mathbf{p}}_c + 2\overline{\mathbf{C}}_c\overline{\mathbf{v}}_c + \overline{\mathbf{E}}_{cc}\overline{\mathbf{p}} + \overline{\mathbf{C}}_{cc}\overline{\mathbf{v}}) + \overline{\mathbf{p}}_{\mathbf{I}cc} \tag{15f}$$
$$\overline{\mathbf{E}}\overline{\mathbf{p}}_{\omega\omega} = \overline{\mathbf{C}}\overline{\mathbf{v}}_{\omega\omega} - (2\overline{\mathbf{E}}_\omega\overline{\mathbf{p}}_\omega + 2\overline{\mathbf{C}}_\omega\overline{\mathbf{v}}_\omega + \overline{\mathbf{E}}_{\omega\omega}\overline{\mathbf{p}} + \overline{\mathbf{C}}_{\omega\omega}\overline{\mathbf{v}}) + \overline{\mathbf{p}}_{\mathbf{I}\omega\omega} \tag{15g}$$
$$\overline{\mathbf{E}}\overline{\mathbf{p}}_{\rho c} = \overline{\mathbf{C}}\overline{\mathbf{v}}_{\rho c} - \frac{1}{2}(\overline{\mathbf{E}}_\rho\overline{\mathbf{p}}_c + \overline{\mathbf{E}}_c\overline{\mathbf{p}}_\rho + \overline{\mathbf{C}}_\rho\overline{\mathbf{v}}_c + \overline{\mathbf{C}}_c\dot{\overline{\mathbf{v}}}_\rho + \overline{\mathbf{E}}_{\rho c}\overline{\mathbf{p}} + \overline{\mathbf{C}}_{\rho c}\overline{\mathbf{v}}) + \overline{\mathbf{p}}_{\mathbf{I}\rho c} \tag{15h}$$
$$\overline{\mathbf{E}}\overline{\mathbf{p}}_{\rho\omega} = \overline{\mathbf{C}}\overline{\mathbf{v}}_{\rho\omega} - \frac{1}{2}(\overline{\mathbf{E}}_\rho\overline{\mathbf{p}}_\omega + \overline{\mathbf{E}}_\omega\overline{\mathbf{p}}_\rho + \overline{\mathbf{C}}_\rho\overline{\mathbf{v}}_\omega + \overline{\mathbf{C}}_\omega\overline{\mathbf{v}}_\rho + \overline{\mathbf{E}}_{\rho\omega}\overline{\mathbf{p}} + \overline{\mathbf{C}}_{\rho\omega}\overline{\mathbf{v}}) + \overline{\mathbf{p}}_{\mathbf{I}\rho\omega} \tag{15i}$$
$$\overline{\mathbf{E}}\overline{\mathbf{p}}_{c\omega} = \overline{\mathbf{C}}\overline{\mathbf{v}}_{c\omega} - \frac{1}{2}(\overline{\mathbf{E}}_c\overline{\mathbf{p}}_\omega + \overline{\mathbf{E}}_\omega\overline{\mathbf{p}}_c + \overline{\mathbf{C}}_c\overline{\mathbf{v}}_\omega + \overline{\mathbf{C}}_\omega\overline{\mathbf{v}}_c + \overline{\mathbf{E}}_{c\omega}\overline{\mathbf{p}} + \overline{\mathbf{C}}_{c\omega}\overline{\mathbf{v}}) + \overline{\mathbf{p}}_{\mathbf{I}c\omega} \tag{15j}$$

For second order perturbation analysis, equations (15) are solved recursively for the unknown components of $\overline{\mathbf{p}}$ and $\overline{\mathbf{v}}$, $\overline{\mathbf{p}}_\rho$ and $\overline{\mathbf{v}}_\rho$, $\overline{\mathbf{p}}_c$ and $\overline{\mathbf{v}}_c$, $\overline{\mathbf{p}}_\omega$ and $\overline{\mathbf{v}}_\omega$, $\overline{\mathbf{p}}_{\rho\rho}$ and $\overline{\mathbf{v}}_{\rho\rho}$, $\overline{\mathbf{p}}_{cc}$ and $\overline{\mathbf{v}}_{cc}$, $\overline{\mathbf{p}}_{\omega\omega}$ and $\overline{\mathbf{v}}_{\omega\omega}$, $\overline{\mathbf{p}}_{\rho c}$ and $\overline{\mathbf{v}}_{\rho c}$, $\overline{\mathbf{p}}_{\rho\omega}$ and $\overline{\mathbf{v}}_{\rho\omega}$, $\overline{\mathbf{p}}_{c\omega}$ and $\overline{\mathbf{v}}_{c\omega}$, and substituted appropriately into equations (14a,b), which are used below to evaluate the first and second order moments (mean and standard deviation) of $\mathbf{p}$ and $\mathbf{v}$. However, only first order perturbation terms were used for the purpose of this study, since the present paper is intended for illustration purposes only.

<u>Expectatons of a Model with Three Random Variables</u>

Equations (14a,b) provide the basis for the probabilistic analysis of the response of acoustic systems. The mean values of the pressures at any point can be approximated by

$$E\{p(c,\rho,\omega)\} \cong \overline{p} + E\{\varepsilon_\rho\}\overline{p}_\rho + E\{\varepsilon_\omega\}\overline{p}_\omega + E\{\varepsilon_c\}\overline{p}_c + \frac{1}{2}E\{\varepsilon_\rho^2\}\overline{p}_{\rho\rho} + \frac{1}{2}E\{\varepsilon_\omega^2\}\overline{p}_{\omega\omega} + \frac{1}{2}E\{\varepsilon_c^2\}\overline{p}_{cc} + E\{\varepsilon_c\varepsilon_\rho\}\overline{p}_{c\rho} + E\{\varepsilon_\rho\varepsilon_\omega\}\overline{p}_{\rho\omega} + E\{\varepsilon_c\varepsilon_\omega\}\overline{p}_{c\omega} \tag{16}$$

and the mean-square values, correct to second order, by

$$E\{p(c,\rho,\omega)^2\} \cong \overline{p}^2 + 2E\{\varepsilon_c\}\overline{p}\,\overline{p}_c + 2E\{\varepsilon_\rho\}\overline{p}\,\overline{p}_\rho + 2E\{\varepsilon_\omega\}\overline{p}\,\overline{p}_\omega + E\{\varepsilon_c^2\}(\overline{p}\,\overline{p}_{cc} + \overline{p}_c^2) + E\{\varepsilon_\rho^2\}(\overline{p}\,\overline{p}_{\rho\rho} + \overline{p}_\rho^2) + E\{\varepsilon_\omega^2\}(\overline{p}\,\overline{p}_{\omega\omega} + \overline{p}_\omega^2) + 2E\{\varepsilon_c\varepsilon_\rho\}(\overline{p}\,\overline{p}_{c\rho} + \overline{p}_c\,\overline{p}_\rho) + 2E\{\varepsilon_\omega\varepsilon_\rho\}(\overline{p}\,\overline{p}_{\omega\rho} + \overline{p}_\omega\,\overline{p}_\rho) + 2E\{\varepsilon_c\varepsilon_\omega\}(\overline{p}\,\overline{p}_{c\omega} + \overline{p}_c\,\overline{p}_\omega) \tag{17}$$

In order to evaluate these expressions, we need to make probabilistic statements regarding the random variables c, ρ and ω, i.e., probability densities $f_c(c)$, $f_\rho(\rho)$ and $f_\omega(\omega)$. For the purpose of this study, a simple uniform probability density distribution was assumed for all the random variables. Thus the evaluation of $E\{\varepsilon_c\}$, $E\{\varepsilon_\rho\}$, $E\{\varepsilon_\omega\}$, $E\{\varepsilon_c^2\}$, $E\{\varepsilon_\rho^2\}$, $E\{\varepsilon_\omega^2\}$ $E\{\varepsilon_c\varepsilon_\rho\}$, $E\{\varepsilon_\omega\varepsilon_\rho\}$ and $E\{\varepsilon_c\varepsilon_\omega\}$ can easily be performed using the well known statistical relations for the uniform distribution case, Papoulis, 1965.

Once equations (16) and (17), and similar expressions for the velocity terms, are used to approximate $E\{p(c,\rho,\omega)\}$, $E\{v(c,\rho,\omega)\}$, $E\{p(c,\rho,\omega)^2\}$, and $E\{v(c,\rho,\omega)^2\}$, it is useful to evaluate the respective variances (σ^2) and standard deviations (σ) of pressures and velocities:

$$\sigma_p = +\sqrt{\sigma_p^2} = \sqrt{E\{p^2\} - E^2\{p\}} \tag{18a}$$
$$\sigma_v = +\sqrt{\sigma_v^2} = \sqrt{E\{v^2\} - E^2\{v\}}. \tag{18b}$$

Given the mean and variance, it is possible to establish confidence bounds on specific realizations of p and v. One possible approach, for a random variable x with mean value η, is to use the Chebyshev inequality

$$P\{\eta - a < \mathrm{x} < \eta + a\} \geq 1 - \frac{\sigma^2}{a^2}, \tag{19}$$

or the one-sided versions

$$P\{\mathrm{x} < \eta + a\} \geq \frac{a^2}{\sigma^2 + a^2} \tag{20a}$$
$$P\{\mathrm{x} > \eta - a\} \geq \frac{a^2}{\sigma^2 + a^2}. \tag{20b}$$

Specific applications are introduced and discussed in the following section.

APPLICATION

Current Practices

Engineers, analysts and designers are well aware of the uncertainties of material properties which were discussed earlier. One way to account for the uncertainties is to perform an expensive and time consuming Monte Carlo simulation of the system at hand, which is not exactly a practical solution of the problem. Another approach is to assess the uncertainties at three levels using deterministic approaches. First the system response is evaluated using the "best guess" of the material properties. It is then evaluated using a combination of upper, and/or, lower limits of the material properties in order to produce upper and lower bound values of the response quantities of interest. The space between the upper and lower bounds (which will contain the "best guess" point) can be called the "deterministic zone". The deterministic zone can then be used for any design or analysis decision.

The deterministic zone approach, although practical, has several drawbacks. First, it does not provide any information about the behavior of the response parameters of interest outside the zone. It can be at times too conservative, at other times not, and the designer/analyst has no way of knowing which. Confidence estimates are out of the question, i.e., what are the probabilities that response–parameter values will fall outside the deterministic zone? The PBEM, and similar methods, can solve this dilemma for the practicing engineer in a straightforward and accurate way.

Case Study

The theoretical PBEM equations were introduced in a previous section. We will now apply the outlined methodology to the solution of the problem of the acoustic radiation from a pulsating sphere in a random media. The sphere has a center at $R = 0$ and a radius at $R = a$. There is a prescribed uniform radial velocity of $v(a) = Ve^{-i\omega t}$ and therefore complete spherical symmetry exists. The solution to the surface pressure for this case is (Junger and Feit, 1986)

$$p(a) = \frac{i\omega\rho aV}{1 + ika} \tag{21}$$

Following Junger and Feit, the surface impedance, defined as $z_0(a) = p(a)/v(a)$, can be expanded in its real and imaginary components r_0 and m_0. The term r_0 is called the resistance and represents the energy lost in the form of radiated acoustic energy. The term m_0 is called the inertance and represents the entrained mass of the fluid set into motion by the pulsating sphere. We have

$$r_0 = \frac{\rho c(ka)^2}{1 + (ka)^2} \tag{22}$$

$$m_0 = \frac{\rho a}{1 + (ka)^2} \tag{23}$$

Both r_0 and m_0 are frequency dependent. At low frequencies, the pressure associated with the inertance is dominant while at high frequencies, the effect of the resistance dominates and the surface impedance approaches that of a plane wave.

Analysts and systems designers in the field of structural acoustics are aware of the uncertainties in the material properties of the exterior fluid. If the acoustic medium is seawater, the density and sound speed vary as a function of temperature, salinity and pressure (Del Grosso, 1974). In our study, we will assume that the only random variable is the sound speed, c. The sound speed is taken to be a uniformly distributed random variable with a range of $0.95\bar{c} \leq c \leq 1.05\bar{c}$. This range of variability is a typical realistic situation. The mean value of the sound speed is $\bar{c} = 59350\,\text{in/sec}$.

A boundary element model for the pulsating sphere has been developed. The accuracy of the boundary element method in the solution of this problem has been established (Seybert et al. 1988). The problem was solved for a range of frequencies from $0 < ka \leq .5$. At each frequency, the maximum and minimum values for the real and imaginary part of the surface impedance, $r_0/\rho c$ and $m_0/\rho a$, with a probability of non–exceedance of 84% and 97% are calculated using the Chebyshev inequalities which were described previously. These non–exceedance levels are chosen since they correspond to one and two standard deviation from the mean in a normal density function, hence they are widely used in engineering applications.

Figures 2a through 2d shows the results of the probabilistic analysis. Figures 2a and 2b indicate the deterministic solution for the quantities $r_0/\rho c$ and $m_0/\rho a$ along with the probabilities of non–exceedance. In this frequency range, the deterministic value of $r_0/\rho c$ varies between $0. \leq r_0/\rho c \leq .2$ and $m_0/\rho a$ varies from $1. \leq m_0/\rho a \leq .8$. Hence, the effect of the inertance is dominant. The accuracy of the discretization is judged to be adequate for the purposes of this study which is to illustrate the implementation of the methodology . Perhaps more revealing is to look at the results of the probabilistic analysis *relative* to the deterministic solution. This is shown in Figures 2c and 2d. Figure 2c indicates that for a non-exceedence probabilty of 97% , there is a constant uncertainty of about $\pm 20\%$ in the resistance over the frequency range of interest. As was previously mentioned however, the overall effect of the resistance is $\leq 20\%$ in this range. Figure 2d indicates that the uncertainty in the estimate of inertance is increasing with frequency to a value of $\pm 4\%$ at $ka = .5$.

An important point to be made in the interpretation of these results is that because of the normalization of the terms for the resistance $r_0/\rho c$ and the entrained mass $m_0/\rho a$, the deterministic response, as given by these quantities, is *independent* of the of the parameters c and ρ. Hence, the deterministic zone has zero width (Figure 2). The solution of the problem at an upper and lower limit of c will yield *no* information on the uncertainty in a response parameter to a random value of c. Only probabilistic approaches can yield quantitative measures of the sensitivity of the results (surface pressure in this case) to the variability of the acoustic medium.

It should be reiterated that the purpose of the example problem is to illustrate the application of the probabilistic analysis procedure presented herein. Other factors not addressed here that need be investigated are but are not limited to the effect of multiple random variables (ρ and ω in addition to c) and the correlation between them, the interaction of the random media with an elastic structure, the effects of the random media on the predictions of far field sound pressure levels and the application of this approach to acoustic scattering problems.

SUMMARY OF PRESENT WORK

A probabilistic formulation for acoustic radiation problems using the boundary element method is introduced. The formulation assumes that the sound speed, mass density and the angular frequency of oscillation are random variables. Using a Taylor series technique, a perturbation approach is used to solve the problem. The statistical properties of the response measures, such as the mean and mean square are then evaluated. These statistical measures are used to evaluate the response measures at different confidence levels.

The problem of acoustic radiation from a pulsating sphere was studied in order to demonstrate the use and value of the proposed method. Estimates of the uncertainty in the surface pressure and in the proportions of resistance to inertances as a function of frequency in a fluid with random sound speed, are given. It is established that the probabilistic approach is needed in order to ascertain the degree of accuracy to which the response measures can be determined.

We have demonstrated the feasibility and practicality of this approach. It has been presented as an analysis and design tool, since we believe that an analytical method without close coupling to simple and clear design usage is incomplete. We have tried to demonstrate such usage with our application.

PBEM are still in their infancy. Much remains to be done in order that the tremendous potential of PBEM as a powerful analysis and design tool is fulfilled. This is only a first step in the application of this method to steady–state dynamic problems.

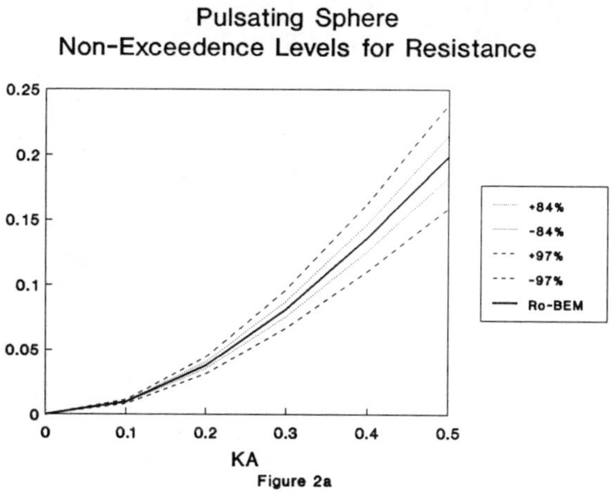

Figure 2a

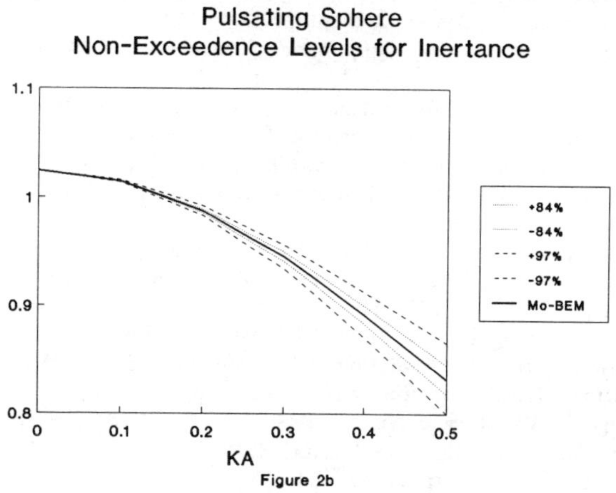

Figure 2b

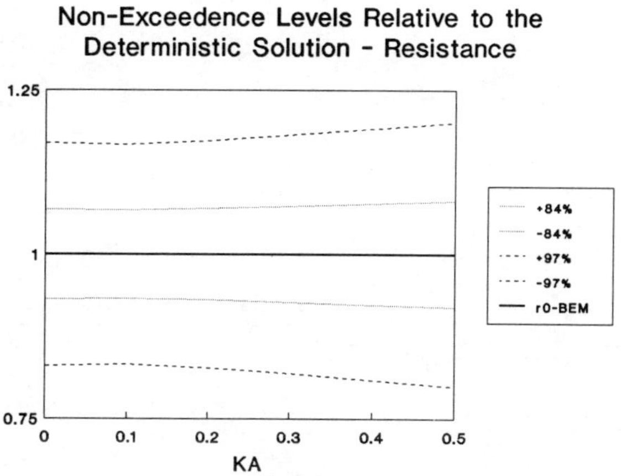

Figure 2c

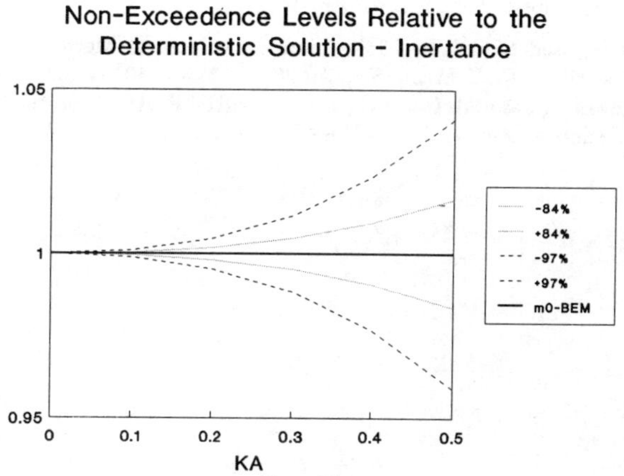

Figure 2d

REFERENCES

Ahmad, S., and Banerjee,P. K., 1988, "Time–Domain Transient Elastodynamic Analysis of 3–D Solids by BEM," *Int J for Num Meth in Eng*, Vol 26, pp. 1709–1728.

Baker, B., and Copson, E., 1987, *The Mathematical Theory of Huygens' Principle*, Chelsea Publishing Company, New York, NY, Third Edition.

Banerjee, P., and Butterfield, R., 1981, *Boundary Element Methods in Engineering Science*, McGraw–Hill UK.

Bellman, R., 1964, *Perturbation Techniques in Mathematics, Physics, and Engineering*, Holt, Rinehart and Winston.

Benaroya, H., and Rehak, M., 1988, "Finite Element Methods and Probabilistic Structural Analysis - A Selective Review," ASME *Applied Mechanics Reviews*, Vol 41, No 5, May 1988.

Benthien, G.W., and Barach, D., 1987, "CHIEF Users Manual," Technical Report, NOSC TD 970, Naval Ocean Systems Center, San Diego, CA, pp. A2–1–A3–4.

Brebbia,C. A., Telles, J., and Wrobel,L., 1984, *Boundary Element Techniques*, Springer–Verlag.

Burton, A.J., and Miller, G.F., 1971, " The Application of Integral Equation Methods to the Numerical Solutions of Some Exterior Boundary Value Problems," *Proc. R. Soc. London A* **323**, pp. 201–210.

Del Grosso, V., 1974, "New Equations for the Speed of Sound in Natural Water (with comparisons to other equations)," *J Acoust Soc Am* **56**(4),pp.1084–1091.

Ettouney, M., Benaroya, H., and Wright, J., 1989a, "Probabilistc Boundary Element Methods (PBEM)," *Computational Mechanics in Probabilistic and Reliability Analysis*, W. K. Liu and T. Belytschko, Editors, ELMPRESS International, 1989.

Ettouney, M., Benaroya, H., and Wright, J., 1989b, "Boundary Element Methods in Probabilistic Structural Analysis (PBEM)," Accepted for publication, *Applied Mathematical Modelling*, 1989.

Everstine, G., Henderson, F., Schroeder, E., and Lipman, R., 1986, "A General Low Frequency Acoustic Radiation Capability for NASTRAN," Technical Report NASA CP–2419, Fourteenth NASTRAN User's Colloquium, National Aeronautics and Space Administration, Washington, D.C., pp. 293–310.

Everstine, G., and Quezon, A., 1988, "User's Guide to the Coupled NASTRAN/Helmholtz Equation Capability (NASHUA) for Acoustic Radiation and Scattering," Technical Report DTRC CMLD–88/03, David Taylor Research Center, Bethesda, MD, pp. 7–9.

Junger M., and Feit, D., 1986, *Sound, Structures, and Their Interaction*, MIT Press, Cambridge, MA, Second Edition.

Meyer, W., Bell, W., Zinn, B., and Stallybrass, M., 1978, "Boundary Integral Solutions of Three Dimensional Acoustic Radiation Problems," *Journal of Sound and Vibration*, **59**(2), pp. 245–262.

Papoulis,A., 1965 *Probability, Random Variables, and Stochastic Processes*, McGraw–Hill.

Rizzo, F., and Shippy,D., 1977, "An Advanced Boundary Integral Equation Method for Three–Dimensional Thermoelasticity," *Int J Engng Sci*, Vol. 11, pp. 1753–1768.

Rizzo, F., Shippy, D., and Rezayat, M., 1985, "A Boundary Integral Equation Method for Radiation and Scattering of Elastic Waves in Three Dimensions," *Int J Engng Sci*, Vol. 21, pp. 115–129.

Seybert, A., Wu, T., and Wu, X., 1988, "Radiation and Scattering of Acoustic Waves from Elastic Solids and Shells Using the Boundary Element Method," *J.Acoust.Soc.Am.* **84**(5) pp. 1906–1912.

Schenck, H., 1968, "Improved Integral Formulation for Acoustic Radiation Problems," *J Acoust Soc Am* **44**(1), pp. 41–58.

Sladek,V., and Sladek,J., 1988, "Boundary Integral Method in Magnetoelasticity", *Int J Engng Sci*, Vol 26, No 5, pp. 401–418.

Vasudevan, R., and McCormick, J., 1988, "HULRAD Theoretical Manual," Technical Report No. WARAD–88/002, Weidlinger Associates, NY.

COMPUTATION OF ACOUSTIC SHAPE DESIGN SENSITIVITY USING A BOUNDARY ELEMENT METHOD

D. C. Smith and R. J. Bernhard
Ray W. Herrick Laboratories
School of Mechanical Engineering
Purdue University
West Lafayette, Indiana

ABSTRACT

Numerical acoustic radiation prediction schemes have developed to the point where they can be used to reliably predict the acoustic response of a vibrating structure. However, the objective of many of the applications of these techniques is the design of a best structural/acoustic configuration for a given application. Thus, the acoustic radiation prediction scheme must be incorporated into a design methodology. Traditionally, the most difficult design variables to incorporate into numerical prediction schemes have been those which define the geometry of the configuration, normally referred to as shape variables. In this investigation, a technique for computing the design sensitivity of the radiated pressure solution to shape variables is formulated and verified. The method uses a boundary element implementation of the Helmholtz integral equation. The potential problems at the characteristic non-uniqueness frequencies of the Helmholtz integral equation are also addressed. The technique is verified for several pulsating sphere examples where analytical solutions are available.

INTRODUCTION

Significant progress is being made in the development and verification of numerical techniques which can be used to accurately predict the acoustic radiation from bodies of general shape with arbitrary boundary conditions (1-8). However, in many applications the numerical radiation prediction techniques are being used as part of a design process to identify the best alternative of a number of different solutions. The acoustic radiation prediction scheme will be used iteratively in conjunction with other models until a satisfactory solution is found. This iterative design process is usually improved significantly if the sensitivity of the performance of a given configuration to the design variables can be determined. Such design sensitivity information is usually useful whether a formal optimization process is implemented or a limited number of design alternatives are selected by the designer.

Bernhard and co-workers have previously investigated the computation and utilization of design sensitivity for applications where the finite element methods are used to model sound fields (9-12). These investigations have considered several types of design variables including the location and properties of noise treatment materials (9) and the geometric shape of enclosures and mufflers (10-12). In addition, Mollo and Bernhard formulated and utilized a boundary element method to find the optimal source strength and location of secondary sources for active noise control (13,14). For both the finite element and boundary element method, the most complicated design variables to incorporate into a design analysis are those associated with the geometry of the system. Both numerical techniques rely on a discretization of the geometry. Changes in geometry generally require changes in the data of the problem.

In this investigation, an acoustical shape design sensitivity analysis procedure based on a boundary element implementation of the Helmholtz integral equation will be developed and analytically verified. The method was developed as part of a structural/acoustic design methodology. Thus, in addition to the development of the shape sensitivity formulation, some aspects of the design methodology will also be discussed.

DESIGN SENSITIVITY ANALYSIS

An example of a simple sensitivity analysis will be presented to provide further explanation of the sensitivity analysis parameters under consideration here. A rectangular plate of dimensions c and d is to be optimized such that the acoustic intensity I is minimized at a given location. The area of the plate is allowed to change within predefined constraints on the plate dimensions. The desired sensitivity information is the partial derivative of the intensity with respect to the area of the plate. The acoustic intensity is defined as

$$I = \frac{1}{2} Re\left(p\bar{v}^{*}\right) \tag{1}$$

where Re denotes the real part of the expression, p is the acoustic pressure, $\bar{v}$ is the acoustic particle velocity and * denotes the complex conjugate. The design variable a is the area of the plate and is defined in terms of the dimensions of the plate, which are the primitive design variables.

$$a = A = cd \tag{2}$$

The sensitivity of the design objective function to the design variable is expanded in terms of the primitive design variables,

$$\frac{\partial I}{\partial a} = \frac{\partial I}{\partial A} = \frac{\partial I}{\partial c}\frac{\partial c}{\partial A} + \frac{\partial I}{\partial d}\frac{\partial d}{\partial A} \tag{3}$$

The expression can be further expanded in terms of the acoustic pressure and velocity as

$$\frac{\partial I}{\partial a} = \frac{1}{2} Re\left[\left(\frac{\partial p}{\partial c}\bar{v}^{*} + p\frac{\partial \bar{v}^{*}}{\partial c}\right)\frac{1}{d} + \left(\frac{\partial p}{\partial d}\bar{v}^{*} + p\frac{\partial \bar{v}^{*}}{\partial d}\right)\frac{1}{c}\right] \tag{4}$$

Thus, the required sensitivity information is obtained entirely in terms of the primitive variables; pressure, velocity and the plate dimensions.

One method of computing sensitivity information is to repeat the previous analysis steps of the design methodology with incremental changes in each of the design variables in turn. Essentially, iterations of the analyses are performed and derivatives computed using finite difference approximations. When the design variable is a shape change and the analysis technique is a numerical procedure such as the finite or boundary element method, the iteration technique becomes very cumbersome due to changes involving the data of model. A more efficient sensitivity analysis procedure for the boundary element method which utilizes the sensitivity of the boundary element matrices to the shape changes has been developed and is demonstrated in this paper.

The predicted sensitivity information can be used to achieve a better configuration using a systematic numerical optimization technique normally from the family of algorithms referred to as gradient search algorithms, or the sensitivity information can be used directly by a human expert. Numerical search algorithms are normally best applied in cases where the design objective function is smooth and regular and a large number of design variables are considered. For the human expert optimizer, the sensitivity information provides insight for determining which changes should be made in the design. Once these changes are identified, iteration of the analysis steps in the design methodology is usually performed until the design objective function meets a predetermined design criterion. Using either optimization procedure, the sensitivity analysis generally improves convergence toward an optimum design, provided accurate results are obtained by prudent selection of design variables and design variable increments and the design objective function is reasonably smooth.

SHAPE DESIGN SENSITIVITY FORMULATION

The direct method of design sensitivity analysis which has been applied by Zeinkiewicz and Campbell (15) for the finite element method will be developed here for the boundary element method acoustical analysis. Sensitivity of the primitive functions of acoustic design objectives (i.e., surface and field pressures) to a given design variable will be demonstrated. In addition to an acoustic sensitivity formulation for the standard boundary element analysis, two other sensitivity formulations will be presented for cases when the CHIEF method (1) of overdetermining the problem is used to resolve the non-uniqueness problem. Verification of the acoustic design sensitivity code developed from these formulations will be shown by using analytical sphere models.

As discussed by a number of previous investigators (1-5), the matrix form of the boundary element approximation to the surface Helmholtz equation (SHE), can be written as:

$$[C]p - [B]v_n = 0 \tag{5}$$

where p is the acoustic pressure and v_n is the normal acoustic velocity on the boundary. The coefficients of the $[C]$ and $[B]$ matrices are obtained from numerical evaluation of the integrals in the SHE formulation. The solution of equation 5 for the surface pressure can be written essentially as

$$p = [C]^{-1}\left\{[B]v_n\right\}. \tag{6}$$

although a matrix decomposition rather than a matrix inversion is usually computed.

The desired sensitivity information is the change of surface pressure due to a change in a given design variable a (i.e., $\partial p/\partial a$). The result of differentiating equation 5 with respect to a is

$$\frac{\partial[C]}{\partial a}p+[C]\frac{\partial p}{\partial a}=\frac{\partial[B]}{\partial a}v_n+[B]\frac{\partial v_n}{\partial a} \quad (7)$$

where $\partial[C]/\partial a$ and $\partial[B]/\partial a$ are the change in the coefficient matrices $[C]$ and $[B]$ due to a change in the design variable and are computed using finite difference approximations. The term $\partial v_n/\partial a$ is the change in the velocity normal to the boundary due to a change in the design variable. Rearranging equation 7 gives

$$\frac{\partial p}{\partial a}=[C]^{-1}\left\{\frac{\partial[B]}{\partial a}v_n+[B]\frac{\partial v_n}{\partial a}-\frac{\partial[C]}{\partial a}p\right\}. \quad (8)$$

The major advantage of using equation 8, as opposed to a simple recomputation of equation 6, to compute design sensitivity is the gain in efficiency from reusing the already decomposed form of the coefficient matrix (i.e., $[C]^{-1}$). Generally for the linear triangular elements used in this investigation, the computational cost of obtaining the matrix coefficients is significantly less than the cost of reducing the coefficient matrix. Therefore, $[C]^{-1}$ is saved in computer memory after initial evaluation of equation 6 and recalled for use in evaluating equation 8. The only extra calculations required are those involved with obtaining the sensitivity of the coefficient matrices to the defined design variable a.

Derivatives of the coefficient matrices are calculated by a first-order forward difference approximation, such that

$$\frac{\partial[C]}{\partial a}=\frac{[C]_{i+1}-[C]_i}{\Delta a} \quad (9a)$$

and

$$\frac{\partial[B]}{\partial a}=\frac{[B]_{i+1}-[B]_i}{\Delta a} \quad (9b)$$

where i denotes the initial matrix and $i+1$ indicates the matrix of the perturbed or incrementally changed state of the structure. Equations 9a and 9b require that new matrices be computed for the perturbed structure (i.e., $[C]_{i+1}$ and $[B]_{i+1}$) and that the matrices $[C]_i$ and $[B]_i$ be recomputed or stored after the initial evaluation of equation 6. The amount of finite difference truncation error and error due to the finite step size, Δa, used to perturb the structure will be determined by the properties of the matrix coefficients. If the matrix coefficients are smooth functions of the design variable, these errors can be minimized by using small incremental changes in the design variable.

SENSITIVITY FORMULATION WITH CHIEF

Equation 8 will produce the surface pressure sensitivities to the defined design variable when the surface Helmholtz equation alone is used. At frequencies where inaccuracies are expected due to the non-uniqueness problem, a different sensitivity formulation based on one of the techniques used to overcome the non-uniqueness problem must be utilized. In this investigation, the CHIEF method (1,17) was used. The matrix equation for the CHIEF method can be written as,

$$[K]p-[G]v_n=0 \quad (10a)$$

where p and v_n are the same variables used in equation 5 and,

$$[K]=\omega_1[C]^{*T}[C]+\omega_2[E]^{*T}[E] \quad (10b)$$

and,

$$[G]=\omega_1[C]^{*T}[B]+\omega_2[E]^{*T}[D] \ . \quad (10c)$$

The matrices $[E]$ and $[D]$ are generally non-square matrices whose coefficients have been computed using the interior Helmholtz equation with field points located inside the structure being analyzed. Weighting functions ω_1 and ω_2 have the effect of assigning confidence levels to the given equations.

Equation 10 can be differentiated with respect to the design variable and an expression for surface pressure sensitivity can be obtained in the same form as equation 8: i.e.,

$$\frac{\partial p}{\partial a}=[K]^{-1}\left\{\frac{\partial[G]}{\partial a}v_n+[G]\frac{\partial v_n}{\partial a}-\frac{\partial[K]}{\partial a}p\right\} \ . \quad (11)$$

Evaluation of equation 11 requires significant computational resources due to the matrix multiplications in equations 10b and 10c. However, because the pressure and velocity vectors (i.e., p and v_n) and the reduced coefficient matrix, $[K]^{-1}$, are known from the initial prediction, evaluation of equation 11 is less expensive than recomputing a new pressure response for each change in the design variable. For example, the initial CHIEF method requires the evaluation of $[C]^{*T}[C]$, which results in n^3 multiplications, where n is the number of elements in the model. For equation 11, the pressure vector $\{p\}$ is known and $[C]^{*T}[C]\{p\}$ is evaluated by first performing the operation $[C]\{p\}$, and then computing $[C]^{*T}\{Cp\}$. This method requires only $2n^2$ multiplications. Consequently, considerable computational savings is realized by performing shape sensitivity with equation 11 as opposed to recomputing the initial prediction. Nevertheless, a significant amount of computer time is needed to evaluate equation 11. Thus, an alternative method was investigated.

ALTERNATE SENSITIVITY FORMULATION WITH CHIEF

Equation 10 will produce a unique solution vector in the vicinity of the characteristic frequencies of the non-uniqueness problem (1,17). Once this unique solution vector is found by using the CHIEF method (i.e., from equation 10), it can be inserted in equation 8 which is formulated directly from the surface Helmholtz equation to obtain,

$$\frac{\partial p}{\partial a} = [C]^{-1}\left\{\frac{\partial [B]}{\partial a}v_n + [B]\frac{\partial v_n}{\partial a} - \frac{\partial [C]}{\partial a}p(\text{unique})\right\} \quad (12)$$

It should be noted that the reduced matrix $[K]^{-1}$ which is stored after completion of the initial CHIEF routine is different than the $[C]^{-1}$ required in equation 12. Thus, a separate matrix decomposition must be performed when evaluating equation 12. Essentially, solving for the sensitivity information in this manner requires about the same amount of computational time as the initial solution of the surface Helmholtz equation (i.e., equation 6) Nevertheless, for many boundary element models it is more efficient than the sensitivity solution of equation 11. The actual computational savings of the method, however, depends on the size of the model and the type of computer multiplication process used, (i.e., vectorized or scalar process.)

FIELD PRESSURE SENSITIVITY FORMULATION

Once the surface pressure sensitivity solution is obtained, the sensitivity of other acoustic parameters can be derived. For instance, if the pressure sensitivity at a field location is desired, the exterior Helmholtz equation can be formulated for computing shape sensitivity in much the same manner as used for the surface Helmholtz equation. The exterior equation can be written as,

$$[C_e]p_s - [B_e]v_{ns} = p_e \quad (13)$$

where the subscript e indicates terms due to the exterior Helmholtz equation and the subscript s designates the surface solution. The exterior pressure sensitivity to the design variable is

$$\frac{\partial p_e}{\partial a} = \frac{\partial [C_e]}{\partial a}p_s + [C_e]\frac{\partial p_s}{\partial a} - \frac{\partial [B_e]}{\partial a}v_{ns} - [B_e]\frac{\partial v_{ns}}{\partial a} \quad (14)$$

where the approximations

$$\frac{\partial [C_e]}{\partial a} = \frac{[C_e]_{i+1} - [C_e]_i}{\Delta a} \quad (15a)$$

and,

$$\frac{\partial [B_e]}{\partial a} = \frac{[B_e]_{i+1} - [B_e]_i}{\Delta a} \quad . \quad (15b)$$

are used. The vector terms in the right hand side of equation 14 are obtained from the surface sensitivity analysis (equations 8 , 11 or 12).

VERIFICATION OF THE SHAPE DESIGN SENSITIVITY PROGRAM

Analytical sphere models were used to verify the boundary element shape design sensitivity formulation. Boundary conditions were specified such that the spheres were modeled to vibrate with a uniform radial velocity. The spherical radius a was designated as the design variable. The partial derivative of the acoustic pressure at a point in the field with respect to the sphere radius was calculated and compared to the theoretical solution. For a sphere vibrating with velocity $u(a)$, the theoretical pressure sensitivity to a radius change is

$$\frac{\partial p}{\partial a} = \frac{e^{-jk(R-a)}}{R}(\rho_o c)\left[\frac{\partial u(a)}{\partial a}\frac{jka^2}{1+jka} + u(a)\left[\frac{-(ka)^2 + 2jka}{1+jka} + \frac{(ka)^2}{(1+jka)^2}\right]\right] \quad (16)$$

where R is a radial distance from the center of the sphere to a location in the field.

The sensitivity solution which is obtained by evaluation of equation 8 will be referred to as the DBEM solution since it is the basic sensitivity formulation of the boundary element method and does not incorporate either of the CHIEF sensitivity formulations. The boundary element program used for these investigations uses a triangular element where pressure and velocity are assumed to be constant on the surface of the element. This element is similar to the one used by Schenck (1), Meyer et al. (3), and Koopman et al. (5). The accuracy of this element is discussed by Bernhard et al. (16).

In figure 1 the theoretical solution (equation 16) for the sensitivity of the pulsating sphere with radius, $a=1$ meter, is compared with the DBEM solution for a 48 element model at a field point were $R=6$ meters. Both the real and imaginary pressure sensitivity solutions are shown. The DBEM solution was obtained by increasing the spherical radius by .01 meters, i.e., $\Delta a = .01m$ and assuming no velocity sensitivity, $\partial v_n/\partial a = 0$. Generally good agreement was found between the theoretical and DBEM solution. In earlier investigations of pulsating sphere models, Gardner et al. showed that the effective radius of these sphere models was the radius of the theoretical model which reproduces the same surface area as the numerical model (16). Figures 2 and 3 show the correlation between the numerical predictions and the theoretical solutions using two smaller radii. The radius $a=.8945$ is the radial location of the centroids of the elements. The theoretical solution for $a=.8945$ is compared with the predicted

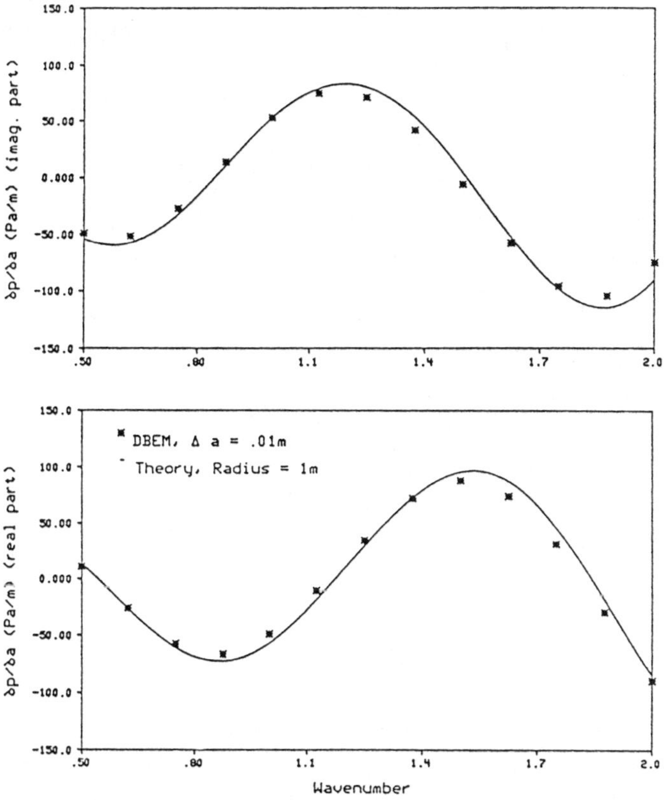

Fig. 1 Real and imaginary parts of the predicted field pressure sensitivity with respect to a radius change at radial distance of 6 meters. 48 element pulsating sphere model.

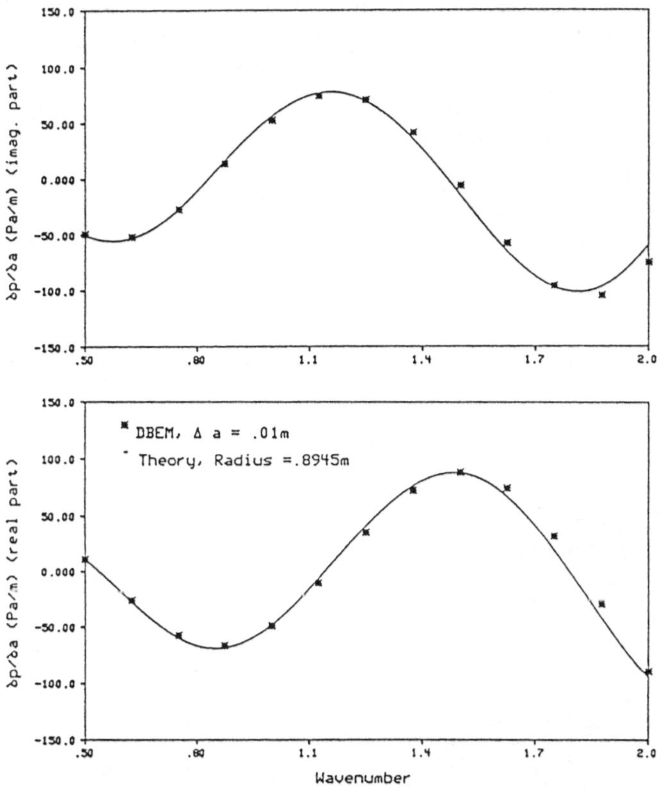

Fig. 2 Real and imaginary parts of the predicted field pressure sensitivity. 48 element pulsating sphere model. Theoretical sphere radius of .8945 meter.

sensitivity solution in figure 2. The results are better than the original results. However, the best results are obtained when the theoretical solution radius is $a = .94$ (figure 3). The radius $a = .94$ is the radius which produces the same surface area as the boundary element model. These sensitivity results are consistent with the results obtained by Gardner et al. for radiation predictions for the crude model used here.

A 256 element sphere model was used as an additional verification study. The 256 element model is expected to more closely approximate the theoretical sphere of 1 meter radius. Figure 4 shows field pressure sensitivity to a radius change ($\Delta a = .01m$) for the 256 element sphere. Also included in this analysis was a velocity design sensitivity, $\partial v/\partial a = (1.,0.)m$. The theoretical solution is obtained with a radius a of 1 meter. Excellent agreement between the numerical and theoretical solution is shown.

Figure 5 shows the magnitude of the sensitivity of field pressure to the change in radius for the 256 element sphere for a broader frequency band. Both DBEM (equation 8) and CHIEF (equation 11) sensitivity predictions are used for analysis with a radius change of .01 meter. For this sphere model the first frequency where the non-uniqueness problem occurs is at a wavenumber of approximately 3.2. It can be seen that the DBEM solution follows the theoretical solution closely except in the vicinity of the non-uniqueness problem, where the predicted solution is beyond the limits of the plot and becomes completely in error. The full CHIEF sensitivity solution is very accurate over the entire frequency range. If indicators of the non-uniqueness problem, such as the interior potential indicator developed by Seybert, et al. (17), can be used to identify where the problem occurs, then the more computationally intensive CHIEF sensitivity analysis need only be utilized for the few analysis frequencies in the vicinity of the non-uniqueness.

Figure 6 compares the alternate CHIEF sensitivity method (equation 12) with the full CHIEF method (equation 11) for the 256 element sphere model. At frequencies very close to the first frequency of the non-uniqueness problem, the alternate CHIEF sensitivity solution deviates slightly from the theoretical solution and from the full CHIEF solution. The deviation is relatively minor and occurs in a much narrower frequency band than the deviation of the DBEM solution shown in figure 5. These solution errors in the alternate CHIEF method are most likely due to the ill-conditioned $[C]$ matrix in the vicinity of the frequencies of the non-uniqueness problem. Significant errors

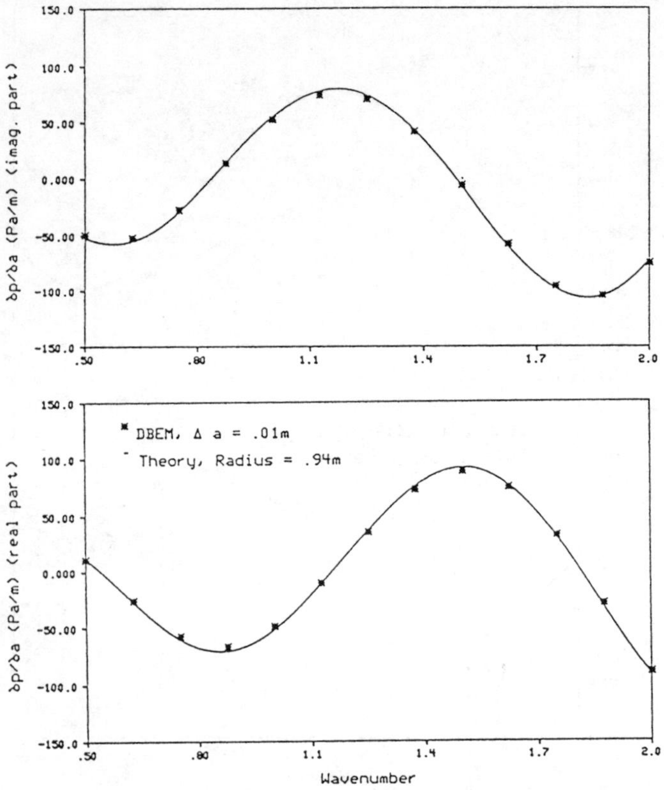

Fig. 3 Real and imaginary parts of the predicted field pressure sensitivity. 48 element pulsating sphere model. Theoretical sphere radius of .94 meter.

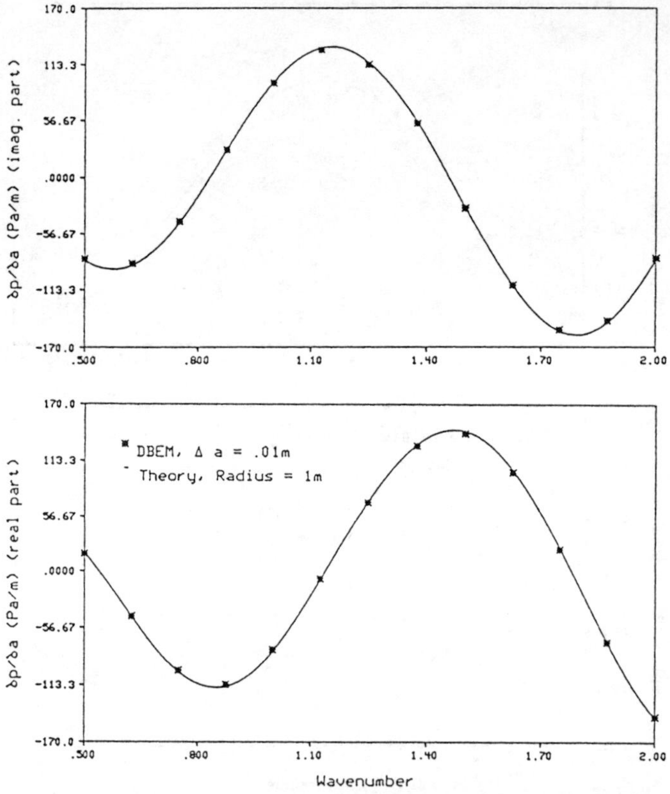

Fig. 4 Real and imaginary parts of the predicted field pressure sensitivity. 256 element pulsating sphere model. $\partial v/\partial a = (1 + j0)\,1/\text{sec}$.

did not occur at these frequencies as is the case with the DBEM solution. Figure 6 indicates that the alternate CHIEF solution is unique even at the characteristic frequencies. However, matrix ill-conditioning results in a poorer solution at these frequencies than the full CHIEF solution (equation 11).

The accuracy of the sensitivity computation methods are also dependent on factors such as step size (i.e., design variable increment), finite difference truncation errors, and numerical and modeling approximation errors. The influence of modeling approximation errors have been demonstrated with the 48 element sphere and were significantly reduced by using a 256 element sphere model. Step size and finite difference errors will be further discussed in the following paragraphs.

An alternative technique for computing the shape design sensitivity is to compute the solution at two values of the design variable, a_1 and a_2, and then compute the sensitivity using the solutions, p_1 and p_2, and a finite difference approximation. For example, the pressure sensitivity could be computed from

$$\frac{\partial p}{\partial a} = \frac{p_2 - p_1}{\Delta a} \tag{17}$$

Figure 7 shows a comparison between field pressure sensitivity (at $k=1$) obtained by the finite difference approximation and the design sensitivity formulation developed here. The finite difference approximation method is referred to here as the iteration technique. Field pressure sensitivity magnitudes are shown for various incremental changes in the radius. The theoretical sensitivity magnitude is 77.3 Pa/m at $k=1$. As Δa decreases, both solutions tend to converge to a value of 76.2 Pa/m. The difference between this value and the theoretical value is most likely due to numerical and modeling approximation errors. The iterated solution diverges more rapidly as Δa increases although both solutions are relatively accurate. The finite difference errors (i.e., errors due to a finite step size, Δa) from equation 17 are probably greater than the approximation errors due to the sensitivity formulation (equation 8). Although finite difference errors are also present in the DBEM sensitivity formulation (from equations 9 and 13) these errors are probably less because the difference equations are applied to the matrix constituents. The magnitude of an appropriate incremental change in the design variable (i.e., Δa) depends on the specific problem under investigation and the associated smoothness of the design objective function.

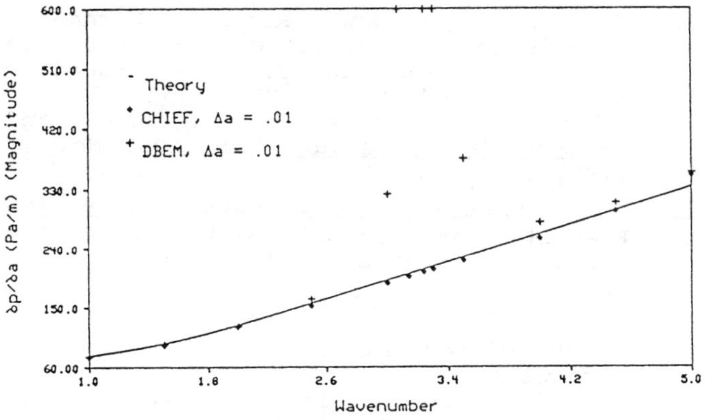

Fig. 5 Magnitude of predicted field pressure sensitivity for a 256 element pulsating sphere at radial distance of 6 meters. Comparison of DBEM and full CHIEF sensitivity solution.

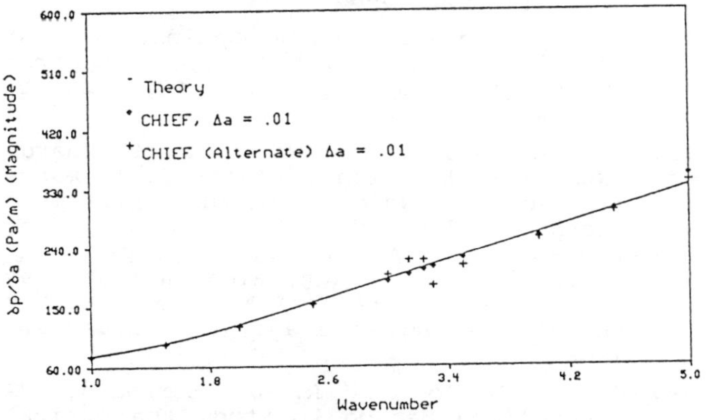

Fig. 6 Magnitude of predicted field pressure sensitivity for a 256 element pulsating sphere. Comparison of full CHIEF sensitivity method with alternate CHIEF method.

Figure 8 shows the field pressure sensitivity magnitudes in the vicinity of the first frequency of the non-uniqueness problem ($k = 3.14$) for increments of radius from .001 meters to .1 meter. The theoretical sensitivity magnitude is 201.2 Pa/m. The iterated solution is compared with the CHIEF sensitivity method and the alternate CHIEF sensitivity method. The iterated solution was obtained by using the CHIEF method for both the initial prediction (to obtain the field pressure p_1) and the prediction for the perturbed sphere due to the change in radius, (to obtain p_2). Equation 17 was then use to compute the field pressure derivative. Almost no difference is shown between the CHIEF method and the iterated solution. This is expected since both methods use the overdetermined CHIEF coefficient matrix $[K]$. For the alternate sensitivity CHIEF method, the coefficient matrix $[C]$ is used and is not consistent with the formulation of the other methods. The alternate CHIEF solution converges to a value of approximately 215 Pa/m which is approximately 7% error. This

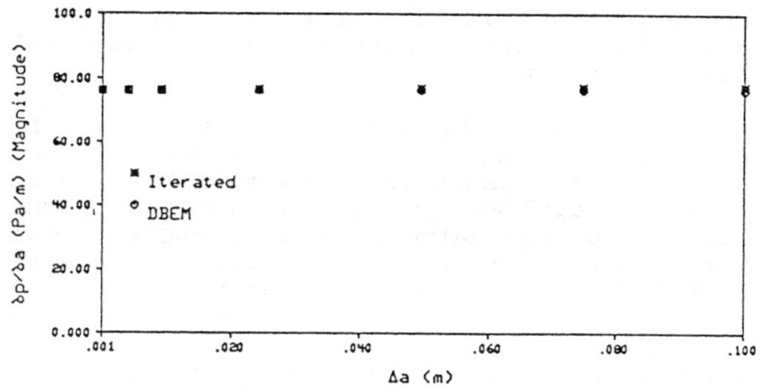

Fig. 7 Magnitude of DBEM field pressure sensitivity for different increments of radius change and comparison with iterated solution. 256 element pulsating sphere model.

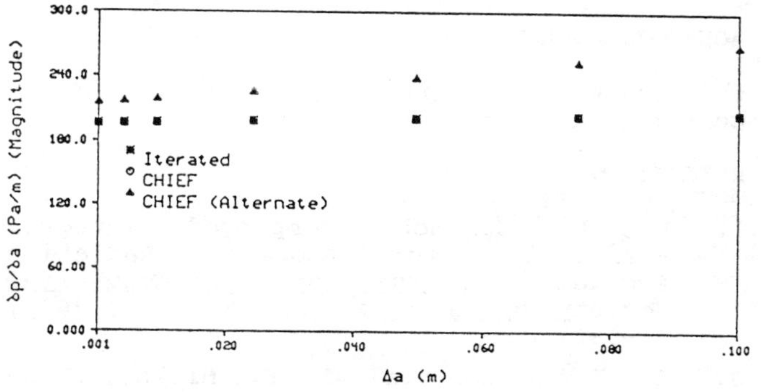

Fig. 8 Magnitude of CHIEF and Alternate CHIEF field pressure sensitivities for different increments of radius change and comparison with iterated solution. 256 element pulsating sphere model.

error is also evident in figure 6. It is apparent that the most accurate of the CHIEF solutions comes from the full CHIEF sensitivity formulation (equation 11). The CHIEF sensitivity formulation is less computationally intensive than the iteration (recomputation) method and provides a viable and accurate method for computing design sensitivities at any frequency.

CONCLUSIONS

Due to the nature of many design problems, optimization procedures can be done more efficiently when information about the sensitivity of parameters in the design objective function to the design variable(s) are known. Thus, a shape sensitivity analysis technique has been formulated for the boundary element method. Several alternative methods of computing the shape sensitivity with boundary element methods have been investigated. The most efficient method investigated here uses a finite difference approximation to compute the boundary element matrix sensitivity. The sensitivity analysis formulation has been verified using analytical pulsating sphere models for which theoretical solutions are available. Several investigations were also included to show the errors which result due to the finite difference approximation and numerical modeling errors.

The non-uniqueness problem of the Helmholtz integral equation necessitates an alternative sensitivity formulation which must be applied in the vicinity of the associated non-uniqueness frequencies. Two formulations which incorporate the CHIEF method have been developed and verified with sphere models. The CHIEF method which utilizes the reduced CHIEF coefficient matrix (stored in computer memory), provides the most accurate results. The method will also consume more computational time than the alternate CHIEF method for many models. The alternate CHIEF method is faster in some instances but produces a relatively small amount of error at frequencies near the frequencies of the non-uniqueness problem.

ACKNOWLEDGMENT

This work was supported by the Ford Motor Company, Dearborn, Mi.

REFERENCES

1. H. A. Schenck, "Improved Integral Formulation for Acoustic Radiation Problems," Journal of the Acoustical Society of America, 44, No. 1 (1968), 41-58.
2. A. J. Burton and G. F. Miller, "The application of integral equation methods to the numerical solution of some exterior boundary-value problems," Proc. Roy. Soc. Lond., A. 323, (1971), 201-210.
3. W. L. Meyer, W. A. Bell, B. T. Zinn and M.P. Stallybrass, "Boundary Integral Solutions of Three Dimensional Acoustic Radiation Problems," Journal of Sound and Vibration, 59(2), (1978), 245-262.
4. A. F. Seybert, B. Soenarko, F. J. Rizzo and D. J. Shippy, "Application of the BIE Method to Sound Radiation Problems Using An Isoparametric Element," ASME paper no. 82-WA/NCA-1, 1982.
5. G. H. Koopmann and H. Benner, "Method for computing the sound power of machines based on the Helmholtz integral," Journal of the Acoustical Society of America, 71, No. 1 (1982), 78-89.
6. P. J. T. Filippi, "Layer Potentials and Acoustic Diffraction," Journal of Sound and Vibration, 54(4), 473-500, (1977).
7. X. F. Wu, A. D. Pierce and J. H. Ginsberg, "Application of a Variational Principle to Acoustic Radiation from a Vibrating Cylinder," Journal of the Acoustical Society of America, 79(S1), S35-6.
8. C. H. Kipp and R. J. Bernhard, "Prediction of Acoustical Behavior in Cavities Using an Indirect Boundary Element Method," ASME Journal of Vibrations, Acoustics, Stress and Reliability in Design, Vol. 109, No. 1, (1987), 22-28.
9. R. J. Bernhard and S. Takeo, "A Finite Element Procedure for Design of Cavity Acoustical Treatments," Journal of the Acoustical Society of America, Vol. 83, No. 6 (June 1988), 2224-2232.
10. J. M. Milner and R. J. Bernhard, "An Investigation of the Modal Characteristics of Non-Rectangular Reverberation Rooms," Journal of the Acoustical Society of America, Vol. 85, No. 2 (Feb. 1989), 772-779.
11. R. J. Bernhard, "Synthesis of Acoustical Shapes Using the Finite Element Method," Journal of Sound and Vibration, Vol. 98, No. 1 (1985), 55-65.
12. R. J. Bernhard, "Shape Optimization of Reactive Mufflers," Noise Control Engineering Journal, Vol. 27, No. 1, July-August, 1986, 10-17.
13. C. G. Mollo and R. J. Bernhard, "Numerical Evaluation of the Performance of Active Noise Control Systems," accepted for the ASME Journal of Vibrations, Acoustics, Stress and Reliability in Design, also published as ASME Paper No. 88-WA/NCA-5, ASME Winter Annual Meeting, Chicago, IL, December, 1988.
14. C. G. Mollo and R. J. Bernhard, "A Generalized Method to Study the Optimal Performance of Active Noise Controllers in Three-Dimensional Enclosures," accepted to the AIAA Journal, also published as AIAA Paper No. AIAA-87-2705, AIAA 11Th. Aeroacoustics Conference, Palo Alto, CA, (October 1987).
15. O. C. Zienkienwicz and J. S. Campbell, "Shape Optimization and Sequential Linear Programming," in R. H. Gallagher and O. C. Zienkiewicz (eds), Optimal Structural Design, Wiley, New York 1973, Chap. 7.
16. R. J. Bernhard, B. K. Gardner, C. B. Mollo and C. R. Kipp, "Prediction of Sound Fields in Cavities Using Boundary Element Methods," AIAA Journal, Vol. 25, No. 9, 1176-1183.
17. A. F. Seybert and T. K. Rengarajan, "The Use Of CHIEF To Obtain Unique Solutions For Acoustic Radiation Using The Boundary Integral Equations," Journal of the Acoustical Society of America, 81, No. 5 (1987), 1299-1306.

TWO-DIMENSIONAL RADIATION AND SCATTERING AT SHORT WAVE LENGTH

W. S. Yoon, J. M. Park and W. Eversman
Mechanical and Aerospace Engineering and Engineering Mechanics
University of Missouri-Rolla
Rolla, Missouri

ABSTRACT

The problem of radiation and scattering by objects in the case when the wave length is much smaller than the characteristic dimension of the radiator or scatterer is investigated. The boundary element method is used with the goal of obtaining accurate results in an efficient computational scheme for wave lengths less than 5% of the object characteristic dimesion. With a systematic application of conventional boundary element techniques it is found that the modelling of such problems leads to excessive computation time due to the large number of elements, demands on Gaussian Quadrature, evaluation of the fundamental solution, and the resulting large non-symmetric matrix equation.

The approach is to use cubic elements, approximate polynomial and asymptotic evaluations of the fundamental solution, and to tailor the order of the Gaussian Quadrature according to the local demands dictated by the distance between sending and receiving points. In addition, out of core equation solvers are investigated.

It is found that results can be obtained which are as accurate as those obtained using conventional Boundary Element techniques, but at greatly reduced cost. The potential application is to problems involving propagation of relatively low frequency sound over large terrain features.

INTRODUCTION

The Boundary Element Method (BEM) has been used widely to predict the sound propagation from an object which is of arbitrary shape and submerged in an infinite acoustic medium. The major advantage is that the complete domain of interest need not be discretized.

Previous work [1] – [4] has presented general methods applicable to acoustic radiation and scattering problems with Dirichlet or Neuman boundary conditions, which use the surface Boundary Integral Equation (BIE) method with an isoparametric element discretization technique in which both the surface acoustic variables and surface geometry are approximated using the same quadratic shape function. However, the BIE method appears to lose some of its attractiveness when it is applied to two dimensional radiation and scattering problems at high frequency. For this problem the surface of the generator must be discretized into moderately sized segments to include an adequate number of waves to approximate the acoustical variables at a given number of integration points. The fundamental solution (Hankel function) is required in the evaluation of the boundary integral equation and due to its form as an infinite polynomial, leads to time consuming computations.

This paper is concerned with radiated and scattered acoustic waves at short wave length from bodies in a two dimensional acoustic wave-conducting medium. The domain to be considered is the space exterior to the surface of the generator including the boundary. Certain conditions are prescribed on the boundary, such as the normal derivative of the velocity potential for radiation and the velocity potential for known incident waves on the rigid obstacle for scattering. The integral on the surface of the generator is performed by dividing the boundary into segments using cubic isoparametric line elements and applying Gaussian Quadrature. While retaining the required order of numerical integration to set up the linear algebraic equation for a given point, asymptotic formulae for Bessel Functions of the first and second kind of order zero replace the fundamental solution for the two dimensional Helmholtz equation. Numerical integration in the case of the sending point close to the receiving point is implemented by using very accurate approximate polynomials for the Bessel Functions and by taking more quadrature points. This is similar in principal to an idea previously introduced by Koopman and Benner [4].

The advantage of this computation scheme is that the asymptotic formulae and approximate polynomials are simple finite polynomials and lead to greatly reduced cost for computation. By making use of these polynomials, which have periodic behavior like the exact fundamental solution (Hankel function) for large arguments, the efficiency of the computer code can be improved while retaining accuracy comparable with the results obtained by using the conventional fundamental solution.

FORMULATION

Consider the linearized acoustic wave equation which, for the time harmonic case, is reduced to the Helmholtz equation,

$$\nabla^2\phi + k^2\phi = 0, \tag{1}$$

within the two-dimensional acoustic medium Ω, as shown in Fig.1, with

$$\nabla\phi \cdot \vec{v} = 0 \tag{2}$$

on sections of the surface which are rigid, and with

$$ik\phi + \frac{Z}{\rho_o c}\frac{\partial\phi}{\partial v} = 0 \tag{3}$$

at sections on the boundary where particle velocity is specified or implemented by specifying the impedence boundary condition. In equation (1)-(3), ϕ is the velocity potential, present in the region Ω exterior to the surface S, k is the wave number, ω/c, where c is the speed of sound, ω is the frequency, and $\vec{v}$ is the unit normal on S directed away from B (Fig.1). The body B is refered to as the generator. Z is the impedence and ρ_o is the ambient density of the exterior fluid medium, here assumed to be air. The BEM formulation is well known [1], however for completeness a brief summary is given here.

The weighted residual method is applied by the introduction of the weighting functions ϕ_o[5] such that

$$\int_\Omega \phi_o(\nabla^2\phi + k^2\phi)d\Omega = 0 \tag{4}$$

By employing integration by parts two times and using the divergence theorem which relates the boundary integral over S to the domain integral over

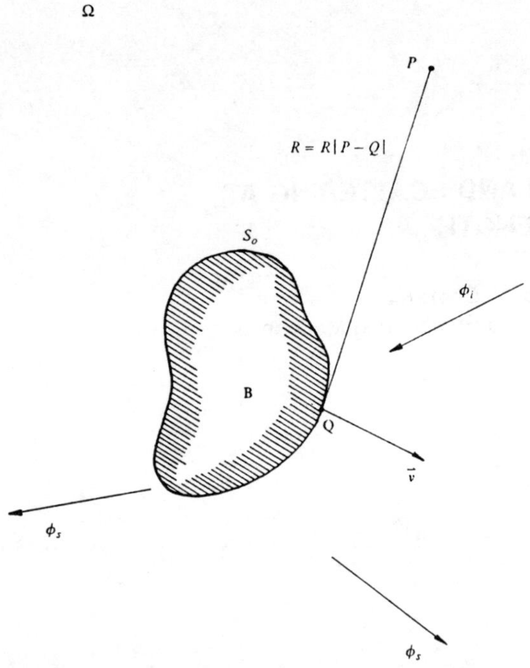

Figure 1. Boundary definition in 2-D acoustic medium

Ω bounded by S for any two functions ϕ and ϕ_o sufficiently smooth and nonsingular in Ω, the basic formulation for the boundary integral equation can be written as

$$\int_S \{\phi\frac{\partial \phi_o}{\partial v} - \phi_o\frac{\partial \phi}{\partial v}\}dS = \int_\Omega \phi(\nabla^2\phi_o + k^2\phi_o)d\Omega, \tag{5}$$

where $\vec{v}$ is the outward normal of S . To reduce the problem to a boundary integral formulation one needs to select the fundamental solution ϕ_o for the Helmholtz equation [5] which satisfies the governing equation in the domain and represents the solution for a concentrated potential at a point P such that

$$\nabla^2\phi_o + k^2\phi_o + \Delta^P = 0, \tag{6}$$

where Δ^p is a Dirac delta function. The solution of Eqn. (6) is the fundamental solution for the two dimensional Helmholtz equation. Due to the sampling property of the Dirac delta function, it follows that

$$\int_\Omega (\nabla^2\phi_o + k^2\phi_o)\phi d\Omega + \phi^P = 0 \tag{7}$$

where ϕ^p represents the value of the unknown function ϕ at the point under consideration. From Eqn. (5)-(7)

$$\phi^P = \int_S \{\frac{\partial \phi}{\partial v}\phi_o - \phi\frac{\partial \phi_o}{\partial v}\}dS \tag{8}$$

The Hankel function of the first kind of order zero

$$H_o^{(1)}(kR) = J_o(kR) + iY_o(kR), \tag{9}$$

where R is the distance between P and Q in Ω, is the fundamental solution for the two-dimensional Helmholtz equation.

By dividing the velocity potential ϕ^p at the sending point P of Eqn.(8) into the scattered velocity potential and the potential created by the prescribed incident wave, the general Helmholtz integral equation can be written [6].

$$C(P)\phi(P) = \int_S (\phi\frac{\partial \phi_o}{\partial v} - \phi_o\frac{\partial \phi}{\partial v})dS + \phi^I(P) \tag{10}$$

The validity of the Helmholtz integral equation is limited to an acoustic medium Ω exterior to a finite generator B in the presence of the prescribed incident wave ϕ^I(*Fig*.1). In Eqn.(10) C(P) is the parameter determined by the location of the point P; 1.0 for the field point outside the body; 0.5 for the field point on the surface of the body; 0 for the field point inside the body [7]. Careful consideration is necessary when the sending point P is located on the surface of the generator. If there is no unique tangent plane at point P, the limit process using a bubblelike hemisphere is no longer valid [5]. In this case, C(P) has the interpretation of the outer solid angle of the two adjacent surfaces. So, if P is at the corner or edge on S, the outer solid angle can be calculated by subtracting the inner solid angle from the value of C(P) for a point P outside the body.

$$C(P) = 1 - C^o(P) \tag{11}$$

where $C^o(P)$ indicates the inner solid angle. Now, since the limit processes depend only on the order of singularity of ϕ_o, which is the same for both the Laplace and Helmholtz operators [6], the inner solid angle can be represented by the counterpart of C(P) in Eqn. (10) for a function satisfying Laplace's equation. Therefore

$$C^o(P) = -\int_S \frac{\partial}{\partial v}(\frac{1}{2\pi}\ln R)dS \tag{12}$$

Now, Eqn. (4) can be rewritten into the complete form of the Helmholtz integral equation

$$\{1 + \int_S \frac{\partial}{\partial v}(\frac{1}{2\pi}\ln R)dS\}\phi(P) = \int_S \{\phi\frac{\partial \phi_o}{\partial v} - \phi_o\frac{\partial \phi}{\partial v}\}dS + \phi^I(P) \tag{13}$$

The method of solution using the Helmholtz integral equation for a specific scattering problem begins with specification of a part of the boundary conditions and corresponding velocity potential prescribed by the incoming wave ϕ^I [8] on the surface of the scatterer. The incoming wave is in the negative direction and is described by $\phi^I = \phi_i\exp(ikx)$. After setting up the linear algebraic matrix equation implied by the discretization of Eqn.(13), the unknown is solved for. Once the potential on the surface of the scatterer is obtained, the source point is placed in Ω to obtain the velocity potential ϕ at any desired exterior field point P. For radiation the approach is identical except that there is no incident wave.

NUMERICAL IMPLEMENTATION

A numerical representation of Eqn. (8) is needed for a body of arbitrary shape. For a boundary in the two-dimensional acoustic medium, the surface is discretized into segments by using line elements. Let four nodes be defined on each element (see Fig.2). Then, actual coordinates X and Y of any receiving point Q on the surface of the generator are assumed to be given in terms of the nodal coordinates X_α and Y_α by using piecewise interpolation functions [9]

$$X(\rho) = \sum_{\alpha=1}^{4} N_\alpha(\rho)X_\alpha \tag{14}$$

$$Y(\rho) = \sum_{\alpha=1}^{4} N_\alpha(\rho)Y_\alpha \tag{15}$$

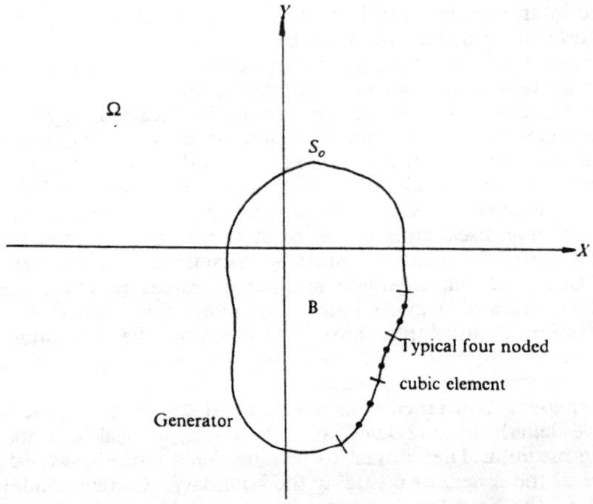

Figure 2. Mesh of the cubic element on the surface

$N_\alpha(\rho)$ are the cubic shape functions given by [18];

$$N_i(\rho) = \prod_{i \neq j} \frac{(\rho - \rho_j)}{(\rho_i - \rho_j)} \qquad i,j = 1,2,3,4 \tag{16}$$

where ρ is the local coordinate, $-1 \leq \rho \leq 1$ and $\rho_1 = -1$, $\rho_2 = -1/3$, $\rho_3 = 1/3$, $\rho_4 = 1$.

Equation (14) and (15) map each element on the generator onto a parent element which is a straight line in the local ρ, η coordinate system. The same procedure is repeated to approximate the boundary variables at the local coordinate ρ

$$\phi_m(\rho) = \sum_{\alpha=1}^{4} N_\alpha(\rho)\phi_{m\alpha} \tag{17}$$

$$\frac{\partial \phi_m(\rho)}{\partial v} = \sum_{\alpha=1}^{4} N_\alpha(\rho)\frac{\partial \phi_{m\alpha}}{\partial v} \tag{18}$$

where $\phi_{m\alpha}$ and $\partial\phi_{m\alpha}/\partial v$ are the values of ϕ and $\partial\phi/\partial v$ respectively, at node α on element m. Application of Eqn. (14)-(18) to Eqn. (13) yields

$$\{1 + \sum_{m=1}^{N} \int_{-1}^{1} \frac{\partial}{\partial v}(\frac{1}{2\pi} \ln R(\rho))J_m(\rho)d\rho\}\phi(P) = \sum_{m=1}^{N}\sum_{\alpha=1}^{4} \phi_{m\alpha} \int_{-1}^{1} \frac{\partial \phi_o}{\partial v} N_\alpha(\rho)J_m(\rho)d\rho$$

$$- \sum_{m=1}^{N}\sum_{\alpha=1}^{4} \frac{\partial \phi_{m\alpha}}{\partial v} \int_{-1}^{1} \phi_o N_\alpha(\rho)J_m(\rho)d\rho + \phi^I(P) \tag{19}$$

where N is the number of elements used to discretize the generator and $J_m(\rho)$ is the Jacobian of the transformation given by Eqn. (14) and (15) for element m [7].

$$J_m(\rho) = [(\frac{dx}{d\rho})^2 + (\frac{dy}{d\rho})^2]^{\frac{1}{2}} \tag{20}$$

A numerical presentation for $\vec{v}$ is needed to describe the curved surface by using the given nodes in each element [10]. Such a representation is

$$\vec{v}(\rho) = (-\frac{dy}{d\rho}\vec{i} + \frac{dx}{d\rho}\vec{j})\frac{1}{J_m(\rho)} \tag{21}$$

While keeping the standard procedure for setting up the linear algebraic equations, special treatment of the fundamental solution is implemented. As was mentioned previously, the fundamental solution for the two dimensional Helmholtz equation is the Hankel function of the first kind and zero order,

$$\phi_o(kR) = \frac{i}{4}H_o^{(1)}(kR). \tag{22}$$

The main modification lies in the reduction of the infinite polynomial representation of the Hankel Funtion to a much simpler finite polynomial [4]. Local demand for accuracy is addressed by using very accurate polynomial [11] for the Bessel Functions of the first and second kind. Except at these points the Hankel function is replaced by asymptotic formulae for every evaluation. The asymptotic formula for the Hankel function of the first kind is [12]

$$\phi_o(kR) = \frac{i}{4}(\frac{2}{\pi kR})^{\frac{1}{2}} e^{i\beta_n} f_n(kR) \tag{23}$$

where

$$\beta_n = kR - \frac{1}{2}(n + \frac{1}{2})\pi \tag{24}$$

$$f_n(kR) = P_n(kR) + iQ_n(kR) \tag{25}$$

In Eqn. (22)-(24), n is the order of the Hankel function of the first kind and $f_n(kR)$ in Eqn. (25) is the infinite polynomial which converges to unity as the distance R goes to infinity. Only one or two terms of the polynomial $f_n(kR)$ are taken as the coefficient for the asymptotic evaluation.

To avoid excessive computations for numerical integration, the number of quadrature points is chosen according to the demand for accuracy. Generally more points are needed to describe the rapid variation of the fundamental solution when the sending point P approaches the receiving point Q. That is, when R approaches zero, more quadrature points are needed to implement the singularity of the fundamental solution.

Now let P be a node identified by the global number j, then Eqn. (19) gives

$$\sum_{m=1}^{N}\sum_{\alpha=1}^{4} \phi_{m\alpha} a_{mj}^{\alpha} - \phi_j(1 + \sum_{m=1}^{N} C_{mj}) = \sum_{m=1}^{N}\sum_{\alpha=1}^{4} \frac{\partial \phi_{m\alpha}}{\partial v} b_{mj}^{\alpha} - \phi_j^I \tag{26}$$

where

$$a_{mj}^{\alpha} = -\sum_{l=1}^{L} \frac{i}{4} k(\frac{2}{\pi kR(\rho_l)})^{\frac{1}{2}} e^{i\beta_1} f_1(kR(\rho_l)) \frac{\partial R(\rho_l)}{\partial v} N_\alpha(\rho_l)J_m(\rho_l)w_l \tag{27}$$

$$b_{mj}^{\alpha} = \sum_{l=1}^{L} \frac{i}{4}(\frac{2}{\pi kR(\rho_l)})^{\frac{1}{2}} e^{i\beta_o} f_o(kR(\rho_l))N_\alpha(\rho_l)J_m(\rho_l)w_l \tag{28}$$

$$C_{mj} = \sum_{l=1}^{L} \frac{1}{2\pi R(\rho_l)} \frac{\partial R(\rho_l)}{\partial v} J_m(\rho_l)w_l \tag{29}$$

In Eqn. (27)-(29), L is the number of the sampling points for Gaussian Quadrature according to the demand for accuracy , ρ_l is the local coordinate of the quadrature point and w_l is the appropriate weight corresponding to the quadrature point.

For a total number of nodes M, we can produce M simultaneous linear algebraic equations in terms of the boundary conditions and unknowns by applying Eqn. (26) for every node defined in the discretization scheme. By adopting the global designation for $\phi_{m\alpha}$ and $\partial\phi_{m\alpha}/\partial v$, Eqn. (26) can be written in matrix form as

$$[A]\{\phi\} = [B]\{\frac{\partial \phi}{\partial v}\} + \{\phi^I\} \tag{30}$$

where the elements of matrices A and B are made up of combination of the expressions in Eqn. (27)-(29), and the elements of the matrices $\{\phi\}$, $\{\partial\phi/\partial v\}$ and $\{\phi_I\}$ are the values of ϕ, $\partial\phi/\partial v$ and ϕ_I, respectively at each node.

OUT-OF-CORE EQUATION SOLVER

The solution of the full non-symmetric matrix systems encountered in the present application of the Boundary Element Method requires a large amount of computer storage. Thus, an out-of-core equation solver was developed to solve the large full unsymmetric matrix equation by using the block elimination technique [13]and [14]

The algebraic equation can be expressed in matrix form composed of the submatrices A_{ij} and subvectors U_i and F_i

$$\begin{bmatrix} A_{11} & A_{12} & \cdot & \cdot & \cdot & \cdot & A_{1M} \\ A_{21} & A_{22} & \cdot & \cdot & \cdot & \cdot & A_{2M} \\ \cdot & \cdot & \cdot & \cdot & \cdot & \cdot & \cdot \\ A_{M1} & A_{M2} & \cdot & \cdot & \cdot & \cdot & A_{MM} \end{bmatrix} \begin{Bmatrix} U_1 \\ U_2 \\ \cdot \\ U_M \end{Bmatrix} = \begin{Bmatrix} F_1 \\ F_2 \\ \cdot \\ F_M \end{Bmatrix} \tag{31}$$

where the dimension of A_{ij} is $N \times N$ and the dimension of U_i and F_i is $N \times 1$. The dimension of the matrix A is $(N \times M) \times (N \times M)$ where M is the number of submatrices in rows or columnns. In this technique, the submatrices A_{ij} are operated on in a manner similar to elimination of coefficients in the Gaussian elimination technique. The submatrices in the Mth row or column may have zero elements, because the actual size of the matrix is less than or equal to $N \times M$ and greater than $N \times (M-1)$. Thus, when the submatrix $A_{M,M}$ is used as the pivotal submatrix, the dimension of $A_{M,M}$ should be reduced to remove the zero elements for obtaining $A_{M,M}^{-1}$. The equation solvers in Reference [13] – [14] are used after all the submatrices and subvectors of the system are prepared and stored in the external file. These solvers need to retrieve the sbmatrices for use from the external file and store the output submatrices except for the null and indentity submatrices in the external file for every elmination procedure.

For the 2-D problem using BEM, the matrix can be assembled row by row, independently of other rows. Thus, the block elimination technique can be applied when each of the M rows is prepared, instead of when the complete matrix A and vector F is prepared. The in-core storage areas $N \times (N \times M + 1)$ for the submatrices A_{ij} and subvector F_i in every row of A are used as the work area for the block elimination, and an additional work area $2(N \times N)$ is needed. One $N \times N$ is used for the inverse matrix of the pivotal submatrix, and the other $N \times N$ is used for the submatrix multiplication and for reading the data from the external file. The complete block elimination for each row can be performed in this work area without storing the original submatrices and the submatrices produced from the matrix multiplication before the completion of the block elimination.

The block elimination procedure for the 2-D BEM problem is described in what follows; The submatrices A_{1j} and subvector F_1 in the first row of Eqn. (31) are computed by BEM and stored in the work area. The premultiplication of the first row by A_{11}^{-1} gives

$$[I] \quad A_{11}^{-1}A_{12} \quad A_{11}^{-1}A_{13} \quad ... \quad A_{11}^{-1}A_{1M} \qquad A_{11}^{-1}F_1$$

or

$$[I] \qquad \bar{A}_{12} \qquad \bar{A}_{13} \quad ... \quad \bar{A}_{1M} \qquad \bar{F}_1 \tag{32}$$

where $\bar{A}_{ij}$ and $\bar{F}_i$ denote the final submatrices and subvector after the completion of the block elimination. After the results in Eqn. (32) are obtained and stored for the next use in the external file, the A_{2j} and F_2 in the 2nd row of Eqn. (31) are created and stored in the work area. The premultiplication of Eqn. (32) by A_{21} and the substraction from the original 2nd row yields

$$[0] \quad A_{22}-A_{21}\bar{A}_{12} \quad A_{23}-A_{21}\bar{A}_{13} \quad ... \quad A_{2M}-A_{21}\bar{A}_{1M} \qquad F_2-A_{21}\bar{F}_1$$

or

$$[0] \qquad A'_{22} \qquad A'_{23} \quad ... \quad A'_{2M} \qquad F'_2 \tag{33}$$

In this procedure, $\bar{A}_{1j}$ is retrieved from the external file and stored for use in the work area $N \times N$ for obtaining the corresponding A'_{2j}. The premultiplication of Eqn. (33) by A'^{-1}_{2j} yields

$$[0] \quad [I] \qquad A'^{-1}_{22}A'_{23} \quad ... \quad A'^{-1}_{22}A'_{2M} \qquad A'^{-1}_{22}F'_2$$

or

$$[0] \qquad [I] \qquad \bar{A}_{23} \quad ... \quad \bar{A}_{2M} \qquad \bar{F}_2 \tag{34}$$

The final results of the 2nd row shown in Eqn. (34) are stored for the next use in the external file. Carrying out the procedure as described above yields the final form of A after the completion of the block elimination.

$$\begin{bmatrix} [I] & \bar{A}_{12} & \cdots & \bar{A}_{1,M-1} & \bar{A}_{1,M} \\ [0] & [I] & \cdots & \bar{A}_{2,M-1} & \bar{A}_{2,M} \\ \vdots & \vdots & & \vdots & \vdots \\ [0] & [0] & \cdots & [I] & \bar{A}_{M-1,M} \\ [0] & [0] & \cdots & [0] & [I] \end{bmatrix} \begin{Bmatrix} U_1 \\ U_2 \\ \vdots \\ U_{M-1} \\ U_M \end{Bmatrix} = \begin{Bmatrix} \bar{F}_1 \\ \bar{F}_2 \\ \vdots \\ \bar{F}_{M-1} \\ \bar{F}_M \end{Bmatrix} \tag{35}$$

The final submatrices $\bar{A}_{ij}$ and subvectors $\bar{F}_i$ in Eqn. (35) are stored in the external file when the block elemination for each row is comlpeted, and are retrieved to compute the final submatrices and subvector for the next row. For example, the final submatrices and subvector in the 1st row are used to obtain those in the 2nd row, and the 1st and 2nd are used for the 3rd, and so on. Since $\bar{F}_M$ is the solution to U_M, the U_{M-1} can be obtained by back substitution

$$U_{M-1} = \bar{F}_{M-1} - \bar{A}_{M-1,M}\bar{F}_M$$

In general, the solution to the equation can be expressed as

$$U_{i-1} = \bar{F}_{i-1} - \sum_{j=i}^{M} \bar{A}_{j-1,j}\bar{F}_j \qquad i = 2,3, ... , M$$

The out-of-core equation solver needs more computation time than in-core solver, because the computation time is influenced by reading of data from and writing data to the external file. The out-of-core solvers in Reference [13] and [14] are based on the conventional Gaussian elimination technique, in which the 1st submatrices from the 2nd row to the Mth are made null by using the 1st submatrix of the 1st row as a pivotal submatrix, and the 2nd submatrices from the 3rd row to Mth are made null by using the 2nd submatrix of the 2nd row as a pivot, and so on. This technique needs a much larger amount of reading of data from and writing to the external file than the technique used in this paper, even though the same work area as used in this paper is used. The block elimination technique used in this paper can be applied for solving any kind of large algebraic equation after the matrix is computed. For the 2-D BEM problem, this technique reduces the conventional external storage to less than 50 %, because only the final effective submatrices in Eqn. (35), not the full matrix in Eqn. (31), are stored in the external file. Also, if this is used for an in-core job, the conventional in-core storage for solving the equation is reduced to about 60 %.

RESULTS AND DISCUSSION

In this section we examine the BEM formulation, Eqn. (13), by solving two dimensional radiation and scattering at high frequency in which the object is a rigid cylinder. A number of test cases were run to demonstrate the advantages of using the approximate polynomials and asymptotic formulae, and of tailoring the order of the Gaussian Quadrature.

Each case was run in four different computer codes and yielded results about the computation time for setting up the linear algebraic matrix equation (30). The accuracy compared with the analytical result for a given problem was plotted for comparison. An odd number of sampling points were taken for the Gaussian Quadrature, since the coordinates of an even number of Quadrature points have the possibility of being very close to coordinates of the nodal points in the cubic element, and therefore produce unpredictable errors. While taking a relatively small number of Quadrature points, a large number of elements were selected to include the proper number of waves in one element. Integration orders were chosen according to the demands of the accuracy required to describe the variations of the fundamental solution in one element. This computational scheme is based on the fact that the asymptotic form of the fundamental solution becomes progressively more simple as kR increases. Numerical experiments show that the integrations for the elements which contain the singular point or points very close to it, are critical to successfully computing the Helmholtz integral due to the singularity inherent in the imaginary part of the fundamental solution (Bessel functions of the second kind). Thus, more quadrature points are taken to predict the singular behavior of $\phi_o(kR)$ and $\partial\phi(kR)/\partial v$ in Eqn. (13) when the sending point approaches very close to the receiving point.

Test cases are as follows; (A) ϕ_o is taken as the exact Hankel function expansion. Numerical integration is performed by taking 7 Quadrature points for all elements on the surface; (B) ϕ_o is represented by the asymptotic form of Bessel functions, and therefore the Hankel function, except when kR in Eqn.(21) is less than 3, in which case ϕ_o is replaced by approximate polynomials for the Bessel functions. Seven quadrature points in each element were used. ; (C) Same as case (B) except that 5 points are sampled for Gaussian Quadrature when the receiving point is more than one element apart from sending point; (D) Same as (C) except that 3 Quadrature points were taken when the receiving point is more than one element apart from the sending point. By taking advantage of the symmetry of the object, the discretization scheme covers only half of the boundary for the problem considered.

The first test case is the scattering of a plane acoustic wave from a rigid cylinder. The problem was run for ka = 5,10,25,50 and 100, where "a" is the radius of the cylinder. The results are shown in Fig. (3)-(7) and Tables (1)-(5). In Fig.(3)-(7) the distance from the origin represents the magnitude of the normalized scattered potential ϕ_s/ϕ_i on the sending point at a distance of five radii and a given polar angle. Here, ϕ_s and ϕ_i are the scattered and the incident potentials and the direction of the incoming wave is shown by the arrow on a small inset in the figure. That is, the vertical axis Y is $|\phi_s/\phi_i| \sin(\theta)$ and the horizontal axis X is $|\phi_s/\phi_i| \cos(\theta)$, where θ is the given angle of the sending point from the plane which is parallel to the direction of the incoming incident wave. The results of cases (A)-(D) plotted as solid lines, which are nearly indistinguishable, are compared with the analytical results plotted as dotted lines. The discretization schemes are summarized in Table (1)-(5). The number of elements was chosen to limit the error to around 5%. The numerical results were compared with the analytical results given by [15];

$$\frac{\phi_s}{\phi_i} = \sum_{m=1}^{\infty} A_m \cos(m\theta)[J_m(kR) + iY_m(kR)] \tag{36}$$

where

$$A_m = -\,\varepsilon_m i^{m+1} e^{-i\gamma_m} \sin(\gamma_m), \tag{37}$$

$$\gamma_o = \tan^{-1}\left(-\frac{J_1(ka)}{Y_1(ka)}\right) \tag{38}$$

$$\gamma_m = \tan^{-1}\left(\frac{J_{m-1}(ka) - J_{m+1}(ka)}{Y_{m+1}(ka) - Y_{m-1}(ka)}\right) \tag{39}$$

and $\varepsilon_o = 1$, and $\varepsilon_m = 2$ for all values of m lager than unity. The notation in Eqn. (36)-(39) follows that of Morse [15].

As was mentioned in the previous section, $f_n(kR)$ in the asymptotic formulae in Eqn. (23)-(25) is an infinite polynomial, therefore proper representation of $f_n(kR)$ is needed. The choice of the terms retained depends on the requirement to limit errors and expedite the operation of the computer code. When the distance R becomes large, only a few terms of the real and imaginary part of $f_n(kR)$ are required to describe the actual behavior of the fundamental solution (Hankel function). For the scattering problem only one term each of the real and imaginary parts was retained, so that

$$f_n(kR) = 1 + i\left(\frac{n^2 - 0.25}{2kR}\right) \tag{40}$$

To demonstrate the success of the computational scheme, the computation time for setting up Eqn. (30) for each case (A)-(D) is compared in Tables 1-5.

Table 1. Scattering around the rigid cylinder (ka = 5.0, 25nodes)

case	cpu time (sec)	error (%)	reduction rate
A	20.60	0.21	1.00
B	1.66	1.63	12.41
C	1.34	1.64	15.37
D	1.01	1.66	20.40

Table 2. Scattering around the rigid cylinder (ka = 10.0, 31nodes)

case	cpu time (sec)	error (%)	reduction rate
A	26.90	3.06	1.00
B	2.62	3.06	10.27
C	2.02	3.01	13.32
D	1.41	3.65	19.08

Table 3. Scattering around the rigid cylinder (ka = 25.0, 91nodes)

case	cpu time (sec)	error (%)	reduction rate
A	152.0	0.82	1.00
B	23.0	1.15	6.61
C	17.5	1.15	8.69
D	11.7	1.04	12.99

Table 4. Scattering around the rigid cylinder (ka = 50.0, 181nodes)

case	cpu time (sec)	error (%)	reduction rate
A	515.0	5.02	1.00
B	92.5	5.24	5.57
C	68.0	5.25	7.57
D	43.8	4.52	11.76

Table 5. Scattering around the rigid cylinder (ka = 100.0, 445nodes)

case	cpu time (sec)	error (%)	reduction rate
A	2623.3	6.56	1.00
B	564.0	6.85	4.65
C	413.0	6.90	6.35
D	264.0	6.15	9.94

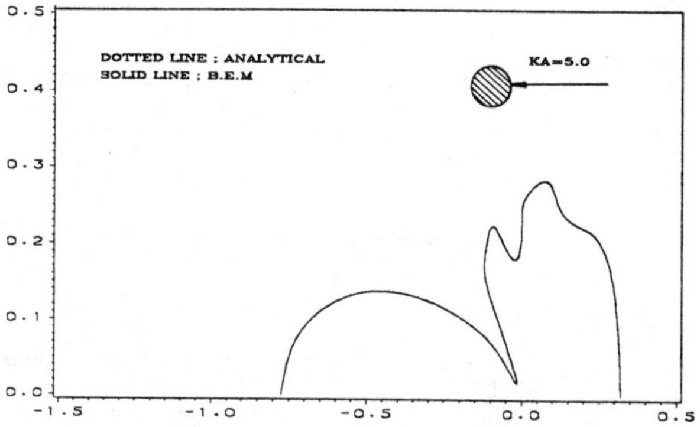

Figure 3. Plot of scattered potential versus polar angle at a distance of five radii for a rigid cylinder (ka = 5.0, 25nodes, 8elements)

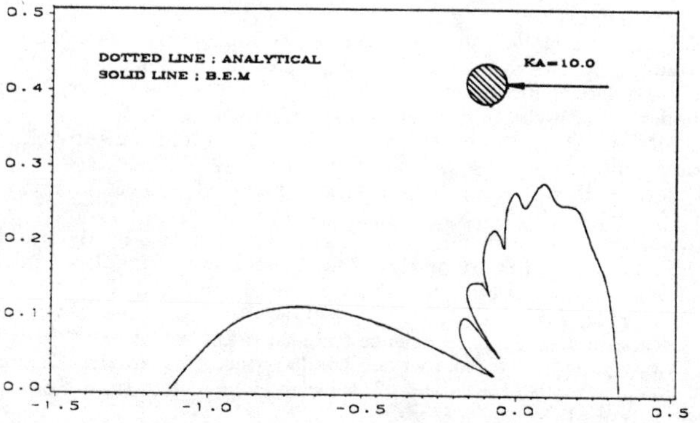

Figure 4. Plot of scattered potential versus polar angle at a distance of five radii for a rigid cylinder (ka = 10.0, 31nodes, 10elements).

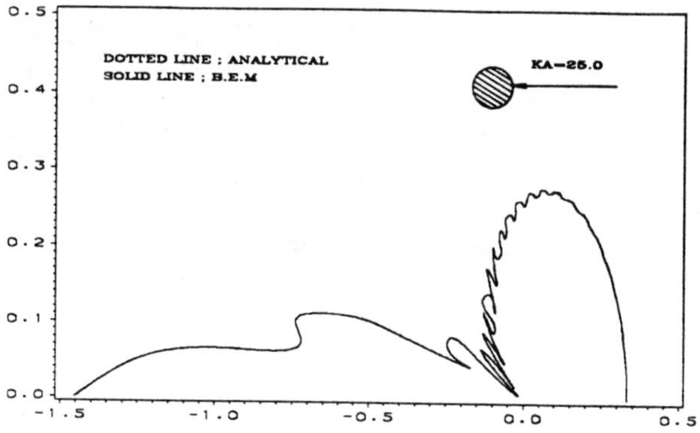

Figure 5. Plot of scattered potential versus polar angle at a distance of five radii for a rigid cylinder (ka = 25.0, 91nodes, 30elements)

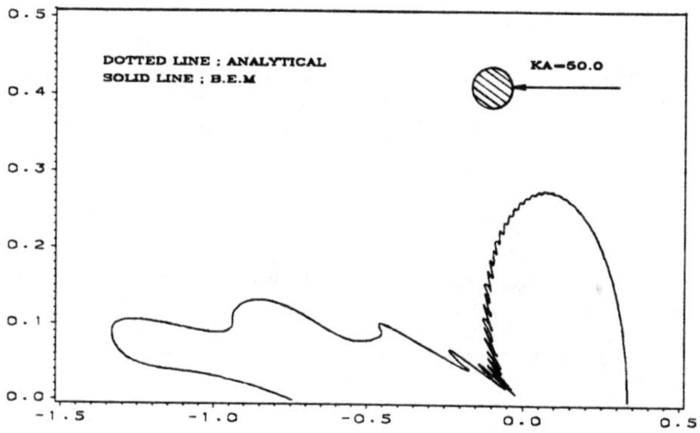

Figure 6. Plot of scattered potential versus polar angle at a distance of five radii for a rigid cylinder (ka = 50.0, 181nodes, 60elements)

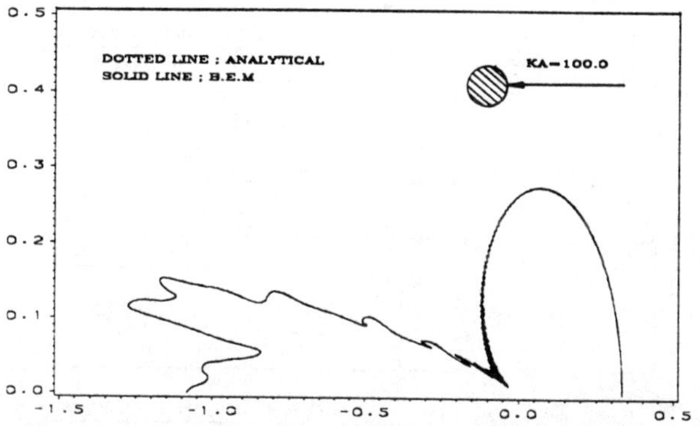

Figure 7. Plot of scattered potential versus polar angle at a distance of five radii for a rigid cylinder (ka = 100.0, 445nodes, 148elements)

Table 1-5 shows greatly reduced cost for the set up process in which the reduction ratio is at least 5 for every case (B)-(D) while keeping the accuracy close to that of case (A) or even better as was shown in Fig. 3-7. For relatively low frequency, the application of asymptotic formulae is not as practical, especially with regard to the accuracy as shown on Table 1. As the frequency becomes higher, the usefulness of the asymptotic formulae and the approximate polynomial is proved for both accuracy and computation time. According to the results shown on Fig. 3-7 and Tables (1)-(5) the most efficient computational scheme is case (D) for each problem. The cost for computation was saved by a ratio of approximately 10 compared with case (A) (Tables 1-5). The accuracy is clearly not compromised, as also shown in Tables 1-5.

The second test case considered is radiation from a circular cylinder of radius "a". The boundary condition on the cylinder is a uniform $U_a = \partial\phi/\partial v$ (pulsating cylinder), and the normalized frequency of vibration of the cylinder is ka = 10,100,150,152 and 200. The discretization procedure follows the scattering case. Shown for comparison in Tables 6-10 are the computation times for setting up Eqn. (30) and the accuracy of the potential on the surface of the cylinder in comparison with the analytical results given by [15];

$$\phi_o = A_o[J_o(ka) - iY_o(ka)] \tag{41}$$

where,

$$A_o = \frac{4}{k}\frac{U_a}{C_o^2}(J_1(ka) + iY_1(ka)) \tag{42}$$

$$C_o = \frac{2J_1(ka)}{\sin(\gamma_o)} \tag{43}$$

In the case of the radiation problem with homogeneous boundary conditions less terms of $f_n(kR)$ in Eqn. (25) are needed to approximate the fundamental solution with reasonably small errors. For the pulsating cylinder only one term of the real part of Eqn.(25) was taken, that is $f_n(kR) = 1$ was included in the asymptotic formulae in every step of the evaluations.

Tables 6-10 show the cost for computation to be reduced greatly by the modifications, just as in the scattering problem. Accuracy was not compromised as shown in table 7-10. As in the scattering problem, the application of asymptotic formulae and approximate polynomials are not as effective for the case of low frequency (Table 6). Table 8 shows the example of the failure to yield a unique solution when the BIE method is applied to exterior Neuman problem [1] – [2] with characteristic frequency. The given value ka = 150 is extremely close to the eigenfrequency of the homogeneous interior Dirichlet problem for a rigid cylinder [16]. Table (9) shows how rapidly the error reduces as the frequency passes the eigenfrequency. In the present investigation we have made no attempt to eliminate this well known problem. Other investigators have offered methods of eliminating the difficulty [16] which will ultimately be implemented in the code developed here.

Table 6. Radiation from pulsating cylinder (ka = 10.0, 13nodes)

case	cpu time (sec)	error (%)	reduction rate
A	4.63	0.62	1.00
B	0.40	0.62	11.58
C	0.36	0.62	12.86
D	0.32	2.15	14.47

Table 7. Radiation from pulsating cylinder (ka = 100.0, 241nodes)

case	cpu time (sec)	error (%)	reduction rate
A	767.72	5.77	1.00
B	152.00	2.09	5.05
C	112.80	2.09	6.81
D	74.33	2.22	10.33

Table 8. Radiation from pulsating cylinder (ka = 150.0, 445nodes)

case	cpu time (sec)	error (%)	reduction rate
A	2428.10	102.31	1.00
B	503.00	23.14	54.83
C	372.65	23.14	6.52
D	243.01	23.77	10.00

Table 9. Radiation from pulsating cylinder (ka = 152.0, 445nodes)

case	cpu time (sec)	error (%)	reduction rate
A	2446.40	0.55	1.00
B	510.00	0.33	4.80
C	374.59	0.33	6.53
D	247.11	0.34	9.9

Table 10. Radiation from pulsating cylinder (ka = 200.0, 445nodes)

case	cpu time (sec)	error (%)	reduction rate
A	2353.40	5.69	1.00
B	502.00	3.45	4.69
C	372.34	3.45	6.32
D	242.22	3.33	9.72

OBSERVATIONS AND CONCLUSIONS

Many of the problems associated with two dimensional scattering and radiation problems at high frequency have been addressed in the context of creating an efficient and accurate Boundary Element scheme for computing the radiated or scattered acoustic fields.

Cubic isoparametric elements have been used and offer a good compromise between accuracy and efficiency. The use of asymptotic and polynomial approximations for the Hankel function provides a substantial benefit in reducing the time required to set up the algebraic equations. At high frequency, the resulting large number of algebraic equations requires the implementation of an out of core linear equation solver and this has been successfully implemented at only a modest penalty in computation time. It has been found that the singularity which is inherent in any representation of the fundamental solution of the Helmholtz equation is a principal source of error. The only effective way to minimize this source of error is by a suitable choice of the order of the quadrature. However, the use of efficient and accurate approximations for the fundamental solution minimizes the cost of the quadrature scheme.

The results obtained in the present investigation support all of the innovations introduced in the application of the BEM to the scattering and radiation problems investigated. The cost of computation is substantially reduced with virtually no impact on accuracy.

ACKNOWLEDGEMENT

This work was supported under a grant from the Acoustic Branch in the Structures Directorate at NASA Langley Research Center

REFERENCES

1. Chertock,G.,"Sound Radiation from Vibrating Surfaces",Journal of the Acoustical Society of America, Vol.36, No.7,1964, pp.1305-1313

2. Schenk, H.A.,"Improved Integral Formulation for Acoustic Radiation Problems." Journal of the Acoustical Society of America, Vol.44, No.1, 1968, pp.41-58

3. Copley, L.G.,"Integral Equation Method for Radiation from Vibrating Bodies." Journal of the Acoustical Society of America, Vol.41, No.4, 1967, pp.807-816

4. Koopman, G.H., and Benner, H.,"Method for Computing the Sound Power of Machines Based on the Helmholtz Integral." Journal of the Acoustical Society of America, Vol.71, 1982, pp.77-89

5. Banerjee, P.K. and Butterfield, Editor,"Developements in Boundary Element methods-1",Applied Science Publishers Ltd, London, (1979)

6. Seybert, A.F., Soenarco, B., Rizzo, F.J. and Shippy, D.J., "An advanced Computational Method for Radiation and Scattering of Acoustic Waves in Three Dimensions," Journal of the Acoustical Society of America, Vol.77, 1985, pp.362-368

7. Seybert, A.F. and Soenarco, B., "Application of BIE Method to Sound Radiation Problems using an Isoparametric Element," ASME trans. J. Vib. Acoust. Stress Reliabil. Des. 106, 1984, pp.414-420 ,pp.362-368

8. Rschevkin, S.N., Editor, "The Theory of Sound", The Macmillan company,New York,(1963)

9. Hughs, J.R. Thomas, Editor,"The Finite Element Method", Prentice Hall, Inc., Englewood cliff,(1987)

10.Baradari, F.,"Comparative study of the Boundary Technique and the Finite Element Method in Two Dimensional Eigenvalue Problem",Thesis, University of Missouri-Rolla, (1980)

11.Abramowitz, M. and Stegan, I.A.,Editor, "Handbook of mathematical functions", Dover Publications, Inc.,(1965)

12.Relton, F.E.,Editor,"Applied Bessel functions", Dover Publications, Inc., New York, (1965)

13.Das, P. C. (1978), "A disk based block elimination technique used for the solution of non-symmetrical fully populated matrix systems encountered in the boundary element methods," Recent Advances in Boundary Element Methods (C. A. Brebia, Ed.), Pentech Press, London.

14.Hofmeister, L. D. (1978), "An out-of-core solver for full, unsymmetric matrices," Int. J. Num. Meth. Eng., Vol. 12, pp. 721-731.

15.Morse, P.H., Ingard, K.U.,Editor,"Thoeritical Acoustics", Princeton University Press, (1986)

16.Seybert, A.F. and Rengrajan, T.K., "The use of CHIEF to obtain unique solutions for Acoustic Radiation using Boundary Integral Equations." Journal of the Acoustical Society of America, 1987

17.Banerjee, P.K. and Butterfield, Editor,"Developements in Boundary Element methods-2",Applied Science Publishers Ltd, London, (1979)

18.Copley, L.G., "Fundamental Results concerning Integral Representations in Acoustic Radiation," Journal of the Acoustical Society of America, Vol.44, No.1, 1968, pp.1626-1632

ITERATIVE SOLUTION OF BOUNDARY ELEMENT EQUATIONS FOR THE EXTERIOR HELMHOLTZ PROBLEM

S. Amini, C. Ke and P. J. Harris
Department of Mathematics and Statistics
Polytechnic South West
Plymouth, United Kingdom

In this paper we study an efficient boundary element method for the determination of the acoustic field around arbitrary-shaped finite structures immersed in an infinite homogeneous acoustic medium.

The direct boundary integral equation due to Burton and Miller is employed to overcome the non-uniqueness of solution associated with the classical boundary integral formulations of the exterior Helmholtz equation. The choice of the coupling parameter in the Burton and Miller formulation is discussed in order to minimise the condition number of the boundary integral equation.

Numerical results are presented for the problem of acoustic radiation from several structures of practical interest.

NOMENCLATURE

c	Sound speed in D_+
D_+	Unbounded region exterior to S
D_-	Bounded region interior to S
G_k	Free-space Green's function
k	Acoustic wavenumber
L_k	Single-layer Helmholtz potential
M_k	Double-layer Helmholtz potential
M_k^T	Derivative of L_k along n
n	Outward normal to S
N_k	Derivative of M_k along n
S	Surface of the structure
t	Time
v_n	Normal component of structure velocity
α	Solid angle
$\triangle$	Reference right angled triangle
η	Coupling parameter
ρ	Density of the medium D_+
κ	A general integral operator
σ	Surface density function
Φ	Sound pressure
ϕ	Sound pressure (harmonic)
ω	Angular frequency of sound source

1 INTRODUCTION

Many linear boundary value problems of mathematical physics and engineering, traditionally solved by domain methods, may be re-formulated as integral equations over the boundary of the domain of interest [8,9,10,11,16]. In cases where the domain of interest is of infinite extent boundary integral equations and their subsequent discretisation by finite element type methods (known as boundary element methods) can offer an obvious advantage over the domain methods, as the dimensionality of the problem is effectively reduced by one.

In this paper we are concerned with the determination of the acoustic field radiated (or scattered) by an arbitrary shaped, finite, vibrating structure immersed in an infinite homogeneous acoustic medium. Such a problem is of considerable interest in many areas including underwater acoustics and aeronautics. [5,28,31].

In section 2 we consider some "direct" boundary integral formulations of the exterior Neumann problem for the Helmholtz equation. We discuss briefly the problem of existence and uniqueness of solutions as well as the conditioning of these formulations. In section 3 we present an overview of our numerical methods based on piecewise constant collocation over linear triangular boundary elements.

The solution of the resulting large, full, non-hermitian boundary element equations by iterative methods is considered in section 4. In particular we propose and implement a modification to the multi-grid method, based on "diagonal splitting", which results in the convergence of these methods, even when dealing with "hyper-singular" operators, namely the derivative of the double layer potential. Finally in section 5, we present the results of our numerical method for the acoustic radiation from a sphere, an ellipsoid, and a cylinder.

2 FORMULATION

Let S be a closed, bounded, piecewise smooth surface is $\mathcal{R}^3$. We denote the interior and exterior of S by D_- and D_+ respectively. The propagation of small amplitude acoustic waves through a homogeneous medium is modelled by the linear wave equation

$$\nabla^2\Phi(p,t) - \frac{1}{c^2}\frac{\partial^2}{\partial t^2}\Phi(p,t) = 0 \quad p \in D_+ \tag{1}$$

where the scalar complex-valued function $\Phi(p,t)$ is the acoustic pressure at time t and c is the speed of sound in the medium. Sound pressure is related to particle velocity through

$$\nabla\Phi(p,t) = -\rho\frac{\partial v}{\partial t}(p,t) \tag{2}$$

where ρ denotes the density of the medium. If we assume a harmonic time dependence of the form $e^{-i\omega t}$, where ω is the angular frequency of the sound source, equation (1) becomes the Helmholtz (reduced wave) equation

$$(\nabla^2 + k^2)\phi(p) = 0 \quad p \in D_+ \tag{3}$$

where $\Phi(p,t) = e^{-i\omega t}\phi(p)$ and $k = \omega/c$ is the acoustic wavenumber. It now follows from (2) that

$$\frac{\partial\phi(p)}{\partial n} = i\omega\rho v_n(p) \quad p \in S \tag{4}$$

where n denotes the unit outward normal to S and $v_n(p)$ denotes the normal particle velocity, which we assume is known. To ensure the existence of a unique solution to (3) with the Neumann boundary condition (4) on the finite boundary, we also need to impose on ϕ the Sommerfeld radiation condition (or a similar condition)

$$\lim_{r_p\to\infty} r_p\left[\frac{\partial\phi(p)}{\partial r_p} - ik\phi(p)\right] = 0 \tag{5}$$

where $p = (r_p, \theta_p, \phi_p) \in \mathcal{R}^3$.

Here we consider a direct boundary integral formulation of equations (3). It follows from Green's second identity that $\phi(p)$ a radiating wave function, (i.e. one satisfying (3) and (5)), satisfies the so called Helmholtz integral formulae for the exterior region, given by

$$\int_S\{\phi(q)\frac{\partial G_k}{\partial n_q}(p,q) - G_k(p,q)\frac{\partial\phi}{\partial n_q}(q)\}dS_q \begin{cases} \phi(p) & p \in D_+ \\ \frac{\alpha(p)}{4\pi}\phi(p) & p \in S \\ 0 & p \in D_- \end{cases} \tag{6}$$

where $\alpha(p)$ is the solid angle subtended by D_+ at $p \in S$ (for p on a smooth part of the surface $\alpha(p) \equiv 2\pi$), and

$$G_k(p,q) = \frac{e^{ikr}}{4\pi r}, \quad r = |p-q| \tag{7}$$

is the free-space Green's function, or the fundamental solution for (3). Assuming for the moment that S is a smooth surface, the equation (6) for $p \in S$, known as the surface Helmholtz equation, can be written in the operator form

$$(-\tfrac{1}{2}I + M_k)\phi(p) = (L_k\tfrac{\partial\phi}{\partial n})(p) \quad p \in S \tag{8}$$

where

$$(L_k\sigma)(p) = \int_S \sigma(q)G_k(p,q)dS_q \tag{9}$$

and

$$(M_k\sigma)(p) = \int_S \sigma(q)\frac{\partial G_k}{\partial n_q}(p,q)dS_q \tag{10}$$

are the single layer and double layer Helmholtz potentials respectively. If the second kind integral equation (8) could be solved for $\phi(p)$, $p \in S$, then it follows from (6) for $p \in D_+$ that the solution of (3) is given by

$$\phi(p) = (M_k\phi)(p) - (L_k\tfrac{\partial\phi}{\partial n})(p) \quad p \in D_+ \tag{11}$$

which involves a a simple integration.

It is well known, however, that for a countable set, $I_1(S)$, of values of k referred to as 'characteristic wavenumbers', the operator $-\frac{1}{2}I + M_k$ is singular. Many formulations valid for all wavenumbers have been proposed since 1965, viz [10,17,21,24]. Here we employ one such formulation, which is the direct formulation counterpart of Panich's [21] indirect formulation due to Burton and Miller [10]. Briefly, on formally differentiating (8) in the direction of n_p, the outward normal at p, we obtain

$$(N_k\phi)(p) = (\frac{1}{2}I + M_k^T)\frac{\partial\phi}{\partial n} \tag{12}$$

where $N_k = \frac{\partial}{\partial n}M_k$ and $M_k^T = \frac{\partial}{\partial n}L_k$. The operator N_k is also singular at a countable set, $I_2(S)$, of values of k. If we couple equations (8) and (12), we obtain

$$\{(-\frac{1}{2}I + M_k) + i\eta N_k\}\phi(p) = \{L_k + i\eta(\frac{1}{2}I + M_k^T)\}\frac{\partial\phi}{\partial n} \tag{13}$$

with $i^2 = -1$, which can be shown to possess a unique solution for all wavenumbers k, provided the coupling parameter η satisfies $Re(k)\eta > 0$.

The operator N_k, the derivative of the double layer potential

$$(N_k\sigma)(p) = \frac{\partial}{\partial n_p}\int_S \sigma(q)\frac{\partial}{\partial n_q}G_k(p,q)dS_q \tag{14}$$

is a hyper-singular operator. In general some mathematical analysis should be carried out before employing a numerical quadrature rule to approximate this operator [19,30]. Here we adopt the formulation of N_k used in [19] which follows from the pointwise variational principle introduced in [29]

$$\begin{aligned} &\int_S \sigma(q)\frac{\partial^2 G_k}{\partial n_p\partial n_q}(p,q)dS_q = \\ &k^2\sigma(p)\int_S n_p\cdot n_q G_k(p,q)dS_q + \int_S\{\sigma(q)-\sigma(p)\}\frac{\partial^2 G_k}{\partial n_p\partial n_q}(p,q)dS_q \end{aligned} \tag{15}$$

The operators on the right hand side of (15) are now more amenable to numerical approximation; see section 3.

It is possible to regularise the operator N_k to yield a modified equation (10) before coupling it with (8). It can be shown that [9,21]

$$(L_kN_k)\sigma(p) = (M_k + \tfrac{1}{2}I)(M_k - \tfrac{1}{2}I)\sigma(p) \quad p \in S \tag{16}$$

By premultiplying equation (12) by L_0, (i.e. L_k with $k = 0$) we obtain

$$\{L_0(N_k - N_0) + M_0^2 - \tfrac{1}{4}I)\}\phi(p) = L_0(\tfrac{1}{2}I + M_k^T)\tfrac{\partial\phi}{\partial n} \quad p \in S \quad (17)$$

which when coupled with (8), similar to (13), yields the regularised Burton and Miller formulation

$$\{(-\frac{1}{2}I+M_k)+i\eta[L_0(N_k-N_0)+M_0^2-\frac{1}{4}I]\}\phi = [L_k+i\eta L_0(\frac{1}{2}+M_k^T)]\frac{\partial\phi}{\partial n} \quad (18)$$

All the operators in (18) are compact and in general more amenable for accurate discretisations.

Either of the boundary integral formulations for (13) or (18) valid for all wavenumbers, may be solved to obtain $\phi(p), p \in S$. Before attempting to discretise them let us discuss briefly the choice of the coupling parameter η. Ideally one should choose $\eta = \eta(S;k)$ so as minimise the condition number of the integral operators in (13) and (18). For a general surface S, analytical evaluation of the condition number of an integral operator is not possible. However to gain some insight, the case of S being a sphere has been considered [1,18] where it was found that the choice of $\eta = \frac{1}{k}$ is an "almost optimal" choice for (13) whilst $\eta = 1$ was found to be satisfactory [1] for a large range of values of k for (18), as in this case the equations are naturally well-conditioned. For a non-spherical surface S, which is not too elongated, it can be argued that the same choice of η should result in well-conditioned equations.

3 NUMERICAL METHODS

The numerical method used in this paper for discretising (13) and (18) is the collocation method based on piecewise constant approximations over linear triangular boundary elements. Consider the second kind integral equation

$$\phi(p) - \int_S K(p,q)\phi(q)dS_q = f(p) \quad p \in S \quad (19)$$

where S is a piecewise smooth surface in $\mathcal{R}^3$. To be precise $S = S_1 \cup S_2 \cup \cdots \cup S_I$, where each S_i is a smooth surface in $\mathcal{R}^3$. We then divide (and approximate) each S_i into a number of flat triangles $S_{ij}, j = 1,2,\ldots,n_i$ having their corners on S_i. Hence

$$S \approx \bigcup_{i=1}^{I}\bigcup_{j=1}^{n_i} S_{ij} = \bigcup_{i=1}^{n} \triangle_i = S_n \quad (20)$$

where $n = \sum_{i=1}^{n} n_i$ and the $\triangle_i's$ are simply a renumbering of the $S_{ij}'s$. Assuming a piecewise constant approximation for $\phi(p)$, the equation (19) can be approximated by

$$\phi_n(p) - \sum_{j=1}^{n}\phi_j \int_{\triangle_j} K(p,q)dS_q = f(p) \quad p \in S_n \quad (21)$$

where $\phi_n(p) \approx \phi(p)$ and $\phi_j \approx \phi(p), p \in \triangle_j$. It is convenient to write (21) in the operator form

$$(I - \kappa_n)\phi_n = f \quad (22)$$

where

$$(\kappa_n\sigma)(p) = \sum_{j=1}^{n}\sigma(p_j)\int_{\triangle_j} K(p,q)dS_q \quad (23)$$

Now choose n collocation points $p_i, i = 1,2,\ldots,n$ where each p_i is the centroid of $\triangle_i$. By insisting that $\phi_n(p_i) = \phi_i$, we obtain the usual piecewise constant collocation equations

$$\phi_i - \sum_{j=1}^{n}\phi_j \int_{\triangle_j} K(p_i,q)dS_q = f(p_i) \quad i = 1,2,\ldots,n \quad (24)$$

which is a full system of linear equations to be solved for $\underline{\phi} = [\phi_1,\phi_2,\ldots,\phi_n]^T$. The equation (21) is of considerable importance in our analysis. Firstly, once (24) has been solved to yield $\phi_1,\phi_2,\ldots,\phi_n$, we may use (21) to obtain the iterated collocation solution

$$\phi_n(p) = \sum_{j=1}^{n}\phi_j \int_{\triangle_j} K(p,q)dS_q + f(p) \quad p \in S_n \quad (25)$$

which is, in general, a continuous function and known to be globally a more accurate approximation to $\phi(p)$ than the collocation solution (though $\phi_n(p_i) = \phi_i$) [27]. Secondly (25), similar to the Nyström extension, allows us to interpolate (and smooth at the same time) values at non-collocation points, which is to be exploited in the multi-grid solution of (24).

The computational complexity of a boundary element method can be divided into 2 major parts, (i) the set up stage, that of evaluating the elements of the $n \times n$ matrix in (24), and (ii) the solution stage, that of solving the system (24). In general the set up time is of the form c_1n^2 where c_1 is a very large positive constant, as the kernel $K(p,q)$ is rather costly to evaluate. Efficiency of the integration rules play a major role in reducing this cost. The solution time by direct methods is of the form c_2n^3 where c_2 is in general a very small number compared with c_1. Therefore, for small values of n, the total cost is dominated by the set-up time. For practical 3-D problems, the value of n could be large and therefore the solution time by direct methods would dominate the total computation time. In cases where the integral operators are compact, having desirable smoothing and spectral properties (zero as the only point of accumulation of their countable set of eigenvalues [20]), their discrete approximations possess similar properties which can be exploited to obtain an efficient iterative scheme such as the conjugate gradient method [4,3,23] and the multi-grid method [3,22,25,26]. Such schemes can yield solutions to (24) to within the level of the discretisation error in c_3n^2 operations, where c_3 is a moderately small constant. In this case therefore the substantial part of the computation is in the set-up stage and much recent work has been carried out to find methods of reducing this [13,23].

Let us briefly consider the numerical integration of the collocation equations. The integrals in (24) are of the form

$$K_{ij} = \int_{\triangle_j} K(p_i,q)dS_q = \mu_j \int_{\triangle} K(p_i(u,v),q(u,v))dudv \quad (26)$$

where $\triangle$ is the reference right angled unit triangle in the (u,v) plane with corners at $(u,v) = (0,1),(0,0)$ and $(1,0)$. The constant μ_j is the Jacobian of the linear transformation taking $\triangle_j$ to $\triangle$, and is equal to twice the area of $\triangle_j$. Here the integrals are non-singular if $i \neq j$ and singular if $i = j$. In the former case however, the closer the collocation point is to the integration region $\triangle_j$, the more peaked, ("nearly singular") the integrand $K(p_i,q)$ will be for q near to p_i. To retain a given level of accuracy in numerical integration, more accurate rules should be employed for "near-singular" integrals, see [2,7].

Here, away form the singularity , we use the 3-point rule

$$\int_{\triangle} f(u,v)dudv \approx \frac{1}{6}[f(\frac{1}{2},0) + f(\frac{1}{2},\frac{1}{2}) + f(0,\frac{1}{2})] \quad (27)$$

with degree of precision 2, whilst for the near-singular integrals we employ the product 3-point Gauss rule (on the triangle), with degree of precision 5. For the evaluation of the singular integrals

K_{ii}, similar to Atkinson [7], we divide the region of integration $\triangle_j$ into three smaller triangles by joining the collocation point p_i to each corner of $\triangle_j$. Each smaller triangle is then transformed to the reference triangle with the singular point p_i transformed to the $(0,0)$ corner in each case. We are now faced with the evaluation of three integrals of the form

$$I = \int_{\triangle} f(u,v)dudv \tag{28}$$

with a singularity at $(0,0)$. Introducing the change of variable [7,12]

$$u = (1-s)t, \quad v = st \quad 0 \le s,t \le 1 \tag{29}$$

we obtain

$$I = \int_0^1 \int_0^1 tf((1-s)t, st)dsdt \tag{30}$$

The integrand in (30) can be shown to be no longer singular in the case of potential type Green functions. The integral (30) can be adequately approximated by a product m-point Gauss rule. Here we choose m=5.

4 ITERATIVE SOLUTION OF LINEAR SYSTEMS

The iterative solution of the full linear system by has been previously studied for 2-D acoustic problems ([3], [4]) and potential problems ([23]). In [3], both the conjugate gradient method and the multi-grid type methods are applied to integral equations with compact operators. In such cases the convergence analyses of the methods can be carried out for these methods using the well-known theory of compact operators [6]. From our experience, in [3], we concluded that the two grid methods are more efficient than the conjugate gradient method and nearly as efficient as multi-grid methods, yet being easier than the latter to implement; (see [3] and [22]). Here, therefore, we shall concentrate on two-grid methods. Briefly, consider the second kind integral equation

$$(\lambda I + \kappa)u = f \tag{31}$$

with κ a compact operator. Discretising it as in §3 on a coarse grid $G[n]$ and a fine grid $G[m]$ with n and m subdivisions respectively, we obtain two discrete collocation equations similar to (24) in the form

$$(\lambda I + K_l)\underline{u}_l = \underline{f}_l, \quad with\ l = n,\ m. \tag{32}$$

The idea is to solve the equation (32) with $l = m$, using the information from the solution of (32) with $l = n$. To implement the two grid method, we require the following quantities :

$$\begin{array}{llll} K_m & : & m \times m \ matrix\ with\ (K_m)_{ij} & = \int_{\Delta_{mj}} K(p_i, q)dS_q; \\ K_{mn} & : & m \times n \ matrix\ with\ (K_{mn})_{ij'} & = \int_{\Delta_{nj'}} K(p_i, q)dS_q; \\ K_n & : & n \times n \ matrix\ with\ (K_n)_{i'j'} & = \int_{\Delta_{nj'}} K(p_{i'}, q)dS_q; \end{array}$$

and vectors $\underline{f}_m, \underline{f}_n$ with $(\underline{f}_m)_j = f(p_j)$ and $(\underline{f}_n)_{j'} = f(p_{j'})$, for $i,j = 1,\cdots,m$ and $i',j' = 1,\cdots,n$. Then starting from an initial guess $\underline{u}_m = \underline{0}$, we can describe the two-grid algorithm as follows(writing x for a vector $\underline{x}$ from here onwards) :

0) Set $r_m = f_m$, $u_m = 0$ and go to step 3);

1) Find the residual on $G[m]$, $r_m = f_m - \lambda u_m - K_m u_m$;

2) Perform smoothing on $G[m]$, $u_m = u_m + r_m/\lambda$ and $r_m = -K_m r_m/\lambda$; (optional step)

3) Restrict the residual r_m to $G[n]$, $r_n = R^n_m r_m$;

4) Solve exactly on $G[n]$, $(\lambda I + K_n)v_n = r_n$;

5) Interpolate v_n to obtain v_m on $G[m]$, $v_m = (r_m - K_{mn}v_n)/\lambda$;

6) Add on the correction v_m to u_m, $u_m = u_m + v_m$;

7) If $\|v_m\| \le TOL$ exit with solution in u_m; otherwise go to step 1).

Here R^n_m is a restriction operator which can be taken to be the so-called injection operator, provided that we choose the nodes of $G[n]$ to coincide with those of $G[m]$. Note that step 4) is in general carried out by a direct solver where the LU decomposition is kept for use in each iteration. The algorithm given above is similar to Atkinson's method II(based on the residual correction principle, see [3]). Missing out the smoothing step 2) will give rise to the corresponding method I of Atkinson. In general, method I may be preferable if the operator κ is sufficiently smooth, since each iteration of this version requires roughly half the number of operations compared with method II. For example, it is easy to show that the convergence of the iterative scheme above is governed by the quantity ([6])

$$\beta = \|(\lambda I + \kappa_n)^{-1}(\kappa_m - \kappa_n)\kappa_m\| \tag{33}$$

If κ is a compact operator, it can shown that $\beta \to 0$ as $n \to \infty$.

Recall from §2 that the regularised Burton and Miller's formulation (18) involves compact operators only and therefore the above algorithm can be readily applied to it. But the major disadvantage of the regularised formulation, from the practical point of view, is the excessive computational cost and complexity of setting up the various discrete operators [15]. On the other hand, due to the presence of the non-compact operator N_k, the direct application of the above iterative method to the formulation (13) simply does not work. As far as we are aware, no efficient techniques have been developed to deal with the discretization of (13) and therefore the direct solution has been used ([19]).

The equation (13) is of the form (31) with $\kappa = M + N$, where the subscripts k have been dropped and N denotes $i\eta N_k$ for ease of notation. Roughly speaking, the problem here is that because of the non-compact operator N, the quantity $\|(\kappa_m - \kappa_n)\kappa_m\|$ can not be shown to be small for all $m > n$ and n sufficiently large; a crucial factor in the convergence analysis of the 2-grid methods. We shall show that by modifying steps 2) and 5) of the procedure above, to reduce the non-smoothing effect of N, we are able to restore the fast convergence of the algorithm for (13).

To explain our modifications, which are in a sense similar to the operator splitting idea of [26] and [25], let us rewrite the step 2) of the algorithm explicitly, in a form suitable for analysis

$$\bar{u}_m = u_m + r_m/\lambda \ and \ \bar{r}_m = -K_m r_m/\lambda, \tag{34}$$

or eqivalently

$$\bar{u}_m = (f_m + K_m u_m)/\lambda \ and \ \bar{r}_m = f_m - \lambda\bar{u}_m - K_m\bar{u}_m.$$

Concentrating on the second term in (34), the residual smoothing step, we have (with $\lambda = -1/2$ and $K_m = M_m + N_m$)

$$\begin{aligned}\bar{r}_m &= 2(M_m + N_m)r_m \\ &= 2(M_m + \bar{N}_m + D_m)r_m \end{aligned} \tag{35}$$

where D_m is the diagonal matrix of diagonal elements of N_m and $\bar{N}_m = N_m - D_m$. We propose to modify the smoothing step 2) given by (35) in the form

$$(I - 2D_m)\hat{r}_m = 2(M_m + \bar{N}_m)r_m, \tag{36}$$

or

$$\hat{r}_m = 2(I - 2D_m)^{-1}(M_m + \bar{N}_m)r_m.$$

Observe that this step increases the computational cost of the original step 2) by m divisions only. Similar splitting of the operator N_m should be carried out for the interpolation step 5). We shall see in the next section that the above modification of the 2-grid method, based on diagonal splitting of the N_k operator, results in the convergence of the scheme as applied to the discretisation of (13).

5 NUMERICAL EXPERIMENTS

Here we present some numerical results for the iterative solution of the boundary integral equations for the exterior acoustic problem. We consider the problem of acoustic radiation from three structures: (S1) a unit sphere. (S2) an axisymmetric ellipsoid ($\frac{x^2}{a^2} + \frac{y^2}{a^2} + \frac{z^2}{b^2} = 1$) with the ratio of major to minor axis b/a=2; (with a=3/4 and b=6/4). (S3) a right circular cylinder with height=2 and base radius=1.

In each case the surface of the structure is approximated by n=112 linear triangular elements. Each triangle is then subdivided into 4 smaller triangles by joining the mid-points of its sides, giving m=4n=448. The test problem considered in all cases is the radiation problem equivalent to that generated by placing 2 point sources inside the structure at locations $q_1 = (0.0, 0.0, 0.5)$ and $q_2 = (0.25, 0.25, 0.25)$ with the strengths $A_1 = 2 + 3i$ and $A_2 = 4 - i$ respectively; [2,14]. The field generated is then given by

$$\phi(p) = A_1 \frac{e^{ik|p-q_1|}}{|\,p - q_1\,|} + A_2 \frac{e^{ik|p-q_2|}}{|\,p - q_2\,|} \tag{37}$$

from which $\frac{\partial \phi}{\partial n}(p)$, for $p \in S$ can be calculated. The measure of the error chosen here is the mean relative error (MRE) defined as

$$MRE = \frac{1}{m} \sum_{i=1}^{m} \frac{|\,\phi(p_i) - \hat{\phi}(p_i)\,|}{|\,\phi(p_i)\,|} \tag{38}$$

where p_i's are the collocation points and $\hat{\phi}(p_i)$ represents the computed approximation to $\phi(p_i)$.

In table 1 we present the results of discretising the regularised equation (18) for the case of structures (S1) and (S3) near their first characteristic wavenumbers k=3.142 and k=3.04 respectively. The coupling parameter $\eta \equiv 1$ is chosen in both cases. The results are for (method I) the iterative algorithm in section 4 without the smoothing step 2. We observe that the convergence of the 2-grid method is rapid. Indeed, for the iterative solution, the total computational time is greatly reduced. We point out here that in table 1 the difference in set-up time is due to the fact that for the direct solution, the matrix multiplications corresponding to product operators $L_0(N_k - N_0)$, M_0^2, and $L_0(\frac{1}{2}I + M_k^T)$ have to be carried out explicitly, whilst in the latter case this is not required (since $\underline{y} = AB\underline{x}$ can be carried out as $\underline{v} = B\underline{x}$ followed by $\underline{y} = A\underline{v}$). (All programs were run on Prime 750 at Polytechnic South West and the value of TOL in our iterative algorithm was set to 10^{-3}).

We shall now concentrate on the computationally more attractive formulation (13) where the suggested modification to the 2-grid algorithm is essential for its convergence. We also found the smoothing step 2 to improve the convergence rate of the scheme. The results in tables 2-6 below are therefore for the modified algorithm with smoothing step 2 included. In tables 2, 4 and 6 the set-up time is around 600 CPU seconds and the direct solution time is around 920 CPU seconds. We observe that in all cases the modified 2-grid algorithm substantially reduces the solution time. In tables 3 and 5 we present results of the sound-field calculations in the exterior domain using (11).

Table 1: Regularised Burton and Miller, equation (18).

	Direct Method		Iterative Method			
Structure	set up time	solution time	set up time	number of iterations	solution time	MRE (%)
S1 (k=3.142)	7000	920	940	4	330	3.8
S3 (k=3.04)	7000	920	940	4	330	4.0

Table 2: (S1) Sphere, Formulation (13)

k	number of Iterations	2-Grid Solution Time	MRE (%)
3.142	3	64	4.2
5	6	112	5.4

Table 3: (S1) Sphere. Error in the exterior field.

k	% Error at (0,0,2)	% Error at (0,0,3)	% Error at (0,0,5)
3.142	2.0	2.2	2.2
5	2.0	2.2	2.2

Table 4: (S2) Ellipsoid. Formulation (13)

k	number of Iterations	2-Grid Solution Time	MRE (%)
3	3	66	5.9
5	4	82	6.2

Table 5: (S2) Ellipsoid. Error in the exterior field.

k	% Error at (0,0,2)	% Error at (0,0,3)	% Error at (0,0,5)
3	2.4	2.4	2.4
5	4.3	3.9	3.5

Table 6: (S3) Cylinder, Formulation (13)

k	number of Iterations	2-Grid Solution Time	MRE (%)
3.04	4	82	4.7
5	6	111	4.2

6 CONCLUSION

Efficient solution of the discrete boundary element equations are of paramount importance in practical applications. We have shown that the two-grid method can be suitably modified, to remain efficient, even when applied to boundary integral equations with non-compact operators. (The superiority of the iterative methods over direct solution methods can only be fully appreciated when the number of equations is large).

If the non-compactness of the integral operator is due to the non-smoothness of the surface S, then the iterative techniques may need to be further modified [25,26]. To overcome the non-uniqueness in the boundary integral formulation of the Helmholtz equation, many researchers employ, in place of the Burton and Miller formulation, the "CHIEF" method due to Schenck [24]. Whether or not the multi-grid idea can be suitably modified to yield an efficient solver for the resulting over determined equations in the CHIEF method deserves further investigation.

ACKNOWLEDGEMENTS: This research was in part sponsored by Admirlaty Research Establishment, Portland, U.K. The authours would also like to thank the referees for their helpful comments.

References

[1] Amini S (1987). On the choice of the coupling parameter in boundary integral formulations of the exterior acoustic problem. To appear in *Appl. Anal.*

[2] Amini S and Wilton D T (1986). An investigation of boundary element methods for the exterior acoustic problem. *Comp. Meth. Appl. Mech. Engng.* 54, pp 49-65.

[3] Amini S and Chen Ke (1988). Iterative techniques for boundary element methods. In *Boundary Element X,* ed. C A Brebbia, Springer-Verlag.

[4] Amini S and Chen Ke (1989). Conjugate gradient method for second kind integral equations- Applications to the exterior acoustic problem. To appear in *J. Engng. Anal.*

[5] Amini S and Harris P J (1988). Boundary element and finite element methods for the coupled fluid-structure interaction problem. In *Boundary Element X,* ed. C A Brebbia, Springer-Verlag, pp 509-520.

[6] Atkinson K E (1976). *A survey of numerical methods for the solution of Fredholm integral equations of the second kind.* SIAM Philadelphia.

[7] Atkinson K E (1985). Solving integral equations on surfaces in space. In *Constructive methods for the practical treatment of integral equations*, ed. Hammerline G and Hoffman K; Birkhauser, Basel, pp 20-43

[8] Brebbia C A (1980). *The boundary element methods for engineers.* Pentech press, Plymouth.

[9] Burton A J (1973). The solution of Helmholtz equation in exterior domains using integral equations. *NPL report NAC30, National Physical Laboratory,* Teddington, Middlesex, UK.

[10] Burton A J and Miller G F (1971). The application of integral equation methods for the numerical solution of boundary value problems. *Proc. Roy. Soc. London* A232, pp 201-210.

[11] Colton D and Kress R (1983). *Integral equation methods in scattering theory.* Wiley, NY.

[12] Duffy M (1982). Quadrature over a pyramid or cube of integrands with a singularity at the vertex. *SIAM J Numer. Anal.* 19, pp 1260-1262.

[13] Hackbusch W (1987). The panel clustering technique for the boundary element method. *Boundary Element IX*, ed. C A Breabbia et al. CMP Springer Verlag pp 463-474.

[14] Hess J L (1968). Calculation of acoustic fields about arbitrary three-dimensional bodies by a method of surface source distributions based on certain wavenumber expansions. *McDonnell Douglas Rept. No. DAC 66901*, Longbeach, CA.

[15] Ivanov V V (1976). *The theory of approximate methods and their application to numerical solution of singular integral equations.* Noordhoff Int. Publ. Leyden.

[16] Jaswon M A and Symm G T (1977). *Integral equation methods in potential theory and elastostatics.* Academic Press London.

[17] Kleinmann R E and Roach G F (1974). Boundary integral equations for three-dimensional Helmholtz equation. *SIAM rev.*16, pp 214-235.

[18] Kress R (1985). Minimising the condition number of boundary integral operators in acoustic and electomagnetic scattering. *Q. J. Mech. Appl. Math.* 38, pp 323-341.

[19] Meyer W L, Bell W A, Zinn B T and Stallybrass H P (1978). Boundary integral solution of three dimensional acoustic radiation problems. *J. Sound and Vibration* 59(2) pp 245-262.

[20] Naylor W and Sell G R (1982). *Linear operator theory in engineering and science.* Applied Mathematical Science, V 40; Springer Verlag.

[21] Panich I O (1965). On the question of solvability of the exterior boundary value problems for the wave equation and Maxwell's equations. *Russian Math. surv.* 20, pp 221-226.

[22] Reichel L (1987). Parallel iterative methods for the solution of Fredholm integral equations of the second kind. In *Hypercube multiprocessors ed. Heath M T, SIAM Philadelphia* pp 520-529.

[23] Rokhlin V (1985). Rapid solution of integral equations of classical potential theory. *J. Comp. Phys.* 60, pp 187-207.

[24] Schenck H A (1968). Improved integral formulations for acoustic radiation problems. *J. Acoust. Soc. Am.* 44, pp 41-58

[25] Schippers H (1982). Application of multi-grid methods for integral equations to two problems from fluid dynamics. *J. Comp. Phys.* 48, pp 441-461.

[26] Schippers H (1987). Multi-grid methods in boundary element calculations. In *Boundary elements IX*, ed C A Brebbia et al. CMP Springer Verlag, pp 475-492.

[27] Sloan I H (1976). Improvement by iteration for compact operator equations. *Math. Comp.* V 30, No. 136, pp 758-764.

[28] Seybert A F, Wu T W and Wu X F (1988). Radiation and scattering from elastic solids and shells using the boundary element methods. *J. Acoust. Soc. Am. 84(5)*, pp 1906-1912.

[29] Stallybrass M P (1967). On a pointwise variational principle for the approximate solution of linear boundary value problems. *J. Mathematics and Mechanics* 16, pp 1247-1286.

[30] Terai T (1980). On the calculation of sound fields around three-dimensional objects by integral equation methods. *J. Sound Vibration* 69(1), pp 71-100.

[31] Wilton D T (1978). Acoustic radiation and scattering from elastic structures. *Int. J. Numer. Methods Engng.* 13, pp 123-138.

TECHNIQUES FOR SPEEDING UP THE BOUNDARY ELEMENT SOLUTION OF ACOUSTIC RADIATION PROBLEMS

S. M. Kirkup and D. J. Henwood
Department of Mathematical Sciences
Brighton Polytechnic, Brighton, United Kingdom
In Collaboration with Ricardo Consulting Engineers PLC

Abstract A boundary element method (based on the integral equation formulation of Burton and Miller (1971) for the exterior Helmholtz equation) for the computation of the properties of the acoustic field around a set of vibrating bodies is described. Methods for speeding up the computation of the discrete forms of the integral operators and the solution of the linear systems that arise in the boundary element method are considered. A Fortran subroutine AC3GEN, which employs speed-up methods, is described. Results from test problems are given.

I INTRODUCTION

A set of vibrating objects in an acoustic medium such as air or water produces an acoustic field. The problem of determining the properties of the three-dimensional acoustic field from a description of the vibration of the objects, whatever their shape and the nature of their vibration, is the acoustic radiation problem (ARP) we consider in this paper. The overall problem is that of writing a computer program which accepts a description of the surfaces and their vibration and, from that, is able to compute satisfactory approximations to the acoustic properties. The overall approach is illustrated in figure 1.

The ARP is modelled by a set of closed vibrating bodies surrounded by an infinite acoustic medium is illustrated in figure 2; this is known as the acoustic radiation model (ARM). The symbol D is used to represent the points in the interior regions, S to represent the points on the boundaries and E to represent the points in the exterior region.

The ARM can be formulated in terms of a boundary integral equation (BIE). Numerical methods based on the BIE are known as boundary element methods (BEMs). The use of the BEM as a means of obtaining a numerical solution of acoustic radiation and scattering problems have been investigated by researchers over the past two or three decades. Difficulties in the achievement of a robust method and in its implementation have hindered its use. However, with the improved integral equation formulations introduced by Kussmaul (1969) and Burton and Miller (1971) and with the development of a wide range of methods for the discretisation of the integral operators in the BIE, progress has been made in resolving these difficulties (Kirkup and Henwood, 1987 or Kirkup, 1989). Further useful information on these formulations and methods can be found in Burton (1973), Kleinmann and Roach (1974), Burton (1976), Amini (1982, 1983) and Kirkup (1989). Results from the application of BEMs based on these formulations to acoustic problems have been described in Meyer et al (1978), Terai (1980), Sayhi et al (1981), Amini and Wilton (1986) and Kirkup (1989).

Acoustic problems are governed by the wave equation. The decompositon of the time variables into Fourier components reduces the ARM to a sequence of three-dimensional exterior Helmholtz problems with Neumann boundary conditions (**HE3N** problems) of the form

$$\nabla^2\varphi(p) + k^2\varphi(p) = 0 \quad (p\in E) \tag{1}$$

$$\frac{\partial\varphi(p)}{\partial n_p} = v(p) \quad (p\in S);\ (v(p)\text{ given}) \tag{1a}$$

$$\lim_{r\to\infty} r\{\frac{\partial\varphi}{\partial r} - ik\varphi\} = 0 \tag{1b}$$

for k taking the values $k_1, k_2, \ldots, k_n$. In the majority of practical situations the time-dependent variables can only be described approximately by a finite linear sum of Fourier components and in order achieve a satisfactory approximation a large number of distinct wavenumbers are often used (see, for example, Skudrzyk (1971) p390-391). The formulae relating the acoustic properties to the potential φ in (1) are given in section II.

The usual way of solving the problem (1) is to apply the BEM for each value of k individually. This requires the construction of the discrete forms of the integral operators using numerical integration and the solution of a linear system of equations. Clearly, for a surface divided into a large number of elements the use of the BEM in this way is a major computational task. In this paper the ARM as it is posed by (1) - that is the complete problem of solving a large sequence of HE3N problems with different wavenumbers and different boundary conditions corresponding to each wavenumber - is considered in order to obtain a computationally more efficient solution. The techniques that are described are for use in the computation of the discrete form of the BIE and in the solution of the linear systems of equations that arise, they are described with a view to their use in BEMs derived via collocation.

The BIE formulation of the HE3N problem (1) that is considered in this paper is the one introduced by Burton and Miller (1971). The formulation is described in section III and the corresponding BEM is stated in section IV. In section V the various ways of speeding up the evaluation of the discrete forms are considered and in section VI a method for speeding up the solution to the linear system of equations is described.

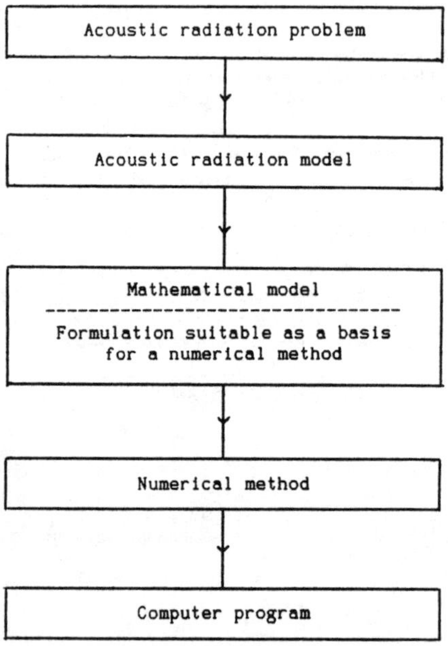

Figure 1. The general approach to obtaining a computational solution to an acoustic radiation problem.

Methods from sections V and VI are implemented in a Fortran subroutine AC3GEN. This subroutine is one of a suite of subroutines for solving acoutic radiation problems (Kirkup and Henwood, 1988) and it is described in section VII. The capabilities of the subroutine for the calculation of sound powers and radiation ratios are demonstrated by applying it to appropriate test problems in section VIII.

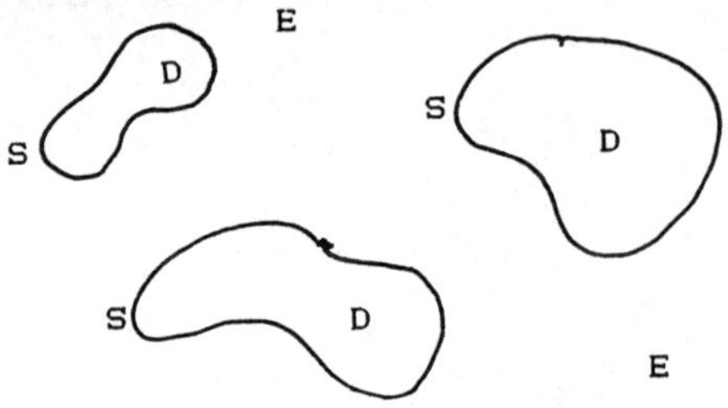

Figure 2. The acoustic radiation model.

II FORMULAE FOR THE ACOUSTIC PROPERTIES

From the solution φ to (1) at each wavenumber k the (time-harmonic) sound pressure is given by:

$$P(\mathbf{p}) = i\rho\omega\varphi(\mathbf{p}) \qquad (\mathbf{p}\in S\cup E), \tag{2}$$

where ρ is the density of the acoustic medium, $\omega=kc$ and c is the speed of sound in the medium. The acoustic intensity on the surface is

$$I(\mathbf{p}) = \tfrac{1}{2}\mathrm{Re}\{P^*(\mathbf{p})v(\mathbf{p})\} \qquad (\mathbf{p}\in S) \tag{3}$$

where * denotes the complex conjugate. The sound power is given by

$$W = \int_S I(\mathbf{q})\, dS_q . \tag{4}$$

The radiation ratio (or radiation efficiency) is given by

$$\sigma_{RAD} = \frac{W}{\tfrac{1}{2}\rho c \int_S v^*(\mathbf{q})v(\mathbf{q})\, dS_q} . \tag{5}$$

III BOUNDARY INTEGRAL EQUATION FORMULATION

The BIE on which we base the BEM is the one introduced in Burton & Miller (1981):

$$\alpha\{M_k\varphi\}(\mathbf{p})-\alpha c(\mathbf{p})\varphi(\mathbf{p})+\beta\{N_k\varphi\}(\mathbf{p}) =$$

$$\alpha\{L_k\frac{\partial\varphi}{\partial n}\}(\mathbf{p})+\beta\{M_k^t\frac{\partial\varphi}{\partial n}\}(\mathbf{p})+\beta c(\mathbf{p})\frac{\partial\varphi}{\partial n}(\mathbf{p}) \tag{6}$$

$$(\mathbf{p}\in D\cup S\cup E)$$

The L_k, M_k, M_k^t, N_k in (6) denote integral operators and are defined as follows:

$$\{L_k\mu\}(p) \equiv \int_S G_k(p,q)\,\mu(q)\,dS_q \;, \qquad (7a)$$

$$\{M_k\mu\}(p) \equiv \int_S \frac{\partial G_k}{\partial n_q}(p,q)\,\mu(q)\,dS_q \;, \qquad (7b)$$

$$\{M_k^t\mu\}(p) \equiv \frac{\partial}{\partial n_p}\int_S G_k(p,q)\,\mu(q)\,dS_q \;, \qquad (7c)$$

$$\{N_k\mu\}(p) \equiv \frac{\partial}{\partial n_p}\int_S \frac{\partial G_k}{\partial n_q}(p,q)\,\mu(q)\,dS_q \;. \qquad (7d)$$

The surface element of integration is at q, n_q is the outward normal at q, $\mu(q)$ is a bounded function defined for $q\in S$, n_p is an outward normal to the boundary passing through p and $p\in D\cup S\cup E$; $G_k(p,q)$ is the free space Green's function for the Helmholtz equation:

$$G_k(p,q) = \frac{e^{ikr}}{4\pi r} \;.$$

The function c(p) is defined as follows:

$$c(p) = \begin{cases} 0 \text{ when } p\in D, \\ \text{solid angle}/4\pi \text{ subtended by the exterior at q when } q\in S, \\ 1 \;\; q\in E \;. \end{cases}$$

The α and β in (6) are complex numbers with $\alpha\neq 0$, $\beta\neq 0$, $\mathrm{Im}(\alpha/\beta)\neq 0$. The conditions on α and β ensure that the integral equation (6) has a unique solution for all k.

IV DERIVATION OF THE BOUNDARY ELEMENT METHOD THROUGH COLLOCATION

The boundary element method is derived by replacing the surface S by an approximate surface and the boundary functions by a linear combination of basis functions. The symbol $\tilde{S}$ is used to represent the approximate surface, the symbols $\tilde{D}$ and $\tilde{E}$ are used to represent the approximate interior and exterior regions. The approximate boundary $\tilde{S}$ can be written as the sum of boundary elements:

$$\tilde{S} = \sum_j \Delta_j \tilde{S} \;.$$

Having replaced S by $\tilde{S}$, the boundary function $\mu(q)$ is replaced by $\tilde{\mu}(q)$ where $q\in\tilde{S}$, and the function c(q) is replaced by $\tilde{c}(q)$.

Similarly, the integral operators L_k, M_k, M_k^t, and N_k defined by equations (7) can be replaced by their analogues $\tilde{L}_k$, $\tilde{M}_k$, $\tilde{M}_k^t$, and $\tilde{N}_k$ with respect to the approximate boundary $\tilde{S}$. For example the $\tilde{L}_k$ operator is defined as follows:

$$\{\tilde{L}_k\tilde{\mu}\}(p) \equiv \int_{\tilde{S}} G_k(p,q)\,\tilde{\mu}(q)\,dS_q \;, \qquad (8)$$

The purpose of the collocation method is to provide a mechanism for describing an approximation to the boundary functions in discrete form. Collocation of the boundary function $\tilde{\mu}(q)$ $(q\in\tilde{S})$ allows the construction of the discrete form of $\{\tilde{L}_k\tilde{\mu}\}(p)$, $\{\tilde{M}_k\tilde{\mu}\}(p)$, $\{\tilde{M}_k^t\tilde{\mu}\}(p)$, and $\{\tilde{N}_k\tilde{\mu}\}(p)$. Let $p_1, p_2, \ldots, p_n \in\tilde{S}$ be the n collocation points and let $\tilde{\chi}_1$, $\tilde{\chi}_2, \ldots, \tilde{\chi}_n$ be basis functions with the usual properties:

$$\text{(i)} \qquad \tilde{\chi}_i(p_j) = \begin{cases} 1 \text{ when } i=j \\ 0 \text{ when } i\neq j \end{cases} \;, \qquad (9a)$$

$$\text{(ii)} \qquad \sum_{j=1}^{n} \tilde{\chi}_j(q) = 1 \qquad (q\in\tilde{S}) \;. \qquad (9b)$$

The boundary function $\tilde{\mu}(q)$ is approximated as follows:

$$\tilde{\mu}(q) \simeq \sum_{j=1}^{n} \tilde{\mu}(p_j)\,\tilde{\chi}_j(q) \qquad (q\in\tilde{S}) \qquad (10)$$

The substitution of the approximation (10) into the definitions of the integral operators allows their expression in discrete form. For example the substitution of (10) into (7) gives:

$$\{\tilde{L}_k\tilde{\mu}\}(p) \simeq \sum_{j=1}^{n} [\;\tilde{\mu}(p_j)\,\{\tilde{L}_k\tilde{\chi}_j\}(p)\;] \;. \qquad (11)$$

The following notation is adopted:

$$[L_k]_{ij} = \{\tilde{L}_k\tilde{\chi}_j\}(p_i) \;, \qquad (12a)$$

$$[M_k]_{ij} = \{\tilde{M}_k\tilde{\chi}_j\}(p_i) \;, \qquad (12b)$$

$$[M_k^t]_{ij} = \{\tilde{M}_k^t\tilde{\chi}_j\}(p_i) \;, \qquad (12c)$$

$$[N_k]_{ij} = \{\tilde{N}_k\tilde{\chi}_j\}(p_i) \;, \qquad (12d)$$

$$\underline{\mu} = [\tilde{\mu}(p_1), \tilde{\mu}(p_2), \ldots, \tilde{\mu}(p_n)]^T \;. \qquad (13)$$

L_k, M_k, M_k^t, and N_k are n×n matrices and $\underline{\mu}$ is an n-vector.

The BEM for the solution of the ARP takes the following form:

■Primary stage. An approximation $\underline{\phi}$ to $\underline{\varphi}$ is obtained by solving the following linear system of equations

$$[\alpha(M_k - C) + \beta N_k]\underline{\phi} = [\alpha L_k + \beta(M_k^t + C)]\underline{v} \quad \square \qquad (14)$$

where $C = \mathrm{diag}\{\tilde{c}(p_1), \tilde{c}(p_2), \ldots, \tilde{c}(p_n)\}$,

$\underline{\varphi} = [\varphi(p_1), \varphi(p_2), \ldots, \varphi(p_n)]^T$,

$\underline{v} = [v(p_1), v(p_2), \ldots, v(p_n)]^T$.

■Secondary stage. An approximation $\phi(p)$ to $\varphi(p)$ $(p\in\tilde{E}\cup\tilde{S})$ is obtained from the summation

$$\phi(p) = \frac{1}{\tilde{c}(p)}\Big(\sum_{j=1}^{n}[\phi_j\{\tilde{M}_k\tilde{\chi}_j\}(p)] \qquad (15)$$

$$+ \sum_{j=1}^{n}[v_j\{\tilde{L}_k\tilde{\chi}_j\}(p)]\Big) \quad (p\in\tilde{E}\cup\tilde{S}) \;\square$$

V THE EVALUATION OF THE DISCRETE FORMS

In this section methods for speeding up the evaluation of the discrete forms of the integral operators are considered. Three different approaches are considered, two based on series approximation and the third based on interpolation. Of the three the one based on interpolation seems the most promising although it requires more storage of values than the other method. The method based on interpolation is the one implemented in the subroutine in section VII.

Taylor expansion of the exp(ikr) term

The exp(ikr) ($r=|\mathbf{r}|$, $\mathbf{r}=\mathbf{p}-\mathbf{q}$) term may be expanded in a Taylor series

$$\exp(ikr) = 1 + ikr + \tfrac{1}{2}(ik)^2r^2 + \ldots \quad . \qquad (16)$$

The substitution of a truncated form of the series (16) into the definition of the integral operators allows the calculation of discrete form of the integral operators by the summation of a finite series. However, the ikr term may not be small and hence a large number of terms may be required to ensure a satisfactory approximation.

An alternative approach, which is applicable only to non-axisymmetric problems and when each basis function is non-zero on a localised set of elements, is derived by Taylor expanding the exp(ikr) term as

$$\exp(ikr)=$$
$$\exp(ikr')(1+ik(r-r')+\tfrac{1}{2}(ik)^2(r-r')^2+\ldots), \qquad (17)$$

where $r' = |\mathbf{p}-\mathbf{q}'|$ is and $\mathbf{q}'$ is a point near to where the basis function is non-zero. It is therefore ensured that the $(r-r')$ term is small. However, for large values of k the $k(r-r')$ term may not be small and hence a fairly large number of terms may be required to ensure a satisfactory approximation. This approach will therefore, perhaps, be feasible only if the values of k of interest are small relative to the size of the elements.

Interpolation of the discrete forms

By evaluating the discrete forms $\{\tilde{L}_k\tilde{\chi}\}(\mathbf{p})$, $\{\tilde{M}_k\tilde{\chi}\}(\mathbf{p})$, $\{\tilde{M}_k{}^t\tilde{\chi}\}(\mathbf{p})$, and $\{\tilde{N}_k\tilde{\chi}\}(\mathbf{p})$ or any linear combination of the discrete forms at some prescribed values of wavenumber k, the values at other wavenumbers can be approximated using interpolation. The success of such a method depends on the number of interpolation points and their position, the nature of the discrete forms as functions of k, the method of interpolation and the range of values over which the interpolant is active. The use of interpolation implies that values of the discrete forms must be stored and hence if this method is used in the primary stage of the boundary element method then several matrices will have to be stored.

The full range of values of wavenumber of interest may be divided into subranges. Interpolation followed by the solution of the **HE3N** problems is then carried out for the values of wavenumber in each subrange individually. In this way the requirement on computer memory is much less than it would be if interpolation is used over the full range directly.

Straightforward polynomial interpolation could be used. Since, in such a case, the interpolation points may be selected by the user, the Chebyshev points should perhaps be chosen, thereby seeking to minimise the maximum error. For example the discrete form $\{L_k\tilde{\chi}\}(\mathbf{p})$ may be approximated for $k\in[k_1,k_2]$ by using the quadratic interpolant.

$$\{\tilde{L}_k\tilde{\chi}\}(\mathbf{p})\simeq \{\tilde{L}_\gamma\tilde{\chi}\}(\mathbf{p})+ \frac{1}{d}(k-\gamma)(\{\tilde{L}_\eta\tilde{\chi}\}(\mathbf{p})-\{\tilde{L}_\gamma\tilde{\chi}\}(\mathbf{p})\}$$

$$+ \frac{1}{2d^2}(k-\gamma)(k-\eta)(\{\tilde{L}_\gamma\tilde{\chi}\}(\mathbf{p})-2\{\tilde{L}_\eta\tilde{\chi}\}(\mathbf{p})\}+\{\tilde{L}_\lambda\tilde{\chi}\}(\mathbf{p})) \qquad (18)$$

where
$$\gamma = k_1+\tfrac{1}{4}(2-\surd 3)(k_2-k_1),$$
$$\eta = \tfrac{1}{2}(k_1+k_2), \qquad (18a)$$
$$\lambda = k_1+\tfrac{1}{4}(2+\surd 3)(k_2-k_1)$$

are the Chebyshev quadratic interpolation points for the interval $[k_1,k_2]$ and

$$d=\eta-\gamma=\lambda-\eta=\tfrac{1}{4}\surd 3(k_2-k_1). \qquad (18b)$$

A similar interpolation can be used on the discrete forms of the other integral operators, or a linear combination of the discrete forms of the integral operators.

The behaviour of the discrete forms $\{\tilde{L}_k\tilde{\chi}\}(\mathbf{p})$, $\{\tilde{M}_k\tilde{\chi}\}(\mathbf{p})$, $\{\tilde{M}_k{}^t\tilde{\chi}\}(\mathbf{p})$, and $\{\tilde{N}_k\tilde{\chi}\}(\mathbf{p})$, as functions of k, is dominated by the exp(ikr) term in the integrand, which is common to the definition of the discrete form of all of the integral operators. For non-axisymmetric problems, and when each $\tilde{\chi}$ is non-zero on a localised set of elements the general success of the interpolation method can be improved by writing the exp(ikr) term in the form of the product of exp(ikr') and exp($ik(r-r')$) where r' is the distance from $\mathbf{p}$ to a point $\mathbf{q}'$ near the elements over which $\tilde{\chi}$ takes non-zero values. If $\tilde{\chi}$ is non zero on a single element then $\mathbf{q}'$ may be chosen to be at the middle of that element. The situation is illustrated in figure 3.

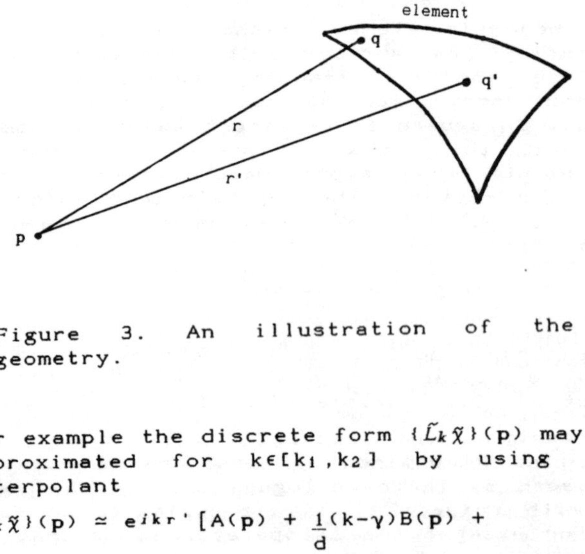

Figure 3. An illustration of the geometry.

For example the discrete form $\{\tilde{L}_k\tilde{\chi}\}(\mathbf{p})$ may be approximated for $k\in[k_1,k_2]$ by using the interpolant

$$\{\tilde{L}_k\tilde{\chi}\}(\mathbf{p}) \simeq e^{ikr'}[A(\mathbf{p}) + \frac{1}{d}(k-\gamma)B(\mathbf{p}) + \frac{1}{2d^2}(k-\gamma)(k-\eta)C(\mathbf{p})]$$

where

$$A(\mathbf{p}) = e^{-i\gamma r'}\{\tilde{L}_\gamma\tilde{\chi}\}(\mathbf{p})$$
$$B(\mathbf{p}) = e^{-i\eta r'}\{\tilde{L}_\eta\tilde{\chi}\}(\mathbf{p}) - e^{-i\gamma r'}\{\tilde{L}_\gamma\tilde{\chi}\}(\mathbf{p})$$
$$C(\mathbf{p}) = e^{-i\gamma r'}\{\tilde{L}_\gamma\tilde{\chi}\}(\mathbf{p}) - 2e^{-i\eta r'}\{\tilde{L}_\eta\tilde{\chi}\}(\mathbf{p})\} + e^{-i\lambda r'}\{\tilde{L}_\lambda\tilde{\chi}\}(\mathbf{p})$$

where γ, η, λ, and d are as defined by (18a,b).

Estimating the values of the discrete forms from previously obtained values

We seek methods for calculating approximations to $\{\tilde{L}_k \tilde{\chi}\}(p)$, $\{\tilde{M}_k \tilde{\chi}\}(p)$, $\{\tilde{M}_k^t \tilde{\chi}\}(p)$, and $\{\tilde{N}_k \tilde{\chi}\}(p)$ for $k=\gamma+\delta$ from their values when $k=\gamma$, where δ may be regarded as being small. Here we consider only a particular example of this approach.

The $\exp(ikr)$ term is regarded as a function of k and r. Let r' be the distance between p and a point q' near the elements over which $\tilde{\chi}$ is non-zero. Refer back to figure 3. We may write

$$\exp(i(\gamma+\delta)r) = \exp(i\gamma r)\exp(i\delta r')\exp(i\delta(r-r'))$$
$$\simeq \exp(i\gamma r)\exp(i\delta r')$$

where in this approximation the $O(\delta(r-r'))$ terms are ignored, hence the methods considered here are applicable only to non-axisymmetric problems and when $\tilde{\chi}$ is non-zero on a localised set of elements.

Using this approximation allows:

(i) $$\{\tilde{L}_{\gamma+\delta}\tilde{\chi}\}(p) = \int_{\tilde{S}} \frac{\exp(i(\gamma+\delta)r)}{4\pi r}\tilde{\chi}\, dS_q$$

$$\simeq \int_{\tilde{S}} \frac{\exp(i\gamma r)\exp(i\delta r')}{4\pi r}\tilde{\chi}\, dS_q$$

$$= \exp(i\delta r')\{\tilde{L}_\gamma \tilde{\chi}\}(p) \ , \qquad (20a)$$

(ii) $\{\tilde{M}_{\gamma+\delta}\tilde{\chi}\}(p)$

$$= \int_{\tilde{S}} \frac{\exp(i(\gamma+\delta)r)}{4\pi r^2}(i(\gamma+\delta)r-1)\frac{\partial r}{\partial n_q}\tilde{\chi}\, dS_q$$

$$\simeq \int_{\tilde{S}} \frac{\exp(i\gamma r)\exp(i\delta r')}{4\pi r^2}(i(\gamma+\delta)r-1)\frac{\partial r}{\partial n_q}\tilde{\chi}\, dS_q$$

$$= \exp(i\delta r')(\{\tilde{M}_\gamma \tilde{\chi}\}(p)+\int_{\tilde{S}}(\frac{\exp(i\gamma r)}{4\pi r^2}\, i\delta r\, \frac{\partial r}{\partial n_q}\tilde{\chi}\, dS_q)$$

$$= \exp(i\delta r')(\{\tilde{M}_\gamma \tilde{\chi}\}(p)+ i\delta\{\tilde{L}_\gamma \tilde{\chi}\}(p)\frac{\partial r'}{\partial n_q'}) , \qquad (20b)$$

(iii) similarly

$$\{\tilde{M}_{\gamma+\delta}^t \tilde{\chi}\}(p) \qquad (20c)$$

$$\simeq \exp(i\delta r')(\{\tilde{M}_\gamma^t \tilde{\chi}\}(p)+ i\delta\{\tilde{L}_\gamma \tilde{\chi}\}(p)\frac{\partial r'}{\partial n_p}) ,$$

(iv) $\{\tilde{N}_{\gamma+\delta}\tilde{\chi}\}(p)$

$$= \int_{\tilde{S}}\frac{\exp(i(\gamma+\delta)r)(2-2i(\gamma+\delta)r-(\gamma+\delta)^2r^2)}{4\pi r^3}\frac{\partial r}{\partial n_p}\frac{\partial r}{\partial n_q}\tilde{\chi}\, dS_q$$

$$+ \int_{\tilde{S}}\frac{\exp(i(\gamma+\delta)r)}{4\pi r^2}(i(\gamma+\delta)r-1)\frac{\partial^2 r}{\partial n_p \partial n_q}\tilde{\chi}\, dS_q$$

$$\simeq \exp(i\delta r')(\{\tilde{N}_\gamma \tilde{\chi}\}(p)+$$

$$+ \int_{\tilde{S}}\frac{\exp(i\gamma r)(-2i\delta r-2\gamma\delta r^2)}{4\pi r^3}\frac{\partial r}{\partial n_p}\frac{\partial r}{\partial n_q}\tilde{\chi}\, dS_q$$

$$+ \int_{\tilde{S}}\frac{\exp(i\gamma r)}{4\pi r^2}\, i\delta r\, \frac{\partial^2 r}{\partial n_p \partial n_q}\tilde{\chi}\, dS_q\ )$$

$$= \exp(i\delta r')(\{\tilde{N}_\gamma \tilde{\chi}\}(p)$$

$$+ 2i\delta\int_{\tilde{S}}\frac{\exp(i\gamma r)}{4\pi r^2}(i\gamma r-1)\frac{\partial r}{\partial n_p}\frac{\partial r}{\partial n_q}\tilde{\chi}\, dS_q$$

$$+ i\delta\int_{\tilde{S}}\frac{\exp(i\gamma r)}{4\pi r}\frac{\partial^2 r}{\partial n_p \partial n_q}\tilde{\chi}\, dS_q\ )$$

$$\simeq \exp(i\delta r')(\{\tilde{N}_\gamma \tilde{\chi}\}(p)+i\delta(2\{\tilde{M}_\gamma \tilde{\chi}\}(p)\frac{\partial r'}{\partial n_p} \qquad (20d)$$

$$+ \{\tilde{L}_\gamma \tilde{\chi}\}(p)\frac{\partial^2 r'}{\partial n_p \partial n_q})) .$$

The approximations (20) suggest methods for the calculation of the approximations to the discrete forms at one wavenumber from their values at another wavenumber.

VI THE SOLUTION OF THE LINEAR SYSTEM

In this section an iterative method for the solution of the linear systems of equations that arise is considered. The use of this type of method on the boundary element solution to acoustic radiation problems was originally considered in Amini (1987).

If an nxn matrix A has known inverse A^{-1} then the solution to the linear system

$$(A+E)\underline{x}=\underline{b} \qquad (21)$$

can be found by using the iterative process

$$\underline{x}_1 = A^{-1}\underline{b} \qquad (22)$$
$$\underline{x}_{j+1} = A^{-1}\underline{b} - A^{-1}E\underline{x}_j \quad j=1,2,\ldots .$$

Alternatively, equivalent information such as the factorisation A=LU may be known, where L is lower triangular and U is upper triangular. The solution to the linear system (21) can then be found by using the iterative process

$$\text{solve for } \underline{x}_1 \quad LU\underline{x}_1 = \underline{b} \qquad (23)$$
$$\text{solve for } \underline{x}_{j+1} \quad LU\underline{x}_{j+1} = \underline{b} - E\underline{x}_j \quad j=1,2,\ldots .$$

Each iteration in methods (22,23) has an $O(n^2)$ computational cost. Convergence in the methods (22,23) is assured if $\|A^{-1}E\|<1$ and, in general, the smaller $\|A^{-1}E\|$ then faster the convergence. The method described is the most elementary iterative method for solving a linear system of equations.

The method develops as follows. The assumption that the respective components of the M_k, M_k^t, and N_k matrices do not vary strongly with k in the range $k\epsilon[k_1,k_2]$ is made (thus the case of axisymmetric problems is perhaps ruled out). The α and β parameters are kept constant (or allowed to vary smoothly) in the range $[k_1,k_2]$. After calculating the inverse (or equivalent) of the matrix that occurs on the left side of the linear system of equations in the primary stage of the BEM for a particular value of $k\epsilon[k_1,k_2]$, this can then be used as an approximate inverse for the iterative solution of the linear systems that arise for any k in the range $[k_1,k_2]$ using methods (22,23).

VII IMPLEMENTATION OF THE BOUNDARY ELEMENT METHOD

In this section a Fortran subroutine 'AC3GEN' for the solution of the ARP is described. The subroutine is based on the **BEM** derived from the BIE of Burton & Miller via collocation. AC3GEN also employs methods for speeding up the evaluation of the discrete forms and the solution to the linear system. An outline of the subroutine is given in section XI. The surfaces S, which may take any shape, are described as the sum of a set of n planar triangles $\Delta\tilde{S}_1, \Delta\tilde{S}_2, \ldots, \Delta\tilde{S}_n$. The normal velocity distribution on the surfaces is described by its value at the collocation points $p_1, p_2, \ldots, p_n$, which are the centroids of the triangles.

The basis functions $\tilde{\chi}_1, \tilde{\chi}_2, \ldots, \tilde{\chi}_n$ are the functions

$$\tilde{\chi}_i(p) = \begin{cases} 1 & \text{if } p \in \Delta\tilde{S}_i \\ 0 & \text{if } p \in \tilde{S} - \Delta\tilde{S}_i \end{cases} \qquad \text{for } i=1,..,n .$$

As input, the subroutine accepts a description of the geometry of the surface (made up of triangles), the wavenumbers under consideration, a description of the normal velocity distribution at each wavenumber, and the coordinates of the exterior points of interest . As output, the subroutine gives, for each wavenumber, the acoustic intensity at the centroid of each triangle, the sound power, the radiation ratio, and the sound pressure at the prescribed exterior points.

Methods for speeding up the boundary element solution to the acoustic radiation problem are employed as follows.

(i) The range of wavenumbers is divided into a set of subranges of approximately equal width. The width of each subrange is chosen to be approximately proportional to the reciprocal of the size of the largest element recieved by the subroutine, although the ratio of the subrange width to the reciprocal of the element size was arrived at only through practical experience with the subroutine.

(ii) For each subrange $[k_1, k_2]$ the weighting parameters are calculated by obtaining approximations to the norms of the matices $M_k - \frac{1}{2}I$ and N_k , so that $\alpha \simeq \|N_k\|_\infty$ and $\beta \simeq i\|M_k - \frac{1}{2}I\|_\infty$ for $k = \frac{1}{2}(k_1 + k_2)$.

(iii) In each subrange $[k_1, k_2]$, the matrices with ij^{th} components

$$[A_k]_{ij} = \exp(-ikr(p_i, p_j))(\alpha\{\tilde{M}_k \tilde{\chi}_j\}(p_i) + \beta\{\tilde{N}_k \tilde{\chi}_j\})(p_i)) \qquad (24a)$$

$$[B_k]_{ij} = \exp(-ikr(p_i, p_j))(\alpha\{\tilde{L}_k \tilde{\chi}_j\}(p_i) + \beta\{\tilde{M}_k{}^t \tilde{\chi}_j\})(p_i)) \qquad (24b)$$

for $k=\gamma, \eta, \lambda$, where γ, η, λ, are as defined by (18a). This gives the matrices A_γ, A_η, A_λ, B_γ, B_η, B_λ.

(iv) The inverse of the matrix $(\alpha M_\eta - \frac{1}{2}\alpha I + \beta N_\eta)$ is evaluated.

(v) Now, for each k in the range $[k_1, k_2]$ approximations to the matrices in (14) can be calculated using

$$[\alpha M_k - \tfrac{1}{2}\alpha I + \beta N_k]_{ij} \simeq e^{ikr'}([A_\gamma]_{ij} + \frac{1}{d}(k-\gamma)([A_\eta]_{ij} - [A_\gamma]_{ij}) + \frac{1}{2d^2}(k-\gamma)(k-\eta)([A_\gamma]_{ij} - 2[A_\eta]_{ij} + [A_\lambda]_{ij})) \qquad (25a)$$
$$- \tfrac{1}{2}\alpha\delta_{ij}$$

$$[\alpha L_k + \beta M_k{}^t + \tfrac{1}{2}\beta I]_{ij} \simeq e^{ikr'}([B_\gamma]_{ij} + \frac{1}{d}(k-\gamma)([B_\eta]_{ij} - [B_\gamma]_{ij}) + \frac{1}{2d^2}(k-\gamma)(k-\eta)([B_\gamma]_{ij} - 2[B_\eta]_{ij} + [B_\lambda]_{ij})) \qquad (25b)$$
$$+ \tfrac{1}{2}\beta\delta_{ij}$$

where $d = \eta - \gamma = \lambda - \eta = \frac{1}{4}\sqrt{3}(k_2 - k_1)$.

(vi) The iterative process (22) is used to solve the linear system (14).

<u>Method</u>

- Calculate normals to elements
- Calculate the diagonal components of L_0 and N_0, as a means of 'subtracting out' the singularity
- Divide the range of wavenumbers into subranges $[k_1, k_2]$.
- Do for each subrange $[k_1, k_2]$
  - Calculate $\alpha \simeq \|N_k\|_\infty$, $\beta \simeq \|M_k - \frac{1}{2}\alpha I\|_\infty$
  - Calculate the A_k, B_k for $k=\gamma, \eta, \lambda$, where γ, η, λ are the Chebyshev interpolation points using equations (24) to give $A_\gamma, A_\eta, A_\lambda, B_\gamma, B_\eta, B_\lambda$ using Gaussian quadrature on a triangle and subtracting out the singularity for the diagonal components.
  - Calculate the inverse of the matrix $[\alpha M_k - \frac{1}{2}\alpha I + \beta N_k]$
  - Do for each point in the exterior p
    - Calculate approximations to $(\tilde{L}_k \tilde{\chi})(p)$, $(\tilde{M}_k \tilde{\chi})(p)$, $(\tilde{M}_k{}^t \tilde{\chi})(p)$, and $(\tilde{N}_k \tilde{\chi})(p)$ for $k=\gamma, \eta, \lambda$.
  - Do for each k in the subrange $[k_1, k_2]$
    - Calculate approximations to the matrices A and B where $A = \alpha M_k - \frac{1}{2}\alpha I + \beta N_k$, $B = \alpha L_k + \beta M_k{}^t + \frac{1}{2}\beta I$ using (25)
    - Use the iterative process (21) to solve the linear system $(\alpha M_k - \frac{1}{2}\alpha I + \beta N_k)\hat{\underline{\phi}} = (\alpha L_k + \beta M_k{}^t + \frac{1}{2}\beta I)\underline{v}$
    - Do for each element $\Delta\tilde{S}_i$
      - Calculate an approximation to the intensity I_i on the element using $\hat{I}_i = \frac{1}{2}\mathrm{Re}\{i\rho\omega v_i \hat{\phi}_i{}^*\}$
    - Calculate an approximation to the sound power W using $\hat{W} = \sum_{i=1}^{n} \mathrm{area}(\Delta\tilde{S}_i)\, \hat{I}_i$
    - Calculate an approximation to the radiation ratio σ_{RAD} using $\hat{\sigma}_{RAD} = \dfrac{\hat{W}}{\frac{1}{2}\rho c \sum_{i=1}^{n} \mathrm{area}(\Delta\tilde{S}_i) v_i v_i{}^*}$
    - Do for each point in the exterior p
      - Calculate approximations to $(\tilde{L}_k \tilde{\chi})(p)$, $(\tilde{M}_k \tilde{\chi})(p)$ using formulae of the form (19).
      - Calculate an approximation to the sound pressure using $\hat{P}(p) = i\rho\omega\hat{\phi}(p)$ with $\hat{\phi}(p)$ calculated using (15).

VIII RESULTS FROM TEST PROBLEMS

Test problems are set up to compute the sound powers and radiation ratios using subroutine AC3GEN and, where appropriate, comparing the results with exact values.

Sound powers for a cube

In the first test problem the surface under consideration is a cube with sides of length 0.2m. The test problem is set up by constructing a substitute problem: the surface normal velocity is prescribed to be the values obtained from the point source at the centre of the cube, as illustrated in figure 4. A point source of unit strength produces a sound power of $k^2\rho c/8\pi$ where k is the wavenumber. Hence a point source of strength $8\pi/(k^2\rho c)$ has unit power and it is used in the test. The wavenumber considered are k=2.0, 2.5, 3.0, ..., 40.0. The cube is divided into 96 uniform elements. The comparison between the results obtained from subroutine AC3GEN and the exact values is given in figure 5.

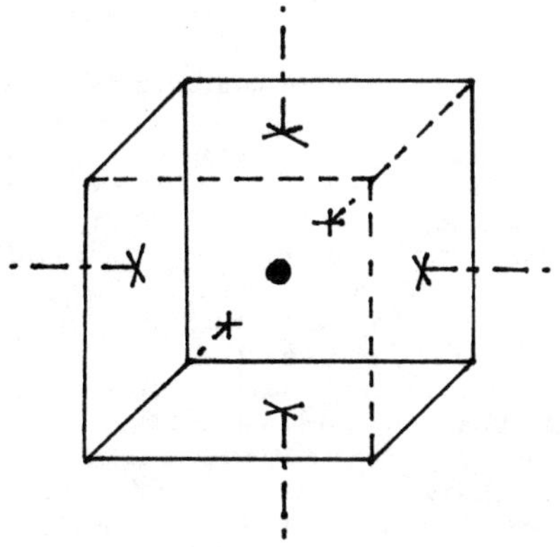

Figure 4. Illustration of the test problem.

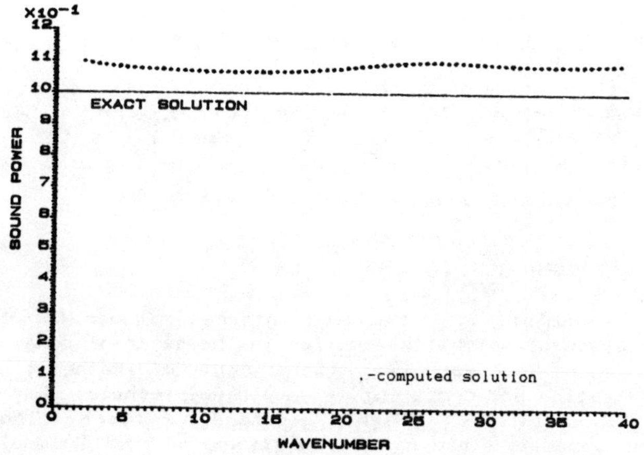

Figure 5. Comparison of computed and exact sound powers for the cube test problem.

Radiation ratio curves for a pulsating and oscillating sphere

In the second test problem the surface under consideration is a sphere of radius 0.5. Two tests are set up in order to compare radiation ratios obtained from subroutine AC3GEN with the exact radiation ratios. The first test is that of a pulsating sphere, the second test is that of an oscillating sphere. The exact radiation ratio for a pulsating sphere (uniform radial motion) is given by the formula

$$\sigma_{RAD} = \frac{(ka)^2}{1+(ka)^2},$$

and the radiation ratio for an oscillating sphere (uniform motion in one direction) is given by the formula

$$\sigma_{RAD} = \frac{(ka)^4}{4+(ka)^4}$$

where k is the wavenumber and a the radius of the sphere, see Morse and Ingard (1968) (p332-336) or Skudrzyk (1971) (p390-391).

The sphere is approximated using 96 triangles of approximately equal size - the approximation to the surface thereby being fairly crude. Figures 6A and 6B compare results obtained from subroutine AC3GEN with exact results for the pulsating and oscillating sphere at the wavenumbers ka=0.1, 0.2, 0.3, ..., 9.9, 10.0 .

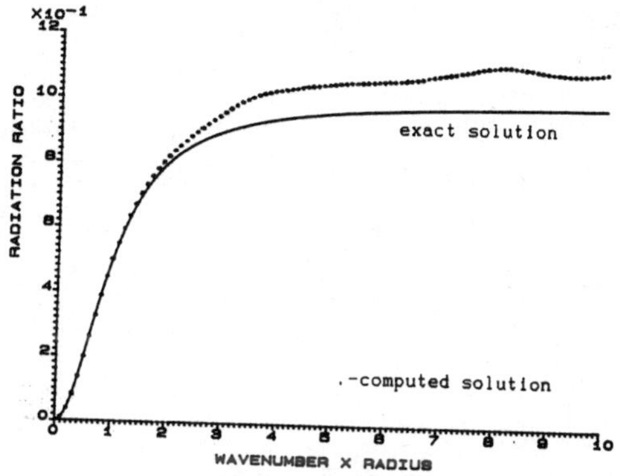

Figure 6A. Comparison of computed and exact sound powers

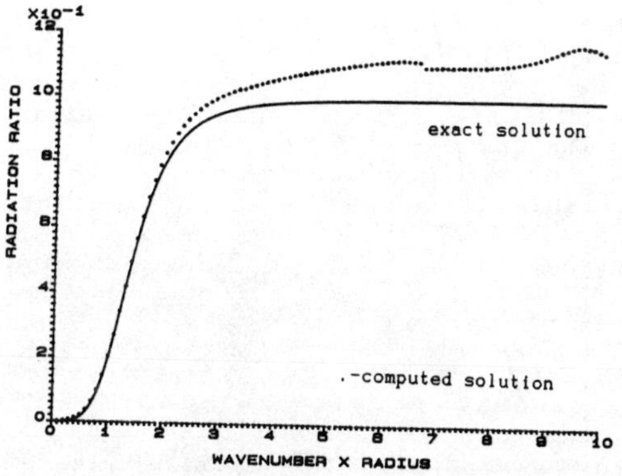

Figure 6B. Comparison of computed and exact sound powers

Radiation ratio curves for a cube with one vibrating side shielded by another cube

This test problem, consists of two cubes with sides of length 0.5. The first cube having a side with uniform vibration, the second cube stands with one side facing the vibrating cube though it is not itself vibrating. Figures 7A and 7B show the radiation ratio curves for the two positions of the cube with vibrating face and shielding face at distances of 0.5 and 5.0 and compare it with the radiation ratio for the situation when there is no second cube.

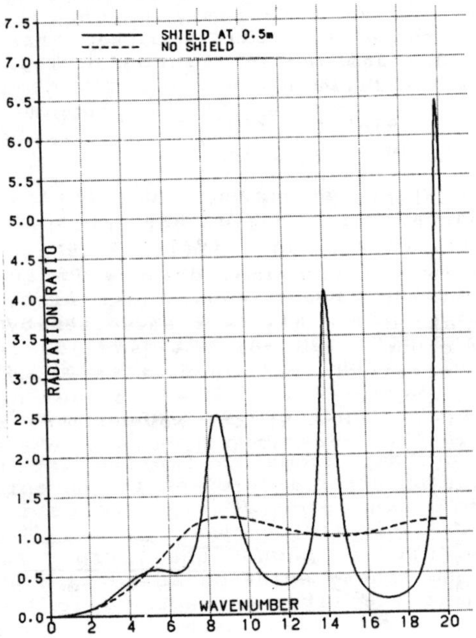

Figure 7A. Radiation ratio curves for cubes distance 0.5 apart and cube alone.

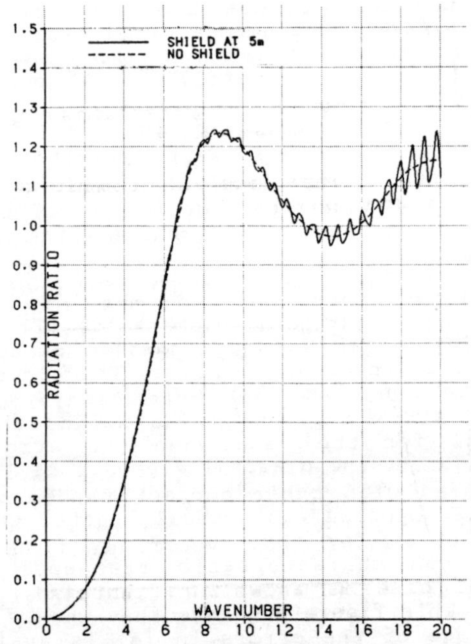

Figure 7B. Radiation ratio curves for cubes distance 5.0 apart and cube alone.

IX CONCLUDING DISCUSSION

In this paper we have described speed-up methods and shown how they may be employed in the BEM based on collocation of the Burton & Miller formulation. The methods have been tested and results given for systems of both one and two separate surfaces.

Of the methods for speeding up the computation of the discrete form the interpolation method is probably the most useful. The method is implemented in the subroutine and the jumps in figures 6A and 6B clearly show the underlying method of partitioning the frequency range into bands. From the evidence of the test problems, the iterative method for solving the linear system seemed to work well. The jumps in figures 6A and 6B indicate that the error from the interpolation is small in comparison with the error from other sources such as the effect of approximating the boundary functions and approximating the boundary.

There are many added difficulties with implementing speed-up methods. The variation of the matrix elements with wavenumber is dependent on the size of the elements. Hence, since the effectiveness of the interpolation rule and the iterative method are dependent on this, the size of the bands should also be closely related to the size of the elements, as they are in subroutine AC3GEN. The use of the speed-up methods also demands that the choice of parameters α and β used remains steady throughout each band of wavenumbers. Otherwise one of the assumptions on which the iterative method is based would not hold and the method would not work. Generally it is found that the iterative method works well near the centre of the band whereas, near the edges of the band, convergence is slow. An analysis of the effectiveness of the iterative method would be useful. It may be possible to construct more successful variants on the iterative method.

Figures 7 show the effect of a shield on the radiation ratio of a vibrating panel on a cube. The figures show that when the second cube is placed at a distance which is of the same order as the wavelengths under consideration acoustic near-resonances can occur. Figure 7B shows that if the distance of the second cube from the first is much greater than the wavelength then the effect of the radiation ratio is small (although weak resonances are evident in the figure).

X REFERENCES

1. Amini, S., "Boundary integral methods for the solution of Helmholtz equation in the exterior domain", MOD Rep 1, Sept 1982, Department of Mathematics, Statistics and Computing, Plymouth Polytechnic, UK.

2. Amini, S., "Highly accurate numerical methods for boundary integral formulations of the Helmholtz equation in the exterior domain", MOD Rep 2, Mar 1983, Department of Mathematics, Statistics and Computing, Plymouth Polytechnic, UK.

3. Amini, S. and Wilton, D.T., "An investigation of boundary element methods for the exterior acoustic problem", Computer methods in Applied Mechanics and Engineering Vol 54, 1986, pp.49-65

4 Amini, S., "An iterative method for the boundary element solution of the exterior acoustic problem", Journal of Computational and Applied Mathematics, Vol 20, Nov 1987, pp.109-117,

5 Burton, A.J. and Miller, G.F., "The application of integral equation methods to the numerical solution of some exterior boundary value problems", Proc. of the Royal Society London A323, 1971, pp.201-210

6 Burton, A.J. "The solution of Helmholtz' equation in exterior domains using integral equations", NPL Report NAC 30, 1973, National Physical Laboratory, Teddington, Middlesex, UK.

7 Burton, A.J. "Numerical solution of acoustic radiation problems", NPL Report OC5/535, 1976, National Physical Laboratory, Teddington, Middlesex, UK.

8 Kirkup S.M. and Henwood D.J., "The discretisation of the integral operators in the application of the boundary element method to Helmholtz problems" Report M12, Oct 1987, Department of Mathematics, Statistics, and O.R, Brighton Polytechnic, UK.

9 Kirkup, S.M. and Henwood, D.J., "Numerical solution of acoustic radiation problems", Report M13, Jan 1988, Department of Mathematics, Statistics, and O.R, Brighton Polytechnic, UK.

10 Kirkup S.M., "Solution of exterior acoustic problems by the boundary element method", PhD thesis, 1989, Brighton Polytechnic, UK.

11 Kleinmann R.E. and Roach, G.F., "Boundary integral equations for the three-dimensional Helmholtz equation", SIAM Review, Vol 16(2), 1974, pp.214-235.

12 Kussmaul, R., "Ein numerisches Verfahren zur Lösung der Neumannschen Außrenraumproblems für die Helmholtzsche Schwingungsgleichung", Computing 4, 1969, pp.246-273.

13 Meyer, W.L., Bell, W.A., Zinn, B.T. and Stallybrass, M.P., "Boundary integral solutions of three dimensional acoustic radiation problems", Journal of Sound and Vibration, Vol 59(2), 1978, pp.245-262

14 Morse P.M. and Ingard, K.U., "Theoretical acoustics", McGraw-Hill, 1968, pp.332-336.

15 Pierce A.D. "Acoustics : an introduction to its physical principles and applications", McGraw-Hill, 1981.

16 Sayhi, M.N., Ousset, Y., Verchery, G., "Solution of radiation problemsby collocation of integral formulations in terms of single and double layer potentials" Journal of Sound and Vibration, Vol 74(2), 1981, pp.187-204.

17 Skudrzyk, E. "The foundations of acoustics", Springer-Verlag, 1971.

18 Terai, T., "On the calculation of sound fields around three-dimensional objects by integral equation methods", Journal of Sound and Vibration, Vol 69(1), 1980, pp.71-100.

XI APPENDIX - SUBROUTINE AC3GEN

```
SUBROUTINE AC3GEN(RHO,C,VERTPTS,ELEMVERT,WAVENUMBERS,
 NORMVELS,EXTPTS,INTENSITIES,POWERS,RADRATIOS,
 EXTPRESSURE)
COMMON/SMKCOM/ MAXELEM,NOELEM,MAXVERT,NOVERT,
 MAXWAVE,NOWAVE,MAXEXTPT,NOEXTPT
COMMON/TGEOME/ NTMAXELEM,NTMAXVERT
COMMON/SURFEL/ NOSURF,NOELS(LMAXSURF)

REAL*4      RHO
REAL*4      C
REAL*4      VERTPTS(MAXVERT,3)
INTEGER*4   ELEMVERT(MAXELEM,3)
REAL*4      WAVENUMBERS(MAXWAVE)
COMPLEX*8   NORMVELS(MAXWAVE,MAXELEM)
REAL*4      EXTPTS(MAXEXTPT,3)
REAL*4      INTENSITIES(MAXWAVE,MAXELEM)
REAL*4      POWERS(MAXWAVE)
REAL*4      RADRATIOS(MAXWAVE)
COMPLEX*8   EXTPRESSURE(MAXWAVE,MAXEXTPT)
```

The COMMON area

The COMMON/SMKCOM/ area contains the eight integer values MAXELEM, NOELEM, MAXVERT, NOVERT, MAXWAVE and NOWAVE.

MAXELEM = the limit on the number of elements,
NOELEM = the actual number of elements,
MAXVERT = the limit on the number of vertices,
NOVERT = the actual number of vertices,
MAXWAVE = the limit on the number of wavenumbers,
NOWAVE = the actual number of wavenumbers,
MAXEXTPT = the limit on the number of prescribed exterior points,
NOEXTPT = the actual number of prescribed exterior points.

The COMMON/SURFEL area states the number of surfaces and the number of elements on each surface.
NOSURF = the actual number of surfaces.
NOELS(LMAXSURF) = the list of the number of elements on each surface. That is NOELS(j) states the number of elements on the jth surface.

The parameters (input)

RHO,C - the density and the speed of sound.
VERTPTS,ELEMVERT describe the geometry of the bodies.
WAVENUMBERS - the list of wavenumbers.
NORMVELS - the velocity normal to each element of the bodies at each wavenumber.
EXTPTS - lists the coordinates of the exterior points.

The parameters (output)

INTENSITIES - the sound intensity on each element at each wavenumber.
POWERS - the sound power at each wavenumber.
RADRATIOS - the radiation ratio at each wavenumber.
EXTPRESSURE - the sound pressure at each exterior point at each wavenumber.

XII ACKNOWLEDGEMENT

The authors would like to thank the Directors of Ricardo Consulting Engineers PLC[1] and Brighton Polytechnic[2] for their sponsorship of the research described in this paper. The authors are grateful to Dr. G. T. Symm and Mr G. F. Miller of the National Physical Laboratory[3] for their helpful advice and to Mr R. J. Tyrrell of Ricardo[1] for his support.

The work described in this paper is taken from the PhD thesis (reference 10).

[1]Ricardo Consulting Engineers PLC,
Bridge Works, Shoreham-by-Sea,
West Sussex, United Kingdom. BN4 5FG.

[2]Brighton Polytechnic,
Mithras House, Lewes Road, Brighton,
East Sussex, United Kingdom. BN2 4AT.

[3]National Physical Laboratory,
Teddington, Middlesex,
United Kingdom. TW11 OIW.

AN IMPROVED FORMULATION FOR ACOUSTIC EIGENMODE EXTRACTION FROM BOUNDARY ELEMENT MODELS

J. P. Coyette and K. R. Fyfe
Dynamic Engineering N.V.
Leuven, Belgium

ABSTRACT

The acoustic boundary element method has been utilized primarily as a direct response analysis technique. Recently, methodologies have been published that enable the formulation of an eigenvalue problem from a boundary element model. These techniques, however, have been very slow due to the inefficient synthesis techniques used in solving the algebraic problems.

In this paper, an alternative procedure is obtained which indirectly calculates the desired acoustic variables. This technique greatly enhances the calculation speed in setting up the eigenvalue problem and in determining the field pressures. A subspace iteration technique is outlined to solve the generalized unsymmetric eigenvalue problem. Examples in both two and three dimensions are presented to show the accuracy of the method.

NOMENCLATURE

A, B	boundary element matrices
$\hat{A}$, $\hat{B}$	matrices of generalized eigenproblem
c	sound speed, coefficient
C, D, T	influence matrices
f	function
G	Green's kernel
J	functional
k	wavenumber
L, U	Crout factorization matrices
n	normal direction
N	interpolation function
p	acoustic pressure
p_a, p_b	pressure components
r	radial distance
S	boundary surface
t	time, frequency shift
V	interior volume
x	field point
y	surface point
ϕ	source strength, eigenvectors
λ	eigenvalue
ω	circular frequency

INTRODUCTION

Amongst the available techniques for numerical acoustic analysis, the finite element and boundary element methods have been the most commonly applied methods to solve problems in the low and intermediate frequency ranges. These frequency restrictions are encountered because it is necessary to construct discrete models which contain a minimum of 5 nodes per smallest acoustic wavelength to be studied. The finite element (FE) method requires an entire field discretization while the boundary element (BE) method needs only to have a model of the surface or boundary of the field. This fact alone makes the boundary element method very attractive due to the much simpler modelling requirements.

The main objectives in the numerical modeling of interior problems lie in the prediction of acoustic eigenmodes and the evaluation of forced response in the frequency domain. FE models are well suited to both types of studies. Acoustic eigenmodes can be computed following procedures similar to those used in structural mechanics. Forced response can be performed by either direct analysis or by modal superposition. Quite recently, new non–modal techniques (Coyette, 1989) have also been developed which greatly enhance the performance obtained from FE models. For BE models however, the only technique available consists of performing a direct analysis, frequency by frequency because the frequency dependent parameters are embedded in the kernel functions. It has thus been difficult to perform an eigenfrequency analysis with the BE method.

This situation has changed recently with the publication of a new BE formulation for acoustic eigenfrequencies, which allow for the derivation of an algebraic eigenfrequency problem (Banerjee et al., 1988). This formulation follows from earlier work of Nardini and Brebbia (1982). In this paper, the method of Banerjee, Ahmad and Wang (1988) is further improved and special attention is devoted to the numerical solutions of the derived unsymmetric generalized eigenvalue problem. Practical problems are solved using both FE and BE approaches. Comparisons of the resulting eigenvalues and eigenmodes demonstrate the versatility and accuracy of the BE eigensolution.

USUAL FINITE ELEMENT AND BOUNDARY ELEMENT FORMULATIONS

With harmonic time dependence like exp $(i\omega t)$, the acoustic problem is governed by the Helmholtz equation :

(1) $$\nabla^2 p + k^2 p = 0 \quad in \quad V$$

where p is the acoustic pressure and k is the acoustic wavenumber ($k = \frac{\omega}{c}$, ω is the circular frequency and c is the speed of sound). The boundary conditions specified are related to pressure velocities and impedances imposed on selected parts of the boundary. When computing eigenfrequencies, it is usually assumed that hard wall conditions are applied on the boundaries. This leads to the constraint :

(2) $$\frac{\partial p}{\partial n} = 0 \quad on \quad S$$

The main problem consists of extracting the eigenfrequencies and modeshapes from a discrete cavity model. This task is usually performed using the FE method which is based on minimization of a functional J (Petyt et al., 1976) :

(3) $$J = \frac{1}{2}\int_V ((\nabla p, \nabla p) - k^2 p^2)\, dV$$

Assuming discretization of V into (volume) finite elements and selecting appropriate (C^0) interpolation function N, both pressure p and pressure gradient ∇p can be expressed in terms of nodal pressure P through relations :

(4) $$p = N\,P$$
$$\nabla p = B\,P$$

The discretized functional takes the form

(5) $$J = \frac{1}{2}(P^T K P - \omega^2 P^T M P)$$

where :

(6) $$K = \int_V B^T B\, dV$$
$$M = \int_V \frac{N^T N}{c^2}\, dV$$

are the acoustic "stiffness" and "mass" matrices.

Minimization with respect to P leads to the generalized eigenvalue problem

(7) $$(K - \omega_i^2 M)\, P_i = 0$$

which enables extraction of eigenvalues ω_i and eigenvectors P_i . Both symmetry and sparseness attributes of K and M matrices can be fully exploited when performing this task (Bathe, 1982).

Another approach to perform acoustic modelling lies in the use of boundary element methods. Both direct and indirect formulations can be selected for this purpose (Bernhard, 1987). In either case, the integral representation for acoustic pressure refers to the frequency dependent kernel G_k . Restricting attention to the direct formulation, the basic integral equation takes the form of :

(8) $$c(x)p(x) = \int_S \left(p(y)\,\frac{\partial G_k(x,\,y)}{\partial n_y} - \frac{\partial p(y)}{\partial n_y}\,G_k(x,\,y) \right) dS(y)$$

where :

(9) $$c(x) = 1 \quad \text{for} \quad x \in V$$
$$c(x) = \frac{1}{2} \quad \text{for} \quad x \in S$$

The G_k kernel is the fundamental solution of (1) for a point source and takes the following forms for 2–D and 3–D problems :

(10) $$G_k(x,\,y) = H_0^{(1)}(kr) \quad (2-D)$$
$$= \frac{\exp\,(-kr)}{4\pi r} \quad (3-D)$$

where $H_0^{(1)}$ is the Hankel function of the first kind and zero–th order while r denotes the distance between field point x and boundary point y.

The usual collocation procedures consists of selecting a boundary element mesh and appropriate interpolation functions for p and $\frac{\partial p}{\partial n}$ on these elements and to write equation (8) at the boundary nodes. The resulting equations can be written in matrix form as :

(11) $$A_k \,.\, P = B_k \,.\, \frac{\partial P}{\partial n}$$

which relates pressures and pressure normal derivatives on the boundary surface. It must be pointed out that both A_k and B_k matrices are full complex (and frequency dependent) matrices. Assuming hard wall boundary condition (2), system (11) reduces to

(12) $$A_k \,.\, P = 0$$

Extraction of eigenfrequencies usually relies on the vanishing condition for the determinant of A_k. This non linear problem can be solved using a secant method (Demey, 1976 and Adeye et al., 1985).

REFINED BOUNDARY ELEMENT FORMULATION

The enhanced procedure proposed in this section is based on the so–called "Dual Reciprocity" method originally developed by Nardini and Brebbia (1982) and successfully applied to acoustic problems by Banerjee, Ahmad and Wang (1988). Basically, this method involves the treatment of two sub–problems followed by some synthesis procedure.

The actual refinement deals with the synthesis procedure together with suitable tools for modes extraction. The solution for the total pressure, p, of the Helmholtz equation can be split up into two components p_a and p_b such that :

(13) $$p = p_a + p_b$$

and satisfying the following field equations

(14) $$\nabla^2 p_a = 0$$

(15) $$\nabla^2 p_b + k^2 p = 0$$

The resulting pressure p satisfies the Helmholtz equation and it must also satisfy the hard wall boundary condition on S, i.e.

(16) $$\frac{\partial p}{\partial n} = \frac{\partial p_a}{\partial n} + \frac{\partial p_b}{\partial n} = 0$$

The solution of these two sub–problems is discussed below together with the improved synthesis procedure.

Problem A

The first sub–problem requires the solution of the Laplace equation $\nabla^2 p_a = 0$ with the boundary condition :

(17) $$\frac{\partial p_a}{\partial n} = \overline{\frac{\partial p_a}{\partial n}} \quad on \quad S$$

where the overbar denotes a fixed value for pressure normal derivative. The integral representation of this problem is widely described in many textbooks about the BE method (Brebbia et al., 1984) :

(18) $$c(X)\; p_a(X) = \int_S \left(p_a(Y)\,\frac{\partial G(X,Y)}{\partial n_Y} - \frac{\partial p_a(Y)}{\partial n_Y}\,G(X,Y) \right) dS(Y)$$

where : $c(X) = 1$ if X is inside the surface S

$c(X) = 1/2$ if X is on a smooth portion of the surface S

$c(X) = 0$ if X is outside the surface S.

The variable, G, is the well known kernel of the Laplace equation given by

(19)
$$G(X,Y) = -\frac{1}{2\pi}\log r \qquad \text{for 2–D problems}$$

$$\text{and } G(X,Y) = -\frac{1}{4\pi r} \qquad \text{for 3–D problems}$$

where r is again the distance between field point X and boundary point Y

Using a boundary element mesh (surface mesh) and selecting appropriate interpolation functions N^e on each element S^e, the discretized form of equation (18) is given by,

(20)
$$c(X_k)\, p_a(X_k) = \sum_e \left[\int_{S^e} [N^e(Y)] \frac{\partial G(X_k,Y)}{\partial n_Y} dS^e(Y) \right] (p_a^e) - \sum_e \left[\int_{S^e} [N^e(Y)]\, G(X_k,Y)\, dS^e(Y) \right] \left(\frac{\partial p_a^e}{\partial n} \right)$$

In this expression, (p_a^e) and $(\frac{\partial p_a^e}{\partial n})$ denote the vectors of nodal pressure and nodal normal derivative respectively related to boundary element e.

Selecting the usual collocation scheme (collocation points being located at the surface nodes), the integral relations (20) can be written in matrix form as :

(21)
$$A \,.\, P_a = B \frac{\partial P_a}{\partial n}$$

where the A matrix is built with contributions from the left hand side and the first component of the right hand side, while the B matrix is composed of the remainder of the right hand side. Explicit computation of the diagonal coefficients of the A matrix can be avoided by application of the rigid mode concept (Brebbia et al., 1984) to equation (21). P_a and $\frac{\partial P_a}{\partial n}$ are vectors containing boundary nodal pressures and pressure gradients. It must be emphasized that equation (21) is frequency independent and that the normal derivative of p_a is mainly related to the normal derivative of p_b through equation (16).

Problem B

For the second problem it is necessary to obtain the solution of

(22)
$$\nabla^2 p_b + k^2 p = 0$$

To be able to analytically solve this equation, one must provide an approximation for the field pressure p. Following from Nardini et al. (1982) and Banerjee et al. (1988), we select the next representation for p,

(23)
$$p(X) = \sum_s c(X,Y_s)\, \phi(Y_s)$$

where $\phi(Y_s)$ is the *strength* of the source located at Y_s, while $c(X,Y_s)$ is the *influence* function related to the source. These sources are assumed to be located at boundary nodes for simplicity. A simple expression for $c(X,Y_s)$ is given by

(24)
$$c(X,Y_s) = R_{max} - R(X,Y_s)$$

where R_{max} is the largest distance between two boundary nodes and R is the usual distance between point X and the nodal source Y_s. This particular choice of the influence function eliminates singularities in the solution procedure. Performing integration of equation (22), the solution p_b can be formulated as

(25)
$$p_b(X) = \sum_s k^2 D(X,Y_s)\, \phi(Y_s)$$

where $D(X,Y_s)$ is given by

(26)
$$D(X,Y_s) = c_1\, R(X,Y_s)^3 - c_2\, R_{max}\, R(X,Y_s)^2$$

with $c_1 = (3\,.\,(d+1))^{-1}$ and $c_2 = (2d)^{-1}$. In these expressions, the variable $d = 2$ for 2–D problems and $d = 3$ for 3–D applications. The validation of expression (26) can be verified by simple substitution of relations (23 – 26) into (22).

Selecting matrix notations (P for total nodal pressures, P_b and $\frac{\partial P_b}{\partial n}$ for nodal pressure and derivative related to problem B and Φ for nodal sources), equations (23) and (25) written at the boundary nodes can be summarized as

(27)
$$P = C\, \Phi$$

(28)
$$P_b = k^2 D\, \Phi$$

(29)
$$\frac{\partial P_b}{\partial n} = k^2\, T\, \Phi$$

where the C and D matrices contain coefficients given by equations (24) and (26) while the T matrix is built from the normal derivative of equation (25) and can easily be computed from the following scalar product :

(30)
$$\frac{\partial P_b}{\partial n}(X) = (\nabla P_b(X),\, n)$$

Synthesis Procedure

This procedure consists of perfoming the coupling of the two previously defined sub–problems. Equations (13) and (16) provide relations between the solution of the a and b sub–problems :

(31)
$$P_a = P - P_b$$

(32)
$$\frac{\partial P_a}{\partial n} = -\frac{\partial P_b}{\partial n}$$

Bringing these relationships into equation (21) allows the elimination of P_a thus,

(33)
$$A\,.\,(P - P_b) = -B\,.\,\frac{\partial P_b}{\partial n}$$

Instead of eliminating the pressure vector P_b (functions of Φ through relation (28) with Φ related to P through equation (27)), we formulate all remaining vectors as a function of Φ and obtain

(34)
$$A\,.\,P = -B\frac{\partial P_b}{\partial n} + A\,.\,P_b$$

(35)
$$A\,.\,C\,.\,\Phi = -k^2 B\,.\,T\,.\,\Phi + k^2 A\,.\,D\,.\,\Phi$$

or

(36)
$$\hat{A}\,\Phi = k^2\,\hat{B}\,\Phi$$

where

(37)
$$\hat{A} = A\,.\,C$$

(38)
$$\hat{B} = A\,.\,D - B\,.\,T$$

The formulation of the above unsymmetric eigenvalue problem requires only elementary matrix manipulations in contrast to the approach of Ahmad and Banerjee (1988) which involves timely matrix inversions. The solution of equation (36) will yield eigenvalues, k_i, and eigenvectors, Φ_i, from which mode shapes can be calculated using equation (23).

SOLUTION PROCEDURE FOR EIGENVALUES

The eigenvalue problem formulated in the previous section is a generalized eigenproblem with real unsymmetric matrices. A large number of solution techniques are available for generalized eigenvalue problems with real symmetric matrices. These techniques enable efficient solution of large structural eigenvalue problems.

Amongst the available techniques, the common subspace iteration method (Bathe, 1982) can be transformed in order to handle unsymmetric problems. Such an extension has been performed recently by Yu (1986) for fluid–structure interaction problems and is summarized below for sake of completeness.

Starting with the general eigenvalue problem,

(39) $$\hat{A}\,\Phi = \lambda\,\hat{B}\,\Phi$$

the analysis proceeds in seven steps summarized below :

Step 1 : Selection of a shift t on eigenvalues and of a set of starting vectors Φ_1. This task can be performed using the same general rules formulated for symmetric problems (Bathe, 1982). Two important parts of this step include determining the number of iteration vectors and the solution of the iteration vectors.

Step 2 : Shifting the eigenvalue problem and Crout factorization of the related left hand side matrix,

(40) $$\hat{A}_t = \hat{A} - t\hat{B}$$

(41) $$\hat{A}_t = LU$$

Step 3 : Computation of the inverse iteration vectors Φ^*_{i+1} through the solution of the unsymmetric system of equations,

(42) $$\hat{A}_t\,\Phi^*_{i+1} = \hat{B}\,\Phi_i$$

Step 4 : Projection of $\hat{A}_t$ and $\hat{B}$ matrices :

(43) $$\hat{A}^*_{t,i+1} = (\Phi^*_{i+1})^T\,\hat{A}_t\,\Phi^*_{i+1}$$

(44) $$\hat{B}^*_{t,i+1} = (\Phi^*_{i+1})^T\,\hat{B}_t\,\Phi^*_{i+1}$$

Step 5 : Solution of the reduced eigenproblem :

(45) $$\hat{A}^*_{t,i+1}\,R_{i+1} = \hat{B}^*_{i+1}\,R_{i+1}\,\lambda_{i+1}$$

where R_{i+1} is the matrix containing eigenvectors while is λ_{i+1} the diagonal matrix with the eigenvalues. This reduced problem may be solved using the real form of either *LZ* (Kaufman, 1975) or *QZ* (Garbow, 1977) methods. Both routines have been used with success here.

Step 6 : Computation of improved eigenvectors,

(46) $$\Phi_{i+1} = \Phi^*_{i+1}\,R_{i+1}$$

Step 7 : Check convergence of the eigenvalues and repeat steps 3 to 6 if necessary. The eigenpaires, $(\lambda + tI,\ \Phi)$, are the subspace eigensolution at the shift t.

APPLICATIONS

The above procedures have been implemented in SYSNOISE (Guisset and Coyette, 1988) which is a general purpose program for acoustic and elasto–acoustic analysis. Boundary element solutions can so be compared with conventional finite element solutions.

To show the accuracy of the above described procedures, two example problems consisting of a three–dimensional rectangular box and a two–dimensional car interior are studied. Both of these studies are compared to finite element calculations using a similar discretization level.

The modeled box measured 1.0 by 0.8 by 0.7 meters. Two finite element meshes and two boundary element meshes where considered in the comparison. The properties of the meshes and the first five eigenfrequencies for each box are shown in table 1 (the exact analytical results are also given in the table). It should be noted the boundary element meshes for this test are obtained by taking the envelope (outer surface) of the corresponding finite element mesh. It is seen that comparable accuracy is obtained between the equivalent FE and BE meshes. It is noted that the FE eigenvalues consistently over–estimate the frequencies, while the BE eigenvalues were higher for the first two or three modes and then below the exact eigenfrequencies for the higher modes.

The modes of a two–dimensional car body were then determined. The finite element is shown in figure 1. The corresponding boundary element mesh consisted of only the surface nodes of this mesh. Table 2 lists a comparison of the first five modal frequencies. Again good agreement is observed. The nodal lines of these modes, resulting from both calculations, are shown overlaid in figure 2. It is seen that there is little difference between these two calculations except for the fifth mode.

To show the nodal lines for the BE calculation, it was necessary to determine the internal field pressures at each resonant frequency once the modes had been calculated. For this purpose, the finite element mesh nodal points were used. It must be stressed that this mesh was only used for the post–processing, and that only the surface mesh was used for the eigenmode and surface pressure determination. The field point calculation serves only to view the spatial variation of the pressure in the enclosed domain, rather than only on the enclosed line boundary.

Another application of the boundary element eigenvalue technique was in the study of the acoustic behavior of a series of car models. Four models were chosen :

1. Sedan
2. Hatchback with cover
3. Hatchback without cover
4. Stationwagen

Table 3 lists the first five modes of the car bodies. Figures 3 and 4 show the nodal contour lines of the first and third modes of these bodies.These results are in good agreement with the general trends observed by Nefske (1985) in his study of car cavity acoustic properties. As expected, the resonant frequencies fall with increasing cavity length and the nodal lines shift towards the rear of the car.

It must be pointed that computation times are greater for the BE models than for the FE models. This result is mainly due to the fact that BE modesl, while dimensionally smaller than FE models, generate full unsymmetric matrices, thus the BE formulation does not have a computational advantage over FE formulation proportional to the size reduction of the model. For pre–design evaluation, the BE model is however very attractive because it enables fast model generation and allows to easily assess effects of geometry modifications.

CONCLUSIONS

An improved boundary element eigen–solution formulation has been presented which avoids costly matrix inversions in the synthesis procedure. A subspace iteration technique has also been outlined to efficiently determine the eigenvalues of the resulting unsymmetric problem. Finally examples have been presented which show the accuracy of the technique in determining both eigenfrequencies and modes of the enclosed fluid medium. The accuracy and relative speed of these BEM eigenfrequency calculations, combined with the greatly simplified meshing requirements, especially for three–dimensional problems, give great prospects for this technique as a design and trouble–shooting tool.

REFERENCES

Adeye, J.O., Bernal, M.J.M, and Pitman, K.E., 1985, "An Improved Boundary Integral Equation Method for Helmholtz Equations," , *International Journal for Numerical Methods in Engineering*, Vol. 21, pp. 779–787.

Banerjee, P.K., Ahmad, S., and Wang, H.C., 1988, "A New BEM formulation for the Acoustic Eigenfrequency Analysis," *International Journal for Numerical Methods in Engineering*, Vol. 26, pp. 1299–1309.

Bathe, K.J., 1982, "Finite Element Procedures in Engineering Analysis," Prentice Hall, Englewood Cliffs.

Bernhard, R.J., Gardner, B.K., and Mollo, C.G., 1987, "Prediction of Sound Fields in Cavities using Boundary Element Methods," *AAIA Journal*, Vol. 25, pp. 1176–1183.

Brebbia, C.A., Telles, J.C.F., and Wrobel, L.C., 1984,"Boundary Element Techniques, Theory and Applications in Engineering," Springer Verlag, Berlin.

Coyette, J.P., 1988, "A New Ritz Vectors Method for Modelling Damped Acoustic Systems," to be published.

Demey, G., 1976, "Calculation of Eigenvalues of the Helmholtz Equation by an Integral Equation", *International Journal for Numerical Methods in Engineering*, Vol. 10, pp. 59–66.

Garbow, B.S., 1977, "The QZ Algorithm to Solve Generalized Eigenvalue Problems for Complex Matrices," *ACM Trans. math. Software*, Vol. 4, pp. 404–410.

Guisset, P., 1988, "SYSNOISE User's Manual," Dynamic Engineering Report, Leuven, Belgium.

Kaufman, L. 1975, "The LZ Algorithm to Solve Generalized Problems for Complex matrices," *ACM Trans. Math. Software*, Vol. 1, pp. 271–281.

Nardini, D., and Brebbia, C.A., 1982, "A New Approach to Free Vibration Analysis using Boundary Elements," *Proceedings, 4th International Conference on Boundary Element Methods*, C.A. Brebbia ed., Springer Verlag, Berlin.

Nefske, D.J., and Sung, S.H., 1985, "Vehicle Interior Acoustic Design Using finite Element Methods," *International Journal of Vehicle Design*, Vol. 6, pp. 24–40.

Petyt, M., Lea, J., and Koopmann, G.H., 1976, "A Finite Element Method for Determining the Acoustic Modes of Irregular Shaped Cavities," *Journal of Sound and Vibration*, Vol. 45, pp. 495–502.

Yu, I.W., 1986, "Solution of Large Unsymmetric Eigensystems for Fluid/Structure Interaction Problems," *Nuclear Science and Engineering*, Vol. 92, pp. 157–161.

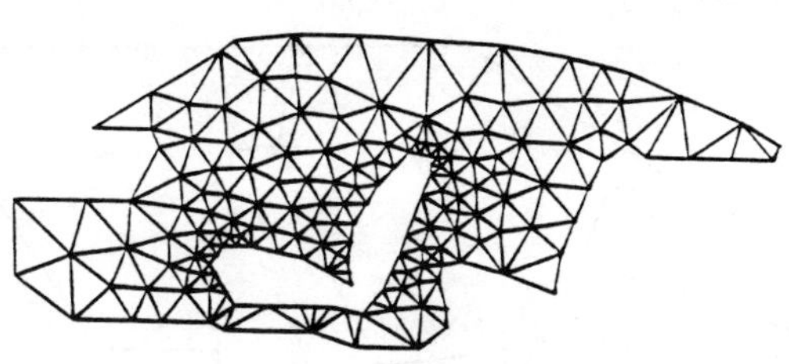

Fig. 1. **FEM mesh of two–dimensional car**

mesh type	FEM mesh 1	FEM mesh 2	BEM mesh 1	BEM mesh 2	analytical
element type	8 node cube	8 node cube	4 node rect	4 node rect	-
# of nodes	36	175	34	130	-
# of elements	12	96	32	128	-
mode 1	179.4 Hz	173.5 Hz	181.4 Hz	173.6 Hz	171.5 Hz
mode 2	236.4 Hz	219.9 Hz	239.2 Hz	221.7 Hz	214.4 Hz
mode 3	270.2 Hz	251.3 Hz	248.3 Hz	225.5 Hz	245.0 Hz
mode 4	296.8 Hz	280.1 Hz	252.4 Hz	268.7 Hz	274.5 Hz
mode 5	324.3 Hz	305.4 Hz	259.7 Hz	293.2 Hz	299.1 Hz

Table 1 : Comparison of FEM and BEM meshes and eigenvalue results

mode	FEM results Hz	BEM results Hz
1	75.4	73.5
2	125.2	117.1
3	133.2	131.2
4	179.6	172.5
5	235.6	233.1

Table 2 : Comparison of FEM and BEM car modal frequencies

mode	sedan Hz	hatch 1 Hz	hatch 2 Hz	stationwagon Hz
1	59.33	57.00	47.76	45.43
2	103.94	103.01	78.07	78.82
3	128.98	112.09	104.41	104.45
4	182.55	157.19	142.43	144.22
5	206.63	205.83	203.41	198.66

Table 3 : Modes of auto cavity styles

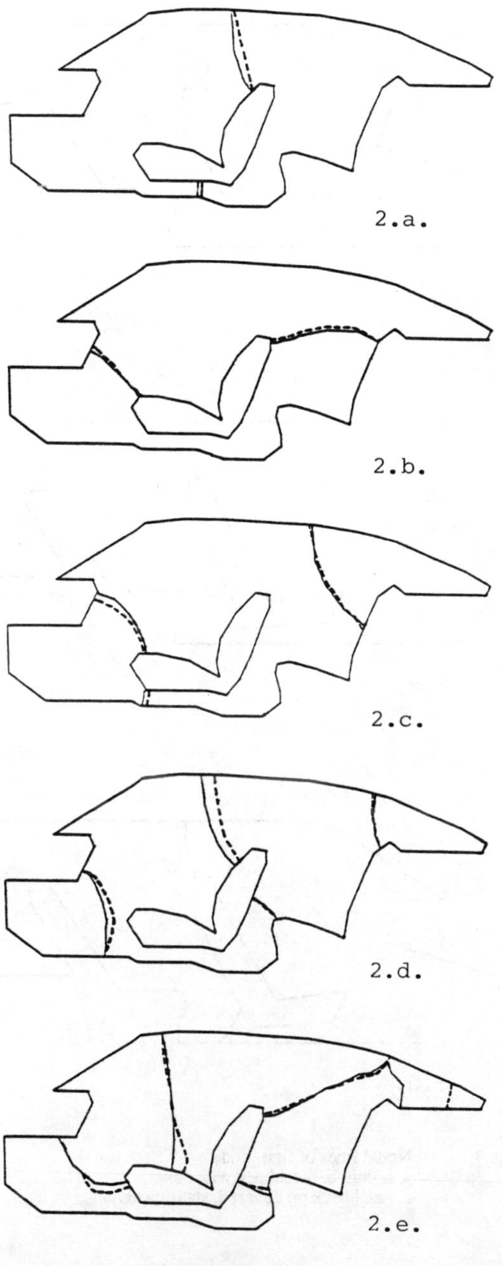

Fig. 2. **Overlay of FEM and BEM nodal lines of first five modes of two–dimensional car**
a. mode 1; b. mode 2.; c. mode 3; d. mode 4; e. mode 5.

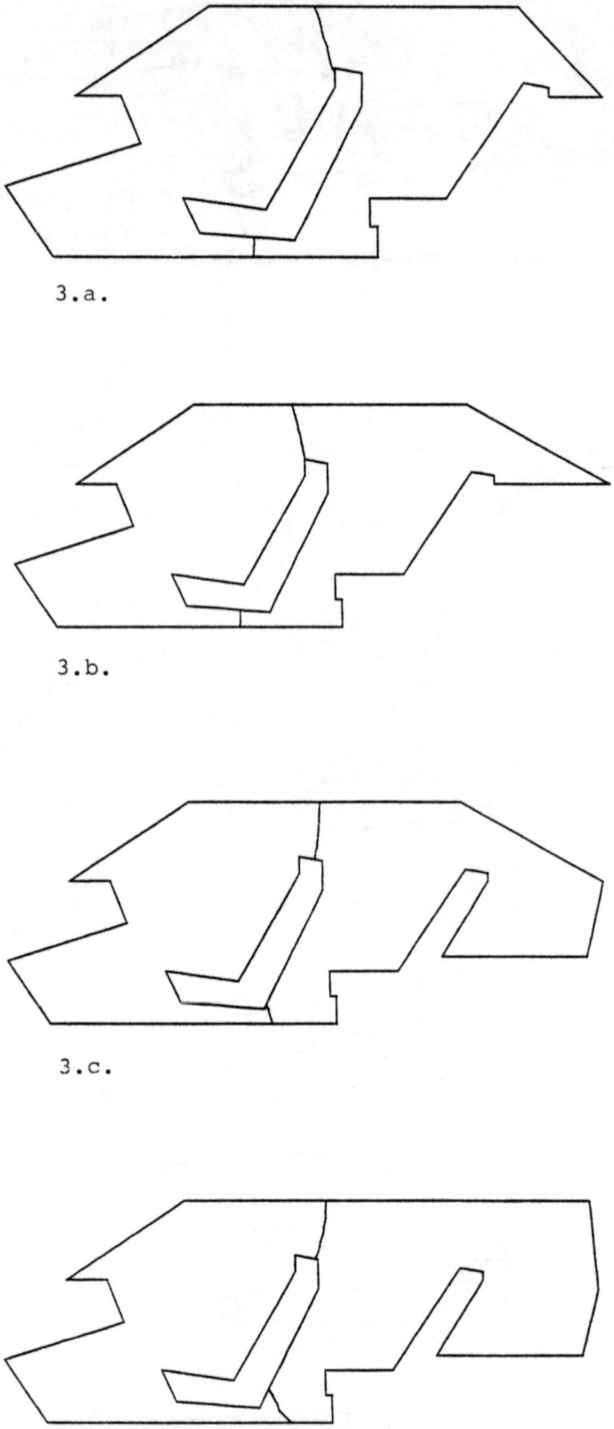

Fig. 3. Nodal lines of first mode
a. sedan; b. hatchback with cover;
c. hatchback no cover; d. stationwagon

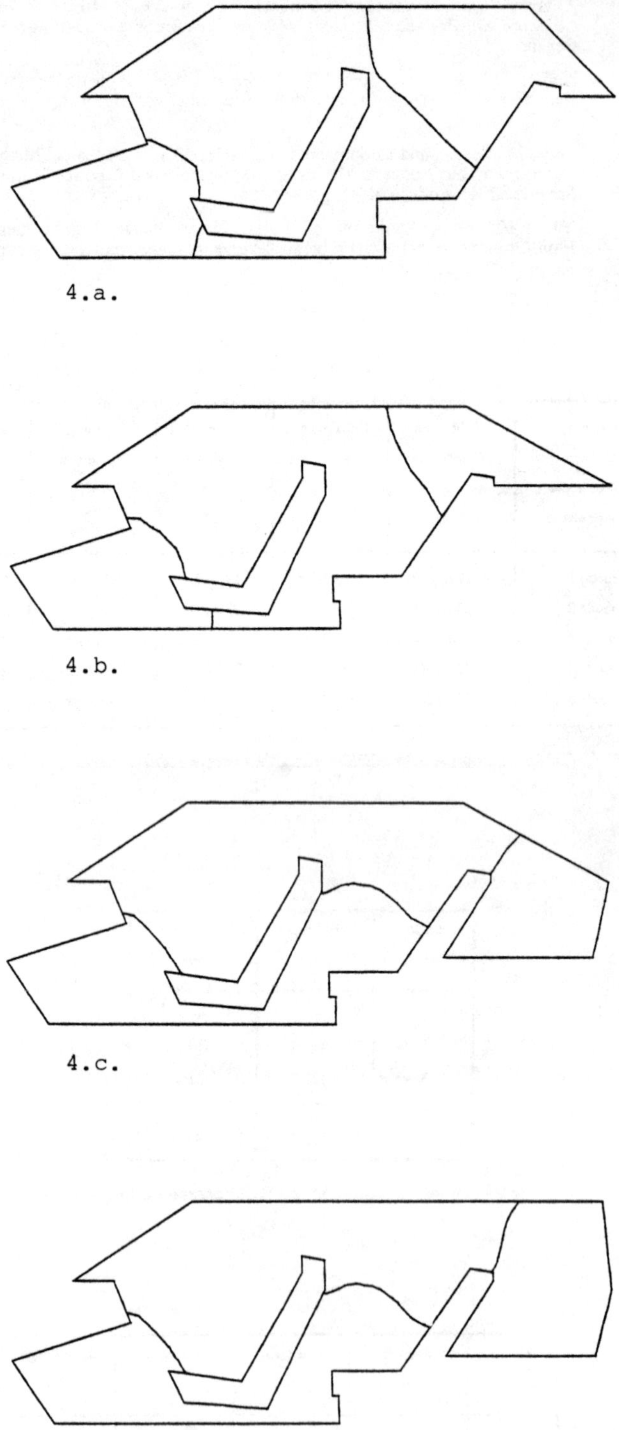

Fig. 4. Nodal lines of third mode
a. sedan; b. hatchback with cover;
c. hatchback no cover; d. stationwagon

TOTAL LEAST SQUARES SOLUTIONS TO ACOUSTIC BOUNDARY ELEMENT MODELS

H. R. Hall and R. J. Bernhard
Ray W. Herrick Laboratories
School of Mechanical Engineering
West Lafayette, Indiana

ABSTRACT

Acoustic radiation problems formulated with boundary element techniques require the solution of a system of linear equations [A][**x**]=[**b**]. The matrix [A] is dense, the vector [**x**] contains the unknowns on the boundary (acoustic pressure, normal motion, impedance, or some combination of the three), and the vector [**b**] is calculated from information known or measured on the boundary. [A], [**x**], and [**b**] are all complex. The coefficients of both [A] and [**b**] include errors due to numerical inaccuracy and data uncertainty. The solution [**x**] is sometimes obtained by least squares techniques which are appropriate when errors are confined to [**b**]. It is shown that the total least squares method can improve the computation of acoustic surface pressure at or near a characteristic frequency where the non-uniqueness problem of the Helmholtz integral equation occurs.

NOMENCLATURE

[A] matrix of coefficients.

$a_{i,j}$ coefficient of [A] at row i, column j.

[**b**] right hand side vector of system of equations [A][**x**]=[**b**].

c speed of sound.

[D] diagonal weighting matrix.

H matrix superscript denoting complex conjugate transpose.

I_m m x m diagonal identity matrix.

k wavenumber, $k = \frac{2\pi}{\lambda}$.

[**r**] residual vector, [**r**] = [A][$\mathbf{x}_k$] - [**b**], where [$\mathbf{x}_k$] is some numerical solution of [A][**x**]=[**b**].

[T] diagonal weighting matrix.

$[U]_{mxm}$ m x m left unitary matrix returned by the singular value decomposition of an m x n matrix: $[A]_{mxn}=[U]_{mxm}[\Sigma]_{mxn}[V]^H_{nxn}$.

[$\mathbf{u}_i$] i^{th} column of [U].

$[V]_{nxn}$ n x n right unitary matrix returned by the singular value decomposition.

[$\mathbf{v}_i$] i^{th} column of [V].

[**x**] vector of unknowns.

[$\mathbf{x}_{LS}$] solution of [A][**x**]=[**b**] obtained by the method of least squares.

[$\mathbf{x}_{TLS}$] solution of [A][**x**]=[**b**] obtained by the method of total least squares.

α $(n+1)^{th}$ element of [$\mathbf{v}_{n+1}$].

λ acoustic wavelength, $c = \lambda f$.

ρ_0 density of the acoustic medium.

$[\Sigma]_{mxn}$ m x n diagonal matrix of singular values.

σ_i i^{th} singular value of $[\Sigma]$.

I. INTRODUCTION

Approximations of the Helmholtz integral equation have been widely utilized to predict acoustic radiation from vibrating sources [1-5]. However, the method is known to have certain problems. At certain frequencies, referred to as characteristic frequencies, there is no unique solution to the Helmholtz integral equation. One of the techniques to overcome non-uniqueness at a characteristic frequency is to over-specify the problem by adding equations for points outside the domain of interest (i.e. inside a closed radiating source). This solution to the non-uniqueness problem was proposed by Schenck and is referred to as the CHIEF method [4]. Helmholtz integral equation methods are also prone to numerical error in the evaluation of the boundary integrals. These integrals are generally singular at some point on the boundary, and thus, sophisticated numerical integration schemes are required to reduce the numerical error to acceptable levels. The methods are also susceptible to data errors due to the geometric approximations that are made to discretize the boundary of the source and the simplifying assumptions made about the variation of the boundary conditions over each element.

Typically, enough data is specified to fully determine the problem. As a result, the errors are incorporated into the solution. In the case of the CHIEF method, the problem is overspecified and so there is some possibility of rejecting some of the error and improving the solution.

Boundary element models which use the CHIEF method produce a system of *m* linear equations in *n* unknowns:

$$[A]_{m \times n}[\mathbf{x}]_n = [\mathbf{b}]_m \, . \qquad (1)$$

The unknown vector $[\mathbf{x}]$ can be the acoustic pressure, the normal motion, impedance, or some combination of the three on the surface of a machine. [A] is a dense matrix whose coefficients are computed from boundary integrals, and $[\mathbf{b}]$ is constructed from boundary integrals and information known on the surface. Frequently, the right hand side, $[\mathbf{b}]$, is constructed from previously calculated or measured values of the normal velocity on the surface of a vibrating machine and $[\mathbf{x}]$ is an estimate of the surface acoustic pressure. The system is solved for $[\mathbf{x}]$, and the product of acoustic pressure and normal velocity at points on the surface gives the surface acoustic intensity, permitting the identification of efficiently radiating portions of the surface [6].

Errors are present in both [A] and $[\mathbf{b}]$. The single largest source of error is that which results from the ill-conditioning of [A] when the frequency of interest approaches a characteristic frequency [4]. This error also affects the right hand side $[\mathbf{b}]$. A discretization error occurs when the boundary values (pressure, normal motion, or impedance) are arbitrarily required to obey some distribution (constant, linear, etc.) over each element. Spatial geometry errors result when the boundary of interest is represented by elements that cannot conform to complex surfaces. These may include such aspects as mistakes in the location and shape of certain elements. If the position of particular elements is incorrectly modeled, the contribution at these elements from other elements will be in error, as will be the contribution from the incorrectly located elements to all other elements. Numerical integration is a source of error that has been extensively investigated for boundary elements and finite elements [7]. When a source point approaches a field point the distance between them approaches zero. Terms in the integrand like $1/|\mathbf{r}|$, where $|\mathbf{r}|$ is the distance between the source and field point, become large and require special consideration to control error. A similar situation exists when one element interacts with an adjacent element for which $|\mathbf{r}|$ is very small throughout the adjacent element's area. This is typical of elements near edges and corners. Solving the equations numerically also introduces error into the solution.

Least squares methods have been used almost exclusively to obtain solutions to acoustic radiation problems that are overdetermined. However, the method of least squares, strictly speaking, only permits the presence of error in the right hand side. It presumes that there are no errors in the equations that result from the modeling process. This paper compares least squares solutions and total least squares solutions. Total least squares is a method derived expressly for the case when errors are present in both sides of the equations generated by the modeling process. Least squares and total least squares may be thought of as 2 different kinds of regression. If a straight line is to be fitted to a set of data points, the least squares fit minimizes the vertical distance between the line and the points. The total least squares fit minimizes the perpendicular distance from the line to each data point. The method of total least squares, or TLS, requires the use of a singular value decomposition algorithm, and the system to be solved must be overdetermined or the TLS solution will not differ from the least squares (LS) solution.

It is the purpose of this paper to show that under certain conditions, which are likely when Helmholtz integral equation methods are applied to "real" problems, TLS solutions can be better than LS solutions. In the following section is a discussion of the TLS method and its implementation. Results are shown for a sphere modeled with 48 triangles.

II. TOTAL LEAST SQUARES

The complete development of the TLS method is found in Golub and Van Loan [8,9] and also in Van Huffel [10]. Both LS and TLS methods will be summarized here to illustrate the fundamental differences in the methods.

Both LS and TLS methods seek a solution that minimizes a particular error norm. To permit certain rows to be weighted more heavily than others, equation (1) is rewritten as

$$[D][A][\mathbf{x}] = [D][\mathbf{b}] \qquad (2)$$

where $[D]_{m \times m}$ is a non-singular diagonal weighting matrix. If all equations are weighted equally, then [D] is the identity matrix.

To develop both LS and TLS methods a residual, $[\mathbf{r}]$, is defined such that

$$[\mathbf{r}] = [A][\mathbf{x}] - [\mathbf{b}] . \quad (3)$$

The least squares problem amounts to finding a solution $[\mathbf{x}]$ that minimizes the norm of $\|[D][\mathbf{r}]\|_2$ where the 2 norm is defined as

$$\|\mathbf{w}\|_2 = \sqrt{\sum_i w_i^* w_i} = \sqrt{\mathbf{w}^H \mathbf{w}} \quad . \quad (4)$$

In other words, the LS solution, $[\mathbf{x}_{LS}]$, seeks a minimum $\|[D][\mathbf{r}]\|_2$ such that $[\mathbf{b}]$ plus $[\mathbf{r}]$ can still be predicted by the columns of [A].

LS solutions of equation (1) may be obtained by a variety of techniques. Two will be discussed here. The LS solution by means of the singular value decomposition (SVD) gives the best results under a wide range of conditions on [A], including rank deficiency. The singular value decomposition is a method for diagonalizing a rectangular (non-square) matrix as

$$[A]_{m\times n} = [U]_{m\times m}[\Sigma]_{m\times n}[V]^H_{n\times n} \quad (5)$$

where [U] and [V] are unitary matrices such that

$$[U]_{m\times m}[U]^H_{m\times m} = I_m , \quad (6)$$

$$[V]_{n\times n}[V]^H_{n\times n} = I_n , \quad (7)$$

and $[\Sigma]_{m\times n}$ is a diagonal matrix with the n singular values σ_i on its diagonal and zeroes elsewhere [11-13]. The superscript H indicates the complex conjugate transpose. The LS solution $[\mathbf{x}_{LS}]$ is then constructed from the SVD of [A] by

$$[\mathbf{x}_{LS}] = \sum_{i=1}^{r} \frac{[\mathbf{u}_i]^H[\mathbf{b}]}{\sigma_i}[\mathbf{v}_i] \quad (8)$$

where $[\mathbf{u}_i]$ and $[\mathbf{v}_i]$ are, respectively, the ith columns of [U] and [V], σ_i is the ith singular value, and r is the rank of [A] [14]. Even if r is relatively small, the time required for the arithmetic operations of equation (8) can be substantial if [A] is large. All results presented in this work used the IMSL subroutine DLSVCR for the SVD of (3) and $[\mathbf{x}_{LS}]$ was constructed as indicated by equation (8). The SVD of an $m \times n$ matrix takes $2mn^2 + 4n^3$ floating point operations, or flops, and the construction of $[\mathbf{x}_{LS}]$ requires the additional flops of equation (8) after the SVD algorithm has returned [U], $[\Sigma]$, and [V] [15].

The second solution procedure for the least squares problem, referred to here as the normal equations, begins by multiplying (2) by the complex conjugate transpose of its left hand side such that

$$\{[D][A]\}^H[D][A][x] = \{[D][A]\}^H[D][b] . \quad (9)$$

To solve equation (9), the results presented in this paper used the IMSL subroutine DLSLCG, which solves the matrix equations by Gaussian elimination after performing an LU decomposition on the left hand side, $\{[D][A]\}^H[D][A]$. The number of flops required for the solution of the normal equations is $(mn^2/2)+(n^3/6)$ [15].

Total least squares seeks a solution that satisfies the equation

$$[A + E][\mathbf{x}_{TLS}] = [\mathbf{b}] + [\mathbf{r}] \quad (10)$$

where [E] is a matrix of error perturbations that may be present in the left hand side of equation (1). Equation (10) can be rewritten as

$$\{[D][A|\mathbf{b}][T]+[D][E|\mathbf{r}][T]\}[T]^{-1}\begin{bmatrix}\mathbf{x}\\-1\end{bmatrix} = 0 \quad (11)$$

where $[D]_{m\times m}$ and $[T]_{(n+1)\times(n+1)}$ are non-singular diagonal weighting matrices. [D] weights the rows and [T] weights the columns of the augmented matrices $[A|\mathbf{b}]$ and $[E|\mathbf{r}]$. TLS seeks a solution that satisfies equation (11) identically and also minimizes the Frobenius norm of $\|[D][E|\mathbf{r}][T]\|_F$. The Frobenius norm is defined as

$$\|A\|_F = \sqrt{\sum_i \sum_j |a_{i,j}|^2} \quad . \quad (12)$$

For the derivation of the TLS method it is convenient to make the substitutions

$$[C] = [D][A|\mathbf{b}][T] , \quad (13)$$

$$[\Delta] = [D][E|\mathbf{r}][T] , \quad (14)$$

$$[\mathbf{w}] = [T]^{-1}\begin{bmatrix}\mathbf{x}\\-1\end{bmatrix} . \quad (15)$$

Thus equation (11) becomes

$$\{[C] + [\Delta]\}[\mathbf{w}] = 0 . \quad (16)$$

The SVD of [C] is required such that:

$$[D]_{m\times m}[A\,|\,\mathbf{b}]_{m\times(n+1)}[T]_{(n+1)\times(n+1)} = [U]_{m\times m}[\Sigma]_{m\times(n+1)}[V]^H_{(n+1)\times(n+1)} . \quad (17)$$

If $[\mathbf{w}]$ is taken as one of the columns of [V], denoted by $[\mathbf{v}_i]$

$$[\mathbf{w}] = [\mathbf{v}_i] \quad (18)$$

and if $[\Delta]$ is formed from the same column such that

$$[\Delta] = -[C]\;[\mathbf{v}_i][\mathbf{v}_i]^H \quad (19)$$

then equation (11) becomes

$$\{[C]-[C][\mathbf{v}_i][\mathbf{v}_i]^H\}[\mathbf{v}_i] = [C][\mathbf{v}_i]-[C][\mathbf{v}_i][\mathbf{v}_i]^H[\mathbf{v}_i] = 0 . \quad (20)$$

Equation (20) will be identically satisfied since the columns of [V] are orthonormal, that is

$$[\mathbf{v}_i]^H[\mathbf{v}_i] = 1 . \quad (21)$$

To select that column of [V] that minimizes $\|\Delta\|_F$, it is convenient to rewrite equation (19) as

$$[\Delta] = -[C][\mathbf{v}_i][\mathbf{v}_i]^H = [\mathbf{q}][\mathbf{v}_i]^H . \quad (22)$$

In equation (22), [**q**] is the vector formed from the multiplication of $[\mathbf{v}_i]$ by [C],

$$[\mathbf{q}] = -[C][\mathbf{v}_i] . \quad (23)$$

Thus $[\Delta]$ is the outer product of two vectors. If an element of $[\Delta]$ located at row r and column s is denoted by $\delta_{r,s}$, equation (22) shows that $\delta_{r,s}$ is given by equation (24):

$$\delta_{r,s} = q_r v_s^* \quad (24)$$

where q_r is the rth element of [**q**] and v_s^* is the sth element of $[\mathbf{v}_i]^H$. The Frobenius norm of $[\Delta]$ is then given by

$$\|\Delta\|_F^2 = \sum_{r=1}^{m}\sum_{s=1}^{n+1} |\delta_{r,s}|^2 = \sum_{r=1}^{m}\sum_{s=1}^{n+1} |q_r v_s^*|^2 . \quad (25)$$

In equation (25), the quantity $|q_r v_s^*|^2$ can be rewritten as

$$|q_r v_s^*|^2 = (q_r v_s^*)(q_r v_s^*)^* = (q_r q_r^*)(v_s v_s^*) . \quad (26)$$

Substituting equation (26) into equation (25), the Frobenius norm of $[\Delta]$ is given by

$$\|\Delta\|_F^2 = \sum_{r=1}^{m} q_r q_r^* \sum_{s=1}^{n+1} v_s v_s^* = [\mathbf{q}]^H[\mathbf{q}]\ [\mathbf{v}_i]^H[\mathbf{v}_i] . \quad (27)$$

Replacing [**q**] as defined by equation (23) yields the desired expression for the Frobenius norm of $[\Delta]$

$$\|\Delta\|_F^2 = \{[C][\mathbf{v}i]\}^H\{[C][\mathbf{v}i]\}[\mathbf{v}i]^H[\mathbf{v}i] \quad (28)$$

or

$$\|\Delta\|_F^2 = [\mathbf{v}i]^H[C]^H[C][\mathbf{v}i] \quad (29)$$

by taking advantage of the orthonormal properties of $[\mathbf{v}_i]$. Equation (29) can be rearranged as

$$[C]^H[C][\mathbf{v}_i] = \|\Delta\|_F^2\ [\mathbf{v}_i] \quad (30)$$

which is in the form of an eigenvalue problem, where $[\mathbf{v}_i]$ is an eigenvector and $\|\Delta\|_F^2$ the corresponding eigenvalue of the matrix $[C]^H[C]$. Since the squares of the singular values returned by the SVD are these same eigenvalues and the singular vectors $[\mathbf{v}_i]$ are the corresponding eigenvectors,

$$[C]^H[C][\mathbf{v}_i] = (\sigma_i)^2[\mathbf{v}_i] . \quad (31)$$

The smallest singular value and the corresponding singular vector solves the TLS problem and gives the smallest Frobenius norm for the matrix $[\Delta] = [D][E|\mathbf{x}][T]$. Since the singular values are ordered,

$$\sigma_1 \geq \sigma_2 \geq \ldots \geq \sigma_n \geq \sigma_{n+1} , \quad (32)$$

this usually will mean that the total least squares solution, $[\mathbf{x}_{TLS}]$, is constructed from $[\mathbf{v}_{n+1}]$. For $[\mathbf{v}_{n+1}]$ of the form

$$[\mathbf{v}_{n+1}] = \begin{bmatrix} \mathbf{y} \\ \alpha \end{bmatrix} \begin{matrix} \}n \\ 1 \end{matrix} \quad (33)$$

where α is not zero, then the solution can be found from equation (15) as

$$[\mathbf{x}_{TLS}] = \frac{-[T_1][\mathbf{y}]}{\alpha t_{n+1}} \quad (34)$$

In equation (34), $[T_1]$ is the diagonal matrix

$$[T_1] = \mathrm{diag}(t_1,\ t_2,\ \ldots,\ t_n) . \quad (35)$$

formed by deleting the $(n+1)^{th}$ element t_{n+1} of [T].

If the SVD returns $[\mathbf{\Sigma}]$ such that the last k singular values are equal to zero (or nearly so), then [**v**] in equation (33) may be chosen from any of the last k columns of [V], all of which will yield a TLS solution. To obtain the minimum norm solution it is necessary to do a Householder transformation such that

$$[Q]_{(n+1)\times(n+1)}[V']_{(n+1)\times k} = \begin{bmatrix} X \mid \mathbf{y} \\ 0 \mid \alpha \end{bmatrix} \begin{matrix} \}n \\ 1 \end{matrix} \quad (36)$$

where [V'] is constructed from the last k columns of [V] and [Q] then transforms them into the form given in equation (36) (i.e., the last row consists of zeroes except for the final term α). The TLS solution constructed from $\begin{bmatrix} \mathbf{y} \\ \alpha \end{bmatrix}$ in equation (36) will be the minimum norm solution.

It should be noted that $[\mathbf{x}_{TLS}]$ will converge to $[\mathbf{x}_{LS}]$ as t_{n+1} is made smaller. If [A] is of rank n and the residual of the LS solution is zero, then $[\mathbf{x}_{TLS}] = [\mathbf{x}_{LS}]$ regardless of the choice of t_{n+1} [8].

Implementing TLS requires an SVD algorithm and a Householder transformation algorithm. The results presented in this paper made use of the IMSL subroutine DLSVCR which performs an SVD of a complex double precision matrix. A Householder algorithm for complex matrices was adapted from existing software for real matrices. Note that as long as $\sigma_n > \sigma_{n+1}$, constructing $[\mathbf{x}_{TLS}]$ is only a matter of multiplying each of the first n components of the last column of [V] by a complex number formed from the diagonal of [T] and the last element of $[\mathbf{v}_{n+1}]$. This is more efficient than the procedure used for constructing $[\mathbf{x}_{LS}]$ from the SVD of [A] given by equation (8).

III. RESULTS FOR THE SPHERE

For a pulsating sphere of radius *a* with a uniform surface velocity $U_0 e^{j\omega t}$, the pressure on the surface is given by

$$p_s = \frac{\rho_0 c U_0 ka(ka+j)}{1+(ka)^2} \tag{37}$$

where ρ_0 is the density of the acoustic medium, c is the speed of sound, and k is the wavenumber,

$$k = \frac{2\pi}{\lambda} \tag{38}$$

where λ is the wavelength. The results for the sphere were obtained from a boundary element program utilizing linear superparametric triangular elements. The coefficients of [A] and [**b**] were evaluated with 3 point Gaussian quadrature [16]. For a sphere of radius a = 1 meter with 48 plane triangular boundary elements, the discretized Helmholtz integral equation results in a linear system of 48 equations in 48 unknowns. Additional equations may be generated by requiring the pressure in the interior of the sphere to be zero, producing an overdetermined system of equations. Over-determination has been shown to be particularly useful at characteristic wavenumbers, those values of k for which the solution of the exact boundary integral equations is not unique [4]. The preceeding system of *m* equations in *n* unknowns ($m \geq 48$, $n = 48$) will be solved by LS and TLS methods and the two methods will be compared.

Figure 1 shows the error of the TLS (solid line) and LS (dashed line) solutions over the surface of the sphere at a wavenumber of k = 7.02. A single CHIEF point at (0, 0, $1/\pi$) is used. This frequency is near the second characteristic frequency of the sphere. Unless specified to the contrary, the LS solution was found by solving the normal equations. If p_i is the magnitude of the calculated pressure (LS or TLS) at element *i* the relative error is computed from

$$\text{relative error} = \frac{p_i - p_s}{p_s} \tag{39}$$

where p_s is the magnitude of the complex pressure given by equation (37). Over some elements the error of the TLS solution is greater by about 1%. Over most of the surface the TLS solution is better than the LS solution.

Of particular interest is the performance of TLS when random error is introduced into the columns of the matrix [A]. The jth column of [A] contains coefficients $[a_{i,j}]$ that are the contribution of the jth element to each of the *i* elements, $i=1,2,\ldots,m$. Thus, introducing random error into the columns of [A] is representative of the situation when elements are incorrectly sized or located. Equation (34), rewritten here for convenience,

$$[\mathbf{x}_{TLS}] = \frac{-[T_1][\mathbf{y}]}{\alpha t_{n+1}} \tag{34}$$

shows how the column weighting matrix [T] enters into the TLS solution. To compare LS and TLS with error present in [A], the diagonal matrix [T] is "randomized" by replacing each element (normally 1) with a number, 1 + (*scale*) X (*random number*), where (*random number*) is a random number between -1 and +1. Thus, if (*scale*) is 0.05, then the elements of [T] will vary randomly between 0.95 and 1.05. In each figure, the average error over the surface is plotted as element 49, and the rms error as element 50.

At wavenumber k = 7.02, for the case where the elements of the diagonal weighting matrix [T] vary randomly between 0.95 and 1.05, the TLS solution is better than the LS solution, as shown in figure 2. The results shown in figures 1 and 2 clearly demonstrate that when random error is present in the matrix [A] the TLS method is preferable to the method of least squares.

At frequencies away from the characteristic frequencies, the TLS method was found to have very little effect. In figure 3 the TLS and LS solutions are shown to be approximately the same at a wavenumber k = 0.75 for the case without random error in the columns of [A]. Figure 4 shows the same case but with error on [A] which varies randomly between 0.95 and 1.05. At k = 0.75, TLS and LS are identical with and without random error. If the random errors are made larger, this situation does not change. Thus, the TLS solution only seems to improve the results when random error is present in the columns of the left hand side at frequencies close to characteristic frequencies.

Figure 5 shows the error for the pulsating sphere solution at k = 3.5 which is very near the first characteristic frequency for this model of the sphere. The columns of [A] do not have random error applied in this case. The CHIEF point is at location (0, 0, $1/\pi$). LS error is less than 3%, TLS less than 4%. Introducing random error into the columns of [A] where the elements of [T] vary between 0.95 and 1.05 at this wave number produces a TLS solution that is still not as good as the LS solution, as shown in figure 6.

As a third test case, for k = 3.5, $[\mathbf{x}_{TLS}]$ is "forced" to equal $[\mathbf{x}_{LS}]$ by setting $t_{n+1} = 1.0 \times 10^{-5}$. The results of this case are shown in figure 7. The error of the TLS solution is indistinguishable from the LS error.

As mentioned previously, random error was introduced into the columns of [A] by adjusting the diagonal elements of [T] so that these varied in a random fashion about their nominal value of 1. However, the elements of $[T_1]$ in equation (34) were always kept at unity, since a solution was sought that was typical for the case when the errors are unknown. If the diagonal elements of $[T_1]$ had been used with their correct random values, the TLS solution was actually better than the LS solution for those cases at frequencies away from characteristic frequencies. A typical case is shown in figure 8 for k = 0.75. This figure suggests that when *a priori* knowledge of the errors in [A] is available, the TLS solution will be better than the LS solution.

Figure 9 compares the TLS solution with the LS solution via SVD when an incomplete set of the singular values and singular vectors are used to construct $[\mathbf{x}_{LS}]$. The dashed lines of figure 9 were made by setting the rank r in equation (18) to values less than n. Away from characteristic frequencies the small singular values and singular vectors, which are neglected, do not contribute as much to the solution as they do near the characteristic frequencies.

The computation time required to solve 4 different models by the two LS methods and the TLS method is shown in figure 10. The models consisted of a six sided cube of 12 triangular elements, and 3 spheres of 48, 60, and 96 triangular elements. The LS methods are the solution of the normal equations and the use of the SVD as given in equation (8). The speed advantage of the LS normal equation solution is apparent. The number of floating point operations (flops) required for the solution of the normal equations is $(mn^2/2) + (n^3/6)$. TLS requires slightly more computational time than the SVD of an $m \times n$ matrix, which takes $2mn^2 + 4n^3$ flops. LS via SVD requires the most time because of the additional flops required to construct the LS solution of equation (8) after the SVD algorithm has returned [U], $[\Sigma]$, and [V]. If an algorithm were used that was written expressly for the LS problem and did not return [U], $[\Sigma]$, and [V], then the time advantage of TLS would be reduced. The potential speed advantage of TLS over LS via SVD is of interest in the solution of large models. The TLS solutions presented here also utilized an SVD algorithm that returned all singular values and singular vectors. Some improvement in time and memory requirements will result when a routine is used that computes the smallest singular values and their corresponding singular vectors, since at most only a few of the smallest singular values and singular vectors are needed to solve the TLS problem. Van Huffel, Vandewalle, and Haegemans [17] have published just such SVD algorithms and report speed increases of between 2.5 and 3 times over the SVD routine used in this work.

IV. CONCLUSION

TLS solutions have been shown to be a viable alternative to LS solutions. The performance of the TLS method is attractive when random error is present in the columns of [A]. In addition, there is a speed advantage for TLS over the SVD solution for LS.

The sphere modeled in this work has characteristic frequencies at approximately $k = 3.5$ and 7.0. Away from these values, at a wavenumber of $k = 0.75$, the TLS and LS solutions were the same in the absence of random error. At this wavenumber TLS showed no improvement over LS as random error was introduced into the columns of [A]. This result was typical of the results at other values of wavenumber away from the characteristic frequencies. At the first charactistic frequency, LS was superior to TLS by about 1 percent, with and without random error. At the second characteristic frequency, TLS was better with and without random error by about 3 percent. TLS was shown to be more robust than LS methods for solving radiation problems at the characteristic frequencies.

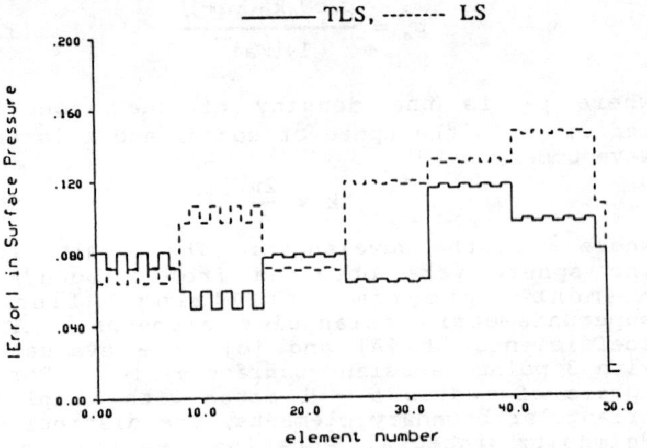

Figure 1. k=7.02, no random error in columns of [A].

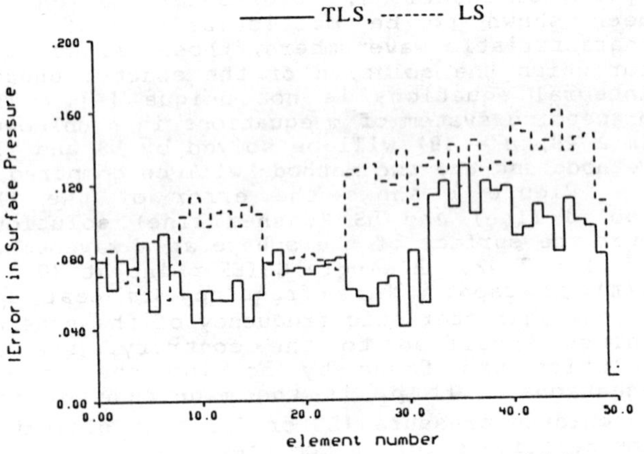

Figure 2. k=7.02, random error in columns of [A].

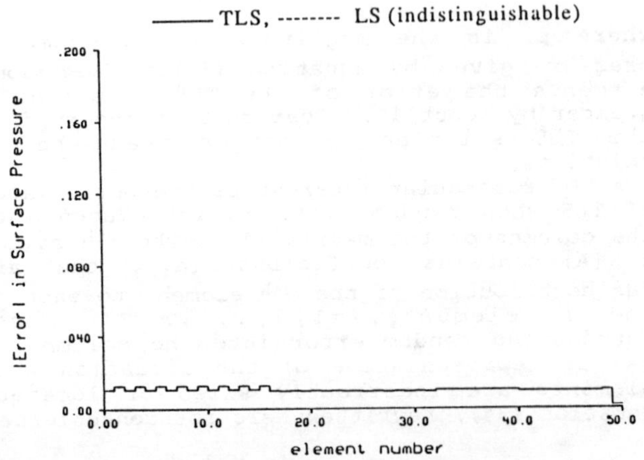

Figure 3. k=0.75, no random error in columns of [A].

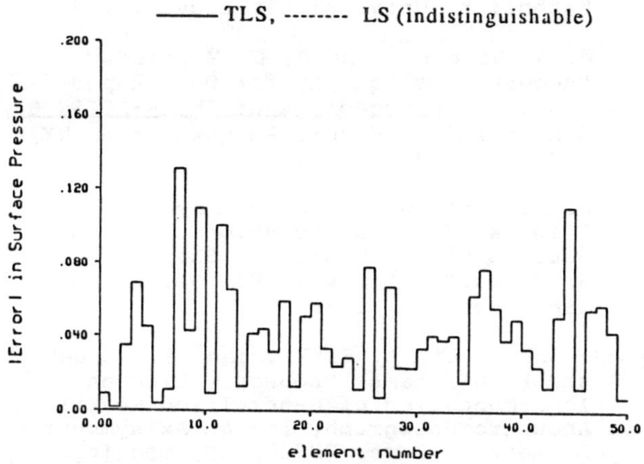

Figure 4. k=0.75, random error in columns of [A].

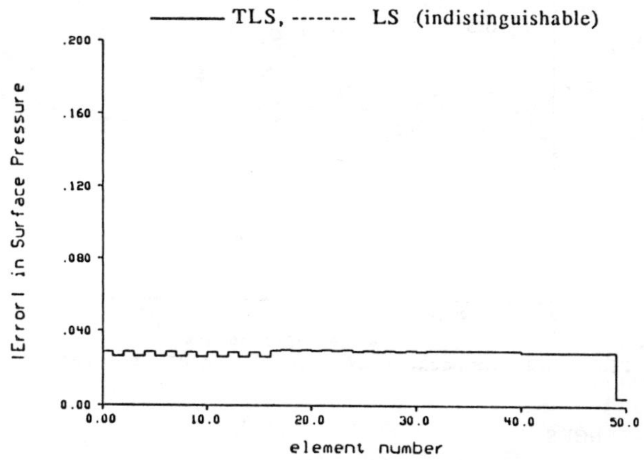

Figure 7. k = 3.5, t_{n+1} = 1.0 X 10^{-5}.

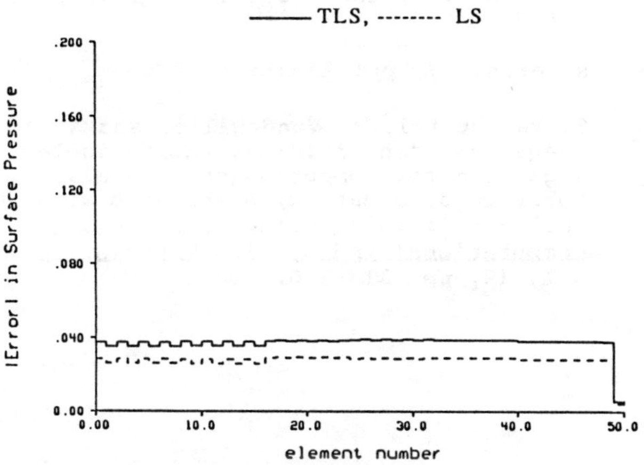

Figure 5. k=3.5, no random error in columns of [A].

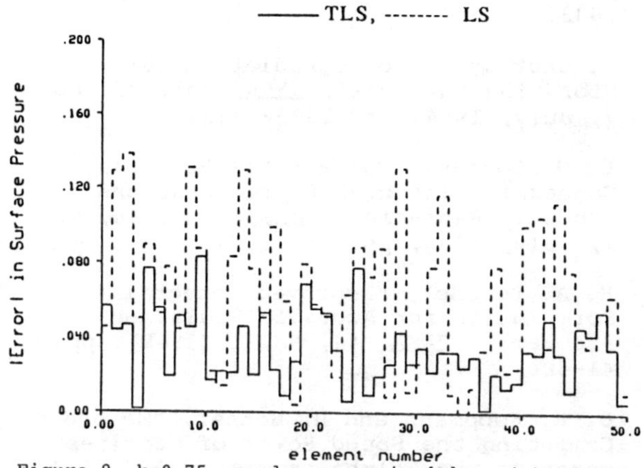

Figure 8. k=0.75, random error in [A], weighted TLS solution.

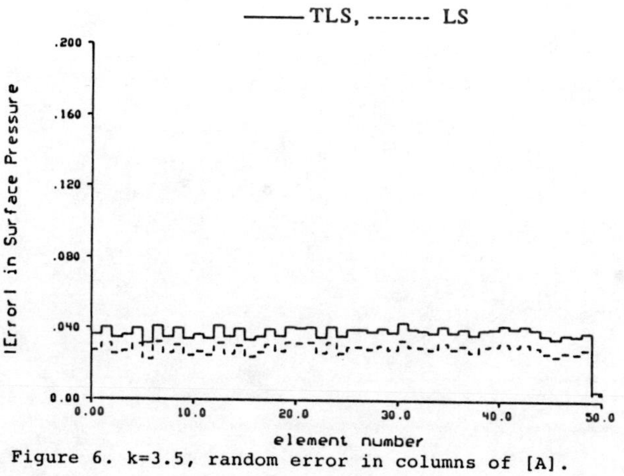

Figure 6. k=3.5, random error in columns of [A].

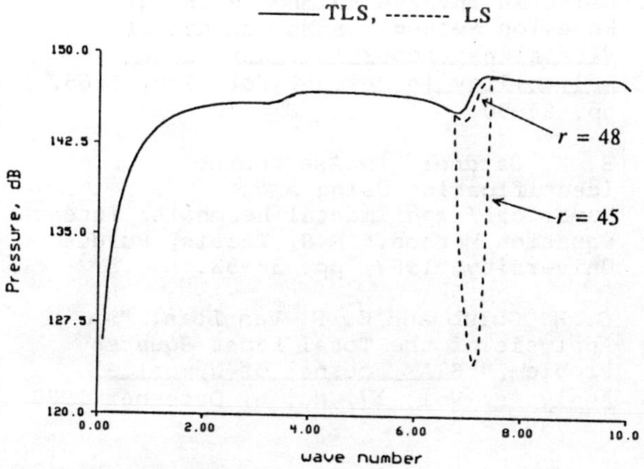

Figure 9. TLS compared to LS solutions for rank r less than n.

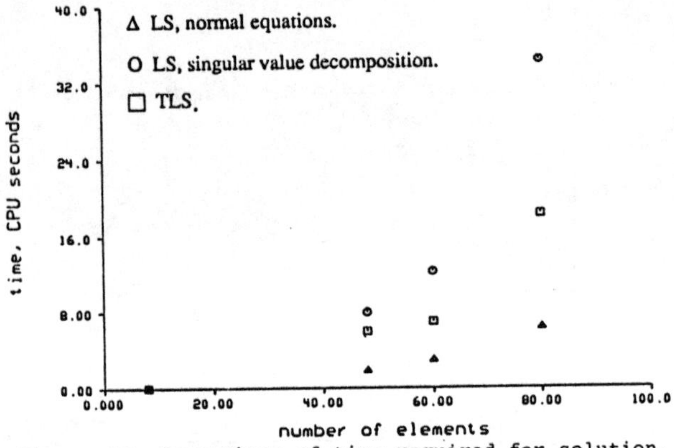

Figure 10. Comparison of time required for solution.

REFERENCES

1. L. H. Chen and D. G. Schweikert, "Sound Radiation from an Arbitrary Body", Journal of Acoustical Society of America, vol. 35, No. 10, October, 1963, pp. 1626-1632.

2. G. Chertock, "Sound Radiation for Vibrating Surfaces", JASA, vol. 35, No. 7, July, 1964, pp. 1305-1313.

3. L. G. Copley, "Fundamental Results Concerning Integral Representations in Acoustic Radiation", JASA, vol. 44, No. 1, July, 1968, pp. 28-32.

4. H. A. Schenck, "Improved Integral Formulation for Acoustic Radiation Problems," JASA, 44, No. 1 (1968), pp. 41-58.

5. G. H. Koopmann and H. Benner, "Method of Computing the Sound Power of Machines Based on the Helmholtz Integral," JASA, Vol. 71, No. 1, 1982, pp. 78-89.

6. B. K. Gardner and R. J. Bernhard, "A Noise Source Identification Technique Using an Inverse Helmholtz Integral Equation Method," ASME Journal of Vibrations, Acoustics, Stress and Reliability in Design, Vol. 110, 1988, pp. 84-90.

7. B. K. Gardner, "Noise Source Identification Using a Numerical/Experimental Helmholtz Integral Equation Method," M.S. Thesis, Purdue University, 1987, pp. 38-53.

8. G. H. Golub and C. F. Van Loan, "An Analysis of the Total Least Squares Problem," SIAM Journal of Numerical Analysis, Vol. 17, No. 6, December 1980, pp. 883 - 893.

9. G. H. Golub and C. F. Van Loan, MATRIX COMPUTATIONS, Johns Hopkins University Press, Baltimore, 1983, pp. 420-425.

10. S. Van Huffel, "Analysis of the Total Least Square Problem and Its Use in Parameter Estimation", doctoral thesis, Dept. of Electrical Engineering, Katholieke Universiteit Leuven, 1987.

11. W. A. Veronesi and J. D. Maynard, "Acoustic Holography for Odd-Shaped Sources," Proceedings of INTER-NOISE 86, Robert Lotz, editor, Poughkeepsie, NY, 1986, pp. 1251-1257.

12. W. A. Veronesi and J. D. Maynard, "Digital Holographic Reconstruction of Sources with Arbitrarliy Shaped Surfaces," JASA, vol. 85, No. 2, pp. 588-598, February, 1989.

13 G. Borgiotti, A. Sarkissian, L. Schuetz, and E. Williams, "Boundary Element Implementation of Generalized Nearfield Acoustic Holography for an Axisymmetric Geometry," Paper BBB10, ASA meeting, December, 1988.

14. Reference 9, p. 138, equation 6.1-5 .

15. Ibid, p. 175, table 6.5-1, p. 177, table 6.5-2 .

16. Reference 7, pp. 23-38.

17. S. Van Huffel, J. Vandewalle, and A. Haegemans, "An Efficient and Reliable Algorithm for Computing the Singular Subspace of a Matrix, Associated with its Smallest Singular Values", Journal of Computational and Applied Mathematics, vol. 19, pp. 313-330, 1987.

A BOUNDARY INTEGRAL FORMULATION FOR THIN-WALLED SHAPES OF REVOLUTION

R. Martinez
Cambridge Acoustical Associates, Incorporated
Cambridge, Massachusetts

ABSTRACT

Much of the recent literature on boundary elements has addressed the analytical and practical limitations of the method. One issue under current debate appears to be the inapplicability of formulations based on the Helmholtz integral to free-flooded shapes characterized by kh<<1, where k and h are the acoustic wavenumber and the object's cross sectional dimension. The present note develops and applies a modal boundary integral technique especially tailored to thin geometries. The problems chosen for its demonstration are cases of acoustic diffraction by an open-ended cylindrical shroud containing a sound source, where the scattering wall's outer and inner surfaces are an infinitesimal distance apart, and are respectively rigid and either rigid or compliant. The study concludes that general shapes with kh<<1 or kh=0 should be treatable by some of the existing numerical approaches, after some analytical re-interpretation.

INTRODUCTION

In 1985 Reut published a letter in the Journal of Sound and Vibration summarizing the state of the art of the then two main boundary-element methodologies and of their ways of treating the eigenfrequencies of a scattering body having a finite volume.[1] The more popular of the pair seems to have been, and apparently still is, Schenck's CHIEF, an approach that supplements the Helmholtz integral with a similar interior formulation that deliberately overdetermines the boundary-value problem and thereby resolves its nonuniqueness difficulties[2] (cf. Seybert & Rengarajan[3], and Tobocman[4], for modern implementations of the method). Reut's alternate technique bears the acronym CONDOR (Composite Outward Normal Derivative Overlap Relation)[1]; it invokes the original ideas of Burton & Miller[5] by applying on the surface of the body the normal derivative of the Helmholtz integral, which combined with the standard form yields a system whose eigenfrequencies are complex and therefore never obstacles in practice (Mathews[6]). Cunefare et al.[7] have instead used the normal derivative equation at interior points to render its otherwise "singular" kernel numerically tractable even by personal computer.

The reason for citing this letter is that its final assessment of the relative merits of these two boundary-element philosophies included a statement on their major common deficiency: neither formulation as at least then configured could tackle scattering or diffracting objects whose cross-sectional dimension was small in terms of the incident wavelength. A primary purpose of the present paper is to show that this failure is somewhat mysterious, because the differentiated Helmholtz equation contained in the CONDOR code should by itself fulfill the mathematical requirements of virtually any thin two-sided geometry.

The problem for which we have chosen to demonstrate this is that of diffraction by a free-flooded cylindrical noise shield having a wall that is by prescription infinitesimally thin (Fig. 1a). Its solutions and predictions are interesting and physically relevant in and of themselves, but another objective of our theoretical development is to introduce a way for checking them based on the interactive energetics of the field source/boundary system. The analysis thus includes the computation of input power at the field point occupied by the insonifying dipole, the power radiated out of the duct's two open ends, the mechanical energy dissipated by the generally damped lining that covers its interior surface; and the numerical proof of the fact that for every frequency examined the sum of the second and third quantities balances the first. This concept of power matching as a diagnostic tool extends that proposed and applied in Ref. 8 for a similar but completely rigid noise shroud.

The last part of the paper outlines the generalization of the procedure, from that applied here for a straight duct geometry, to one of variable cross section obtained by revolving a curved generator about the cylinder's axis.

I DEVELOPMENT

A Formulation

Figs. 1a,b display the diffraction problem and the coordinate system adopted for its solution: a free-flooded cylindrical shroud of length L and normalized radius "a" contains the geometry's (r,z) origin at the axis point that is equidistant from the two open ends;

ϕ is circumferential angle. An axial dipole of strength F_o generates incident field p^{inc} from interior position $\hat{r}, \hat{z}, \hat{\phi}$ (e.g., Pierce[9])

$$p^{inc}(r,\phi,z) = -\frac{F_o}{\pi L^2}\frac{\partial}{\partial z}\left(\frac{e^{ik\sqrt{r^2+\hat{r}^2-2r\hat{r}\cos(\phi-\hat{\phi})+(z-\hat{z})^2}}}{\sqrt{r^2+\hat{r}^2-2r\hat{r}\cos(\phi-\hat{\phi})+(z-\hat{z})^2}}\right) \quad (1a)$$

$$= -\frac{F_o}{\pi L^2}\frac{\partial}{\partial z}\sum_{\nu=0}^{\infty}\left\{\frac{\varepsilon_\nu(-1)^\nu}{\pi}\int_0^\pi \frac{d\beta\cos\nu\beta}{\hat{R}}\, e^{ik\hat{R}}\right\}\cos[\nu(\phi-\hat{\phi})] \quad (1b)$$

where $\hat{R}$ is defined by

$$\hat{R} = \sqrt{r^2+\hat{r}^2+2r\hat{r}\cos\beta+(z-\hat{z})^2} \quad (2)$$

and

$$\varepsilon_\nu = \begin{cases} 1 & \nu=0 \\ 2 & \nu\geq 1 \end{cases} \quad (3)$$

The pipe's outer surface is rigid. A stiffness-controlled, locally reacting material of negligible thickness lines its inner surface between points z_1 and z_2. The theory allows the properties of this internal coating to vary in z but restricts them to be uniform in the circumferential direction ϕ (cf. parameter α below and Fig. 1b). Cylinder halflength L/2 normalizes all spatial variables and wavenumbers in Eqs. (1) and (2), and throughout the analysis; e.g., k is the product of the acoustic wavenumber and L/2. Setting $\hat{\phi}$ to zero for convenience does not compromise the nonaxisymmetric character of the single dipole's insonifying field. The total pressure p, and the corresponding fluid particle displacement w (both of which shall remain dimensional along with p^{inc}), then follow the dependence on ϕ of the circumferential modes $\cos\nu\phi$,

$$p(r,\phi,z) = \sum_{\nu=0}^{\infty} p_\nu(r,z)\cos\nu\phi \quad (4a)$$

$$w(r,\phi,z) = \sum_{\nu=0}^{\infty} w_\nu(r,z)\cos\nu\phi \quad (4b)$$

The equation of radial momentum connects the above at every point in the fluid,

$$\ddot{w}_\nu = -\frac{L}{2\rho}\frac{\partial p_\nu}{\partial r} \quad (5)$$

where ρ is the background density.

The usual approach in diffraction problems is to split the total field into a sum of incident and scattered components. Here p_ν will be p_ν^{inc} plus "inner" ("i" superscripted) pressure

$$p_\nu^i(r,z) = \int_{-\infty}^{\infty} d\bar{k}\, e^{-i\bar{k}z} A(\bar{k}) J_\nu(\gamma r) \quad (6)$$

for r<a, and p_ν^{inc} plus outer pressure

$$p_\nu^o(r,z) = \int_{-\infty}^{\infty} d\bar{k}\, e^{-i\bar{k}z} B(\bar{k}) H_\nu^{(1)}(\gamma r) \quad (7)$$

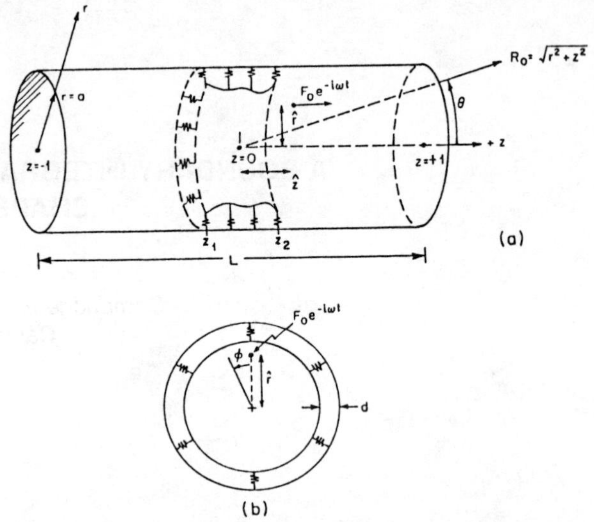

Figs. 1a,b Diffracting geometry and coordinate system. Coating properties do not vary with circumferential angle ϕ.

for r>a. Radial wavenumber γ is

$$\gamma = \sqrt{k^2-\bar{k}^2} \quad (=i\sqrt{-\gamma^2} \text{ for } \bar{k}>k) \quad (8)$$

where the sign of the second equality reflects the $\exp(-i\omega t)$ time factor assumed for all dependent variables. Eqs. (6) and (7) are solutions of the wave equation that behave as they should in their respective spatial regimes: p_ν^i is finite at r=0 and p_ν^o satisfies the radiation condition. J_ν is a Bessel function of order ν; all further references to Hankel function $H_\nu^{(1)}$ will omit its superscript for simplicity of notation.

The boundary condition for the rigid exterior surface of the cylindrical shroud becomes

$$\frac{\partial p_\nu^o}{\partial r} = -\frac{\partial p_\nu^{inc}}{\partial r} \quad ; \quad r=a^+,\ |z|<1 \quad (9)$$

while Hooke's law applies along the interior,

$$p_\nu^{inc} + p_\nu^i = K(z) w_\nu \quad ; \quad r=a^-,\ z_1<z<z_2 \quad (10)$$

where spring stiffness K may have a resistive part to account for boundary losses, through damping factor η:

$$K = K_r(1-i\eta) \quad (11)$$

with subscript r denoting real part [$K_r(1-i\eta)=E_r(1-i\eta)/d$ for a thin elastic coating material of complex Young's modulus E and thickness d, whose extent appears greatly exaggerated in Fig. 1b; the sign of the negative imaginary parts of K and E is consistent with the $\exp(-i\omega t)$ harmonic time function].

Eqs. (5) and (10) yield the boundary condition for the inner surface in terms of incident and scattered pressures, akin to Eq. (9) for the exterior,

$$\frac{\partial p_\nu^{inc}}{\partial r} + \frac{\partial p_\nu^i}{\partial r} = \alpha k^2\left(p_\nu^{inc}+p_\nu^i\right) \quad ; \quad r=a^- \quad (12)$$

with nondimensional compliance parameter α given by

$$\alpha(z) = \frac{2\rho c^2}{K(z)L} \tag{13}$$

where ρc^2 represents the bulk modulus of the fluid medium. Eq. (12) holds both for coated and uncoated sections of the pipe's interior, since with $\alpha=0$ (i.e., for $z<z_1$ and $z>z_2$) it collapses to the proper cancellation of incident and diffracted gradients.

Define now

$$\Delta p_\nu(z) \equiv p_\nu^o(a,z) - p_\nu^i(a,z) \tag{14}$$

$$\Delta p_\nu'(z) \equiv \frac{\partial p_\nu^o}{\partial r}(a,z) - \frac{\partial p_\nu^i}{\partial r}(a,z) \tag{15}$$

by putting r=a in Eqs. (6) and (7); then invert-transform the above relationships and enforce continuity of pressures and fluid particle velocities across r=a for $|z|>1$ to obtain a pair of equations for $A(\bar{k})$ and $B(\bar{k})$:

$$B(\bar{k})H_\nu(\gamma a) - A(\bar{k})J_\nu(\gamma a) = \int_{-1}^{1} \frac{dz}{2\pi} e^{i\bar{k}z} \Delta p_\nu(z) \tag{16}$$

$$\gamma[B(\bar{k})H_\nu'(\gamma a) - A(\bar{k})J_\nu'(\gamma a)] = \int_{-1}^{1} \frac{dz}{2\pi} e^{i\bar{k}z} \Delta p_\nu'(z) \tag{17}$$

Solving for $A(\bar{k})$ and $B(\bar{k})$, and applying Wronskian relationship $J_\nu(\gamma a)H_\nu'(\gamma a) - J_\nu'(\gamma a)H_\nu(\gamma a) = 2i/(\pi\gamma a)$, results in

$$A(\bar{k}) = \frac{i\pi a}{2}\left\{\gamma H_\nu'(\gamma a)\int_{-1}^{1} \frac{d\bar{z}}{2\pi} e^{i\bar{k}\bar{z}} \Delta p_\nu(\bar{z}) - H_\nu(\gamma a)\int_{-1}^{1} \frac{d\bar{z}}{2\pi} e^{i\bar{k}\bar{z}} \Delta p_\nu'(\bar{z})\right\} \tag{18}$$

$$B(\bar{k}) = \frac{i\pi a}{2}\left\{\gamma J_\nu'(\gamma a)\int_{-1}^{1} \frac{d\bar{z}}{2\pi} e^{i\bar{k}\bar{z}} \Delta p_\nu(\bar{z}) - J_\nu(\gamma a)\int_{-1}^{1} \frac{d\bar{z}}{2\pi} e^{i\bar{k}\bar{z}} \Delta p_\nu'(\bar{z})\right\} \tag{19}$$

and thus in expressions for p_ν^o and p_ν^i that are solely in terms of Δp_ν and $\Delta p_\nu'$, e.g.,

$$p_\nu^o(r,z) = \int_{-1}^{1} d\bar{z}\, \Delta p_\nu(\bar{z}) \int_{-\infty}^{\infty} d\bar{k}\, e^{-i\bar{k}(z-\bar{z})} \frac{i\gamma a}{4} J_\nu'(\gamma a)H_\nu(\gamma r) + \int_{-1}^{1} d\bar{z}\, \Delta p_\nu'(\bar{z}) \int_{-\infty}^{\infty} d\bar{k}\, e^{-i\bar{k}(z-\bar{z})} \left(\frac{-ia}{4}\right) J_\nu(\gamma a)H_\nu(\gamma r) \tag{20}$$

These two fundamental components, $\Delta p_\nu'$ and Δp_ν, of the virtual source distribution, stand respectively for monopoles and radial dipoles of yet undetermined strengths; Δp_ν cannot affect the on-axis total noise levels (i.e., at $\theta=0,\pi$ in Fig. 1a), which thus remain equal to p_ν^{inc} for cases when the duct's interior wall is uncompliant (cf. Refs. 8 and 10). The next step is to take the r derivative of Eq. 20 and to let r approach a^+. The procedure generates a kernel for the first term on the right side of the above of the type that is sometimes misleadingly referred to as "highly singular", when in fact a careful analysis shows that its infinities are self cancelling and that its finite part emerges naturally. Since the large $\bar{k}$ behavior of the integrand of this kernel term is

$$\frac{i\gamma a}{4} J_\nu'(\gamma a)H_\nu(\gamma r) \sim \frac{e^{-|\bar{k}|(r-a)}}{4\pi} \tag{21}$$

the logical thing to do is to express its r derivative as

$$\begin{aligned} &\frac{\partial}{\partial r}\int_{-\infty}^{\infty} d\bar{k}\, e^{-i\bar{k}(z-\bar{z})} \frac{i\gamma a}{4} J_\nu'(\gamma a)H_\nu(\gamma r) \\ &= \frac{\partial}{\partial r}\int_{-\infty}^{\infty} d\bar{k}\, e^{-i\bar{k}(z-\bar{z})}\left\{\frac{i\gamma a}{4} J_\nu'(\gamma a)H_\nu(\gamma r) - \frac{e^{-|\bar{k}|(r-a)}}{4\pi}\right\} \\ &+ \frac{\partial}{\partial r}\int_{-\infty}^{\infty} \frac{d\bar{k}}{4\pi} e^{-i\bar{k}(z-\bar{z})-|\bar{k}|(r-a)} \end{aligned} \tag{22}$$

where the first term on the right side turns out now to be at most logarithmically singular at $z=\bar{z}$ and therefore integrable everywhere.[10] The integral on the second term on the right side of Eq. (22) is

$$\int_{-\infty}^{\infty} \frac{d\bar{k}}{4\pi} e^{-i\bar{k}(z-\bar{z})-|\bar{k}|(r-a)} = \frac{1}{2\pi} \cdot \frac{(r-a)}{(r-a)^2+(z-\bar{z})^2} \tag{23}$$

so that the r derivative of the first term on the right side of Eq. (20) becomes, on approach of the surface,

$$\begin{aligned} &\lim_{r\to a^+} \frac{\partial}{\partial r}\int_{-1}^{1} d\bar{z}\, \Delta p_\nu(\bar{z}) \int_{-\infty}^{\infty} d\bar{k}\, e^{-i\bar{k}(z-\bar{z})} \frac{i\gamma a}{4} J_\nu'(\gamma a)H_\nu(\gamma r) \\ &= \lim_{r\to a^+} \frac{\partial}{\partial r}\int_{-1}^{1} d\bar{z}\, \Delta p_\nu(\bar{z}) \int_{-\infty}^{\infty} d\bar{k}\, e^{-i\bar{k}(z-\bar{z})} \cdot \left\{\frac{i\gamma a}{4} J_\nu'(\gamma a)H_\nu(\gamma r) - \frac{e^{-|\bar{k}|(r-a)}}{4\pi}\right\} \\ &+ \frac{1}{2\pi} \lim_{r\to a^+} \frac{\partial}{\partial r}\int_{-1}^{1} d\bar{z}\, \Delta p_\nu(\bar{z}) \cdot \frac{(r-a)}{(r-a)^2+(z-\bar{z})^2} \end{aligned} \tag{24}$$

The last term in the above contains three arbitrary intervals

$$\lim_{r\to a^+} \frac{\partial}{\partial r} \int_{-1}^{1} d\bar{z}\, \Delta p_\nu(\bar{z}) \frac{(r-a)}{(r-a)^2+(z-\bar{z})^2}$$

$$= \lim_{r\to a^+} \frac{\partial}{\partial r} \left(\int_{-1}^{z-\varepsilon} + \int_{z+\varepsilon}^{1} \right) d\bar{z}\, \Delta p_\nu(\bar{z}) \frac{(r-a)}{(r-a)^2+(z-\bar{z})^2}$$

$$+ \lim_{r\to a^+} \frac{\partial}{\partial r} \int_{z-\varepsilon}^{z+\varepsilon} d\bar{z}\, \Delta p_\nu(\bar{z}) \frac{(r-a)}{(r-a)^2+(z-\bar{z})^2} \qquad (25)$$

Passing from Eq. (24) to (25) involves no approximation, though it is now convenient to make the $z-\varepsilon$, $z+\varepsilon$ segment small enough for Δp_ν not to change much within it (the smoothness or continuity of Δp_ν is a condition to be checked a posteriori). Then

$$\frac{\partial}{\partial r} \int_{z-\varepsilon}^{z+\varepsilon} d\bar{z}\, \Delta p_\nu(\bar{z}) \frac{(r-a)}{(r-a)^2+(z-\bar{z})^2}$$

$$\sim \Delta p_\nu(z) \frac{\partial}{\partial r} \int_{z-\varepsilon}^{z+\varepsilon} \frac{d\bar{z}(r-a)}{(r-a)^2+(z-\bar{z})^2}$$

$$= \Delta p_\nu(z) \int_{z-\varepsilon}^{z+\varepsilon} d\bar{z} \left\{ \frac{1}{(r-a)^2+(z-\bar{z})^2} - \frac{2(r-a)^2}{[(r-a)^2+(z-\bar{z})^2]^2} \right\} \qquad (26)$$

Fig. 2 plots the above integrand within brackets for $\varepsilon=.1$ (i.e., for 10% of the cylinder halflength), and for two values of $r-a^+ (\ll\varepsilon)$. As $r\to a^+$ its negative trough and two positive peaks cancel to yield a negative but finite area: $-2/\varepsilon$. This follows either from the direct evaluation of the elementary integrals in the second equality of Eq. (26), or from the first, which gives

$$= \Delta p_\nu(z) \lim_{r\to a^+} 2 \frac{\partial}{\partial r} \tan^{-1}\left(\frac{\varepsilon}{r-a}\right)$$

$$= -2\Delta p_\nu(z) \lim_{r\to a^+} \left\{ \frac{\varepsilon}{(r-a)^2+\varepsilon^2} \right\} = -\frac{2\Delta p_\nu(z)}{\varepsilon} \qquad (27)$$

Eq. (27) states the result of the last term in Eq. (25), which then becomes

$$= \lim_{r\to a^+} \frac{\partial}{\partial r} \int_{-1}^{1} d\bar{z}\, \Delta p_\nu(\bar{z}) \cdot \frac{(r-a)}{(r-a)^2+(z-\bar{z})^2}$$

$$= \lim_{\varepsilon\to 0} \left[\left(\int_{-1}^{z-\varepsilon} + \int_{z+\varepsilon}^{1} \right) \frac{d\bar{z}\, \Delta p_\nu(\bar{z})}{(z-\bar{z})^2} - \frac{2\Delta p_\nu(z)}{\varepsilon} \right]$$

$$\equiv \fint_{-1}^{1} \frac{d\bar{z}\, \Delta p_\nu(\bar{z})}{(z-\bar{z})^2} \qquad (28)$$

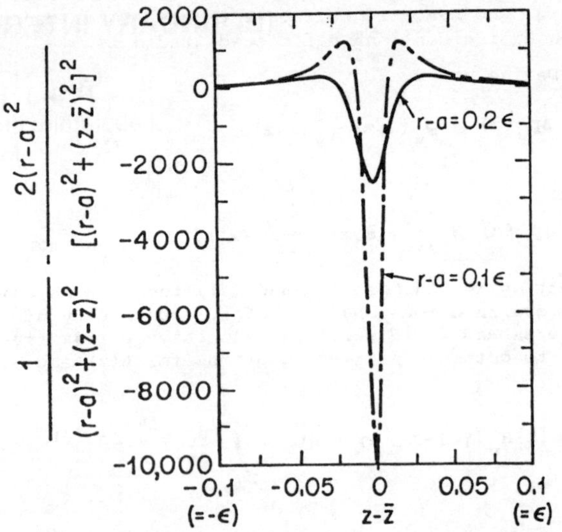

Fig. 2 Plot of integrand of Eq. 26 for $r-a=.2\varepsilon$, $.1\varepsilon$; $\varepsilon=.1$. Top values of twin positive horns are $1/[8(r-a)^2]$ and occur at $z-\bar{z}=\pm\sqrt{3}(r-a)$.

where now ε approaches zero and thereby leads to a similar type of double cancellation and to a perfectly finite result: the external $-2\Delta p_\nu(z)/\varepsilon$ part cancels the two positive infinities within the integral. The crossed integral symbol is the usual nomenclature used in the field of thin-wing theory. Substitution of the conclusions of Eqs. (21-28) into the r. derivative of Eq. (20), and recall of Eq. (9), produces the first of the problem's two coupled integral equations (accordingly labelled "I"):

$$\text{(I)} \quad r=a: \; -\frac{\partial p_\nu^{inc}}{\partial r} = \frac{1}{2\pi} \fint_{-1}^{1} \frac{d\bar{z}\, \Delta p_\nu(\bar{z})}{(z-\bar{z})^2}$$

$$+ \int_{-1}^{1} d\bar{z}\, \Delta p_\nu(\bar{z}) \int_{-\infty}^{\infty} d\bar{k}\, e^{-i\bar{k}(z-\bar{z})} \left\{ \frac{ia\gamma^2}{4} J_\nu'(\gamma a) H_\nu'(\gamma a) + \frac{|\bar{k}|}{4\pi} \right\}$$

$$+ \frac{\Delta p_\nu'(z)}{2}$$

$$+ \int_{-1}^{1} d\bar{z}\, \Delta p_\nu'(\bar{z}) \int_{-\infty}^{\infty} d\bar{k}\, e^{-i\bar{k}(z-\bar{z})} \left\{ \frac{-i\gamma a}{4} J_\nu(\gamma a) H_\nu'(\gamma a) - \frac{1}{4\pi} \right\} \qquad (29)$$

The above is the normal derivative of the Helmholtz integral equation applied to the shroud's exterior surface, although its generation here obviously followed a different path (cf. comments under subheading III, below). The second system equation comes about from a similar set of manipulations on p_ν^i rather than p_ν^o, and from application of Eq. (12):

$$\text{(II)} \quad r=a: \; \alpha k^2 p_\nu^{inc} - \frac{\partial p_\nu^{inc}}{\partial r} = \frac{\alpha k^2}{2}\Delta p_\nu(z)$$

$$+ \alpha k^2 \int_{-1}^{1} d\bar{z}\, \Delta p_\nu(\bar{z}) \int_{-\infty}^{\infty} d\bar{k}\, e^{-i\bar{k}(z-\bar{z})} \left\{ \frac{-i\gamma a}{4} H_\nu'(\gamma a) J_\nu(\gamma a) - \frac{1}{4\pi} \right\}$$

$$+ \frac{i a\alpha k^2}{4} \int_{-1}^{1} d\bar{z}\, \Delta p_\nu'(\bar{z}) \int_{-\infty}^{\infty} d\bar{k}\, e^{-i\bar{k}(z-\bar{z})} H_\nu(\gamma a) J_\nu(\gamma a)$$

$$+ \frac{1}{2\pi} ⨎_{-1}^{1} \frac{d\bar{z}\, \Delta p_\nu(\bar{z})}{(z-\bar{z})^2}$$

$$+ \int_{-1}^{1} d\bar{z}\, \Delta p_\nu(\bar{z}) \int_{-\infty}^{\infty} d\bar{k}\, e^{-i\bar{k}(z-\bar{z})} \left\{ \frac{i\gamma^2 a}{4} J_\nu'(\gamma a) H_\nu'(\gamma a) + \frac{|\bar{k}|}{4\pi} \right\}$$

$$- \frac{\Delta p_\nu'(z)}{2}$$

$$+ \int_{-1}^{1} d\bar{z}\, \Delta p_\nu'(\bar{z}) \int_{-\infty}^{\infty} d\bar{k}\, e^{-i\bar{k}(z-\bar{z})} \left\{ \frac{-i\gamma a}{4} H_\nu'(\gamma a) J_\nu(\gamma a) - \frac{1}{4\pi} \right\} \tag{30}$$

part of which obviously reproduces Eq. (29). An alternate and more compact form of the above therefore results upon subtraction of system equation I:

$$\text{(II')} \quad r=a: \; p_\nu^{inc}(z) = \frac{\Delta p_\nu(z)}{2}$$

$$+ \int_{-1}^{1} d\bar{z}\, \Delta p_\nu(\bar{z}) \int_{-\infty}^{\infty} d\bar{k}\, e^{-i\bar{k}(z-\bar{z})} \left\{ -\frac{i\gamma a}{4} H_\nu'(\gamma a) J_\nu(\gamma a) - \frac{1}{4\pi} \right\}$$

$$+ \frac{ia}{4} \int_{-1}^{1} d\bar{z}\, \Delta p_\nu'(\bar{z}) \int_{-\infty}^{\infty} d\bar{k}\, e^{-i\bar{k}(z-\bar{z})} H_\nu(\gamma a) J_\nu(\gamma a)$$

$$- \frac{\Delta p_\nu'(z)}{\alpha k^2} \tag{31}$$

The last term in Eq. (31) confirms that the strength of the virtual monopoles tends to zero in the limit of vanishing compliance: as $\alpha \to 0$ it becomes arbitrarily large relative to the rest and so yields $\Delta p_\nu' = 0$, which is as should be for a geometry with rigid surfaces that are specified to be an infinitesimal distance apart. The first three terms of Eq. (29) then become the governing integral equation for the dipole distribution Δp_ν (Ref. 8).

The classical analysis of diffraction by a thin semi-infinite rigid barrier deduces that the local behavior of this net "loading" vanishes as the square root of the distance to the edge (cf., for example, Noble[11]). This conclusion applies topologically also to the two ends of the pipe in Fig. 1a as long as the interior lining does not extend all the way to them; i.e., as long as $z_1 \neq -1$ and $z_2 \neq +1$. Our solution technique exploits this piece of information from potential theory and uses the expansion

$$\Delta p_\nu(z) = \sqrt{1-z^2} \sum_{n=0}^{N} a_n U_n(z) \tag{32}$$

where $U_n(z)$ are Chebyshev polynomials of the second kind, defined by (Ref. 12, p. 776, Eq. 22.3.16)

$$U_n(\cos\theta) = \frac{\sin(n+1)\theta}{\sin\theta} \tag{33}$$

They are always finite and nonzero at $z=\pm 1$. Integer N depends on frequency and must be large enough to capture the solution's expected degree of variation between $z=-1$ and $z=1$.

It is also convenient to recall the identity given in Ref. 12, Eq. 22.13.4,

$$\lim_{\varepsilon \to 0} \left(\int_{-1}^{z-\varepsilon} + \int_{z+\varepsilon}^{1} \right) d\bar{z} \frac{\sqrt{1-\bar{z}^2}}{\bar{z}-z} U_n(\bar{z}) = -\pi T_{n+1}(z) \tag{34}$$

[which unfortunately is wrong both in the book's first edition and in the first edition of the Bateman Project Manuscript, to which Ref. 12 refers (Higher Transcendental Functions, Vol. 2, p. 187)]. The T_n are Chebyshev polynomials of the first kind,

$$T_n(\cos\theta) = \cos n\theta \tag{35}$$

Differentiation of Eq. (34) generates four terms, two from the integrand and two from the limits,

$$\lim_{\varepsilon \to 0} \left\{ \frac{U_n(z-\varepsilon)}{-\varepsilon} \sqrt{1-(z-\varepsilon)^2} + \int_{-1}^{z-\varepsilon} \frac{d\bar{z}\sqrt{1-\bar{z}^2}}{(z-\bar{z})^2} U_n(\bar{z}) \right.$$

$$\left. - \frac{U_n(z+\varepsilon)}{\varepsilon} \sqrt{1-(z+\varepsilon)^2} + \int_{z+\varepsilon}^{1} \frac{d\bar{z}\sqrt{1-\bar{z}^2}}{(z-\bar{z})^2} U_n(\bar{z}) \right\}$$

$$= -\pi \frac{\partial}{\partial z} T_{n+1}(z) \tag{36}$$

or

$$⨎_{-1}^{1} d\bar{z} \frac{\sqrt{1-\bar{z}^2}}{(z-\bar{z})^2} U_n(\bar{z}) = -\pi \frac{\partial}{\partial z} T_{n+1}(z) \tag{37}$$

the left side being a consequence of the continuity of $U_n(z)\sqrt{1-z^2}$ everywhere in $|z|<1$. On the right side of the above,

$$\frac{\partial}{\partial z} T_{n+1}(z) = \frac{\partial}{\partial \theta} \cos(n+1)\theta \frac{\partial \theta}{\partial z} \tag{38}$$

becomes

$$\frac{\partial T_{n+1}(z)}{\partial z} = (n+1)\sin(n+1)\theta/\sin\theta = (n+1)U_n(z) \tag{39}$$

because $\theta=\cos^{-1}z$ so that $d\theta/dz=-\csc\theta$. Thus,

$$\oint_{-1}^{1} \frac{d\bar{z}\sqrt{1-\bar{z}^2}}{(z-\bar{z})^2} U_n(\bar{z}) = -\pi(n+1)U_n(z) \tag{40}$$

Ref. 8 stated the above result without intermediate details; the formula fits in nicely with the foregoing analysis by making possible the calculation of the "singular" integral in Eqs. (29) and (30) in closed form,

$$\frac{1}{2\pi} \oint_{-1}^{1} \frac{d\bar{z}\ \Delta p_\nu(\bar{z})}{(z-\bar{z})^2} = -\frac{1}{2} \sum_{n=0}^{N} (n+1)\, a_n U_n(z) \tag{41}$$

Multiplication of Eqs. (29) and (30) (or (31)) by $\sqrt{1-z^2}\ U_m(z)$, and integration from $z=-1$ to $+1$, generates a system of "modal" linear equations with the diagonal containing the contribution of Eq. (41),

$$\int_{-1}^{1} dz \sqrt{1-z^2}\ U_m(z) \left[\frac{1}{2\pi} \oint_{-1}^{1} \frac{d\bar{z}\ \Delta p_\nu(\bar{z})}{(z-\bar{z})^2}\right]$$

$$= -\frac{\pi}{4}(n+1)\, a_n\, \delta_{nm} \tag{42}$$

which has invoked the orthogonality condition for Chebyshev polynomials of the second kind

$$\int_{-1}^{1} dz \sqrt{1-z^2}\ U_m(z)\, U_n(z) = \frac{\pi}{2}\, \delta_{mn} \tag{43}$$

A second expansion, this time in terms of Legendre polynomials $P_n(z)$, describes the as-yet unknown discontinuity in gradient $\Delta p'_\nu$ across the cylinder's wall,

$$\Delta p'_\nu(z) = \sum_{n=0}^{N} b_n P_n(z) \tag{44}$$

The above does not place conditions on the behavior of the gradient solution $\Delta p'_\nu(z)$ at $z=z_1$, z_2, and its role in the solution does not have the special purpose of the $U_n(z)$ series, which was to compute the "singular" part of the diagonal of the integral equation system analytically. It simply produces convenient and quickly convergent integrands for the affected kernel functions. The final result of this combined modal treatment is a $(2[N+1]\times 2[N+1])$ system of linear equations for the $N+1$ a_n coefficients for Δp_ν and of the equally long $N+1$ element vector b_n for $\Delta p'_\nu$:

$$\begin{bmatrix} B_{mn} & | & C_{mn} \\ - & - & - \\ D_{mn} & | & E_{mn} \end{bmatrix} \begin{Bmatrix} a_n \\ - - \\ b_n \end{Bmatrix} = \begin{Bmatrix} F_m \\ - - - \\ G_m \end{Bmatrix} \tag{45}$$

Submatrix B_{mn} and vector F_m are the only ones present when the pipe's interior is rigid (Ref. 8, Eq. 27).

B. Power Balance: A Check on the Self Consistency of the Solution

Ref. 8 has pointed out that the essential drawback of this type of formulation is that the integral equations it produces appear already driven as explicitly as possible by the functional conditions imposed upon them by the object's boundaries. Recall, for example, the special case of rigid exterior and interior surfaces, which yields a Fredholm equation of the first kind with the known term given by the negative of the incident field's gradient at the wall. Also, the construction of the scattered pressure field here is such that it satisfies automatically the remaining "boundary" condition, i.e., that of outbound radiation, through proper choice of radial behavior for p^o_ν in Eq. (7). And so at first there might seem to be nothing more to do but accept whatever set of numbers results for vector $[a_n, b_n]^T$ as the one and only authentic answer from which the total pressure's directivity pattern, its radiated power, etc, are to be calculated.

Ref. 8 has also argued, however, that a numerical match in the computed input and radiated energies for an insonifying dipole/rigid shroud system should certify the solution indirectly by proving that the boundary which the Δp_ν function then straddles behaves reactively, and that the consistent meeting of this power balance over wide frequency bands and combinations of pipe parameters implies that this reactive boundary property must in fact correspond to the correct one of infinite impedance. The idea is that because the matrix system that replaces the integral equation's kernel has complicated, frequency-dependent complex entries, it would be difficult for its inverse to produce wrong solutions that conserve energy at the surface consistently without introducing also spurious resistive effects that violate the proposed power balance.

Here the energy argument becomes more delicate and therefore more persuasive because it demands a balance of not two quantities, but three: (1) the power injected directly into the acoustic medium by the dipole source at its position in the field, (2) the power radiated out of the duct's two open ends, and (3) the mechanical power dissipated by the lossy lining. Economic reasons will here confine the proof of this equilibrium concept to the axisymmetric case of an on-axis dipole; thus $\hat{r}=0$ and $\nu=0$ in what follows.

Refs. 8 and 10 have outlined in detail an analytical procedure for obtaining the real part of the total fluid particle velocity in the axial direction z at the source position $\hat{z}$, and from it have calculated the energy that the freefield dipole puts into the medium while under the feedback influence of the virtual sources that compose the diffracting shroud. The period-averaged input power Π_{inp} turns out to be given by[8]

$$\Pi_{inp} = \left(\frac{2}{L}\right)^2 \frac{F_o^2 k^2}{24\pi\rho c} + \frac{F_o}{2\rho c\cdot k\hat{R}}\, \mathrm{Im}\left\{\tilde{p}^{i}_{\nu=0}\right\} \tag{46}$$

where Im stands for imaginary part and $\tilde{p}^{i}_{\nu=0}$ is defined by

$$\tilde{p}^{i}_{\nu=0} = \frac{1}{\pi} \int_{0}^{2\pi} d\hat{\theta} \cos\hat{\theta}\ p^{i}_{\nu=0}(\hat{R},\hat{\theta}) \tag{47}$$

with integrand pressure $p^{i}_{\nu=0}(\hat{R},\hat{\theta})$ given by Eqs. (6) and (18), with $z=\hat{z}+\hat{R}\cos\hat{\theta}$, $r=\hat{R}\sin\hat{\theta}$. $\hat{R}$ is the radius of a temporary elemental sphere drawn around the position of the insonifying dipole, i.e., $\hat{R}$ is given by Eq. (2) with $\hat{r}=0$; the treatment in Ref. 8 eventually sets $\hat{R}$ to zero after establishing the convergence of the wavenumber integrals that participate in the calculation. By now solution vectors $\Delta p_\nu(z)$ and $\Delta p'_\nu(z)$ are known in terms of coefficients a_m and b_m, respectively. The final form of normalized input power $\tilde{\Pi}_{inp}$ is

$$\tilde{\Pi}_{inp} = \frac{\rho c L^2}{F_o^2}\Pi_{inp} = \frac{k^2}{6\pi} +$$

$$\frac{kaL^2}{4F_o} \text{ Re}\left\{\pi \sum_{m=0}^{N} a_m(m+1)\right.$$

$$\left[\int_0^{\pi/2} d\xi \sin^2\xi\ H_1(ka\sin\xi) J_{m+1}(k\cos\xi)\sin(k\hat{z}\cos\xi - m\pi/2)\right.$$

$$\left. -\frac{2i}{\pi}\int_0^{\infty} d\xi \sinh^2\xi\ K_1(ka\sinh\xi) J_{m+1}(k\cosh\xi)\sin(k\hat{z}\cosh\xi - m\pi/2)\right]$$

$$+\sum_{m=0}^{N} b_m\left[\int_0^{\pi/2} d\xi \sin 2\xi\ H_o(ka\sin\xi) j_m(k\cos\xi)\sin(k\hat{z}\cos\xi - m\pi/2)\right.$$

$$\left.\left. -\frac{2i}{\pi}\int_0^{\infty} d\xi \sinh 2\xi\ K_o(ka\sinh\xi) j_m(k\cosh\xi)\sin(k\hat{z}\cosh\xi - m\pi/2)\right]\right\} \tag{48}$$

where the j_m's are spherical Bessel functions and K_o, K_1 modified Bessel functions of orders o and 1; Re denotes real part.

The normalized period-averaged power radiated by the system to the farfield is

$$\tilde{\Pi}_{rad} = \frac{\rho c L^2}{F_o^2}\Pi_{rad} = \frac{\rho c L^2}{F_o^2}\cdot\frac{1}{2\rho c}\iint_{S_{ff}} dS\,|p^{inc}_{\nu=0} + p^o_{\nu=0}|^2 \tag{49}$$

where $dS=(L/2)^2\ 2\pi R_o^2\sin\theta d\theta$ (cf. Fig. 1a), and the integrand pressures are given by their large distance asymptotic forms.

The period-averaged power dissipated by the lining when its loss factor η is nonzero is defined by, and in normalized form turns out to be,

$$\Pi_{dis} = \frac{1}{2}\iint_{S_{cyl}} dS\ \text{Re}\{p^*\dot{w}\} \tag{50}$$

$$\tilde{\Pi}_{dis} = \frac{\rho c L^2 \Pi_{dis}}{F_o^2}$$

$$= \frac{\pi k a L^4}{4F_o^2}\ \eta \int_{z_1}^{z_2} dz\,|p^i_{\nu=0}(a,z) + p^{inc}_{\nu=0}(a,z)|^2\ \text{Re}[\alpha(z)] \tag{51}$$

C. Benchmark Solution: Finite Cylindrical Cavity with Pressure-Release Terminations

The power balance verifies the self consistency of the model without addressing the expected form of the solution in any particular frequency regime. One possible way to obtain this type of truly external benchmark check is to consider a slender pipe and frequencies that yield low values of ka while sweeping over a fair range of k, i.e., of the product of acoustic wavenumber and halflength L/2. The low ka regime is useful because the duct's two open ends are then known to behave very nearly as pressure-release plugs that trap most of the energy inside, and in the present situation should therefore produce a clear pattern of internal resonances very much like those of a cylindrical room whose two flat ends bear the condition of zero impedance.

This elementary and limiting solution "p_ℓ", that allows no radiation to the exterior is

$$p_\ell(r,z) = -\frac{4\pi F_o}{aL^2}\sum_{m=1}^{\infty}\sum_{\ell=0}^{\infty}\frac{m\cos[\frac{m\pi}{2}(\hat{z}+1)]\sin[\frac{m\pi}{2}(z+1)]J_o(\gamma_\ell r)}{\left\{k^2 - \left[(\frac{m\pi}{2})^2 + \gamma_\ell^2\right]\right\}J_o^2(\gamma_\ell a)} \tag{52}$$

for an on-axis source (i.e., for $\hat{r}=0$). Symbol γ will continue to be the radial wavenumber of the field, except that now it takes on only the "ℓ" discrete-value solutions of the following transcendental equation

$$J_o(\gamma_\ell a)\ \frac{i\rho c}{Z} + \frac{\gamma_\ell}{k}J_1(\gamma_\ell a) = 0 \tag{53}$$

with Z denoting the z-averaged stiffness impedance of the curved interior surface,

$$Z = i\langle K(z)\rangle/\omega \tag{54}$$

K remains generally complex, as stated in Eq. (11).

It is reasonable to expect the fundamental solution of Eq. (53), $\gamma_o a$, to be small if ka and $\rho c/Z$ are low; then

$$\frac{i\rho c}{Z} + \frac{\gamma_o^2 a}{2k} = 0 \tag{55}$$

so that

$$\gamma_o a = \sqrt{-2ka(\frac{i\rho c}{Z})} = \sqrt{-2k^2 a\langle\alpha(z)\rangle} \tag{56}$$

It follows that Eq. (55) is valid when

$$\sqrt{|2k^2 a\langle\alpha\rangle|} \ll 1 \tag{57}$$

The first zero of the denominator in Eq.(52) gives the fundamental resonance $k=k_{m=1,\ell=0}$ of the interior space,

$$k^2_{m=1,\ell=0} - (\frac{\pi}{2})^2 + \frac{2k^2_{m=1,\ell=0}\text{Re}\{\langle\alpha\rangle\}}{a} = 0 \tag{58}$$

i.e.,

$$k_{m=1,\ell=0} = \frac{\pi/2}{\sqrt{1+2\text{Re}\{\langle\alpha\rangle\}/a}} \tag{59}$$

so that the final version of the criterion in Eq. (57) becomes

$$\frac{\pi\sqrt{a|\langle\alpha\rangle|/2}}{\sqrt{1+2\,\mathrm{Re}\{\langle\alpha\rangle\}}} \ll 1 \qquad (60)$$

Again, the purpose of this secondary simplified analysis, and particularly of its final result in Eq. (59), has been to produce rationally an expression against which the eigenstructure of the rigorous calculations presented next can be checked.

II PREDICTIONS OF THE MODEL

All of the results about to be discussed are numerical solutions of the integral equation pair of the open-ended pipe (i.e., the finite-room results will not appear).

Table 1 lists in linear scale the three power components of Eqs. (48), (49) and (51) for a high-aspect ratio shroud (columns 2-4, respectively) vs k (column 1). Column 5 contains the sum of columns 3 and 4; column 6 shows the percent difference between this quantity -- the sum of radiated and dissipated powers -- and the input power in column 2. The least accurate calculation fails to meet the power balance condition by only 1/2 of 1%. The table's title states the geometric and physical parameters of the pipe: $\hat{z}=1/2$ indicates that the axial position of the incident-field dipole is halfway between the pipe's center and the right open end in Fig. 1a. The radial position of the dipole source, $\hat{r}$, is zero. Parameter a=1/4 means that the radius to length ratio for the pipe is 1/8; parameter $\alpha^{-1}=3(1-.3i)$ corresponds to a flat-coating coverage with a material whose product of spring stiffness and cylinder halflength is three times the bulk modulus of the acoustic medium, with loss factor η equal to .3. The lining extends from $z_1=-.8$ to $z_2=.8$. The accuracy of the tabulated power balance is typical of the rest of the calculations.

The results in Table 1 indicate that the maxima of the input, radiated, and dissipated energies occur at the same values of frequency. They also confirm that the radiated levels are below those of input power by an order of magnitude for the ka<<1 regime investigated, in a manner consistent with that expected of the limiting, nonradiating solution in Eq.(52). Fig. 3 shows in logarithmic scale the radiated power component from column 3, normalized by the free-field energy output of the insonifying dipole (which in normalized form is simply the first term on the right side of Eq. (48)). The figure also shows the power radiated by an unlined pipe of equal dimensions, whose peak value occurs at k=1.36, which represents a slight shift down from the limiting value π/2, obtained from Eq.(59) with α=0. The peak power for the lined pipe occurs at k=.75 whereas Eq. (59) would put it at about .8. Fig. 4 repeats the rigid cylinder curve of Fig. 3 but now shows it against a lined-pipe result for a material three times more compliant than that in Fig. 3; i.e., for a normalized stiffness α^{-1} equal to 1*(1-.3i). The peak power shifts down to k=.45 and remains in fairly close agreement with the value from Eq. (59) (=.52), although this second number is now less credible because the criterion that justifies its calculation, i.e., Eq. (60), states that $\pi/(2\sqrt{6})\ll 1$.

III. GENERALIZATION OF THE PROCEDURE TO OPEN-ENDED DUCTS OF VARIABLE CROSS SECTION

We now return to the statement made after Eq. (29) that identified that result as the normal derivative of the Helmholtz integral; more precisely, that identified Eq. (29) as the normal derivative of Kellogg's[13] version of the Helmholtz integral expressed in terms of jump functions for dipoles and sources making up a two-sided surface.

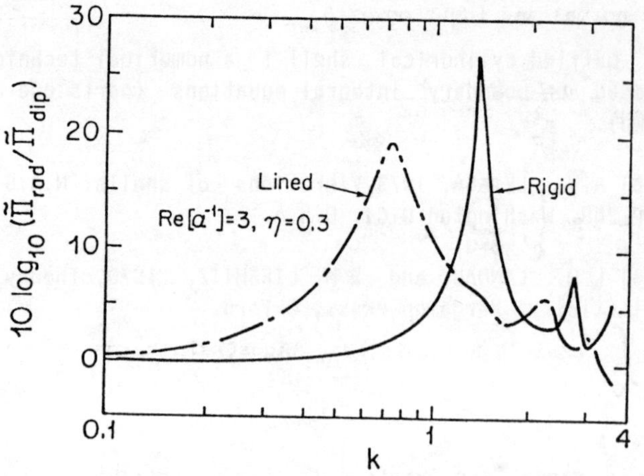

Fig. 3 Calculations of radiated power for open-ended rigid and lined cylindrical shrouds; $\hat{r}=0$, $\hat{z}=1/2$, a=1/4, $z_1=-.8$, $z_2=+.8$; Re$[\alpha^{-1}]=3$.

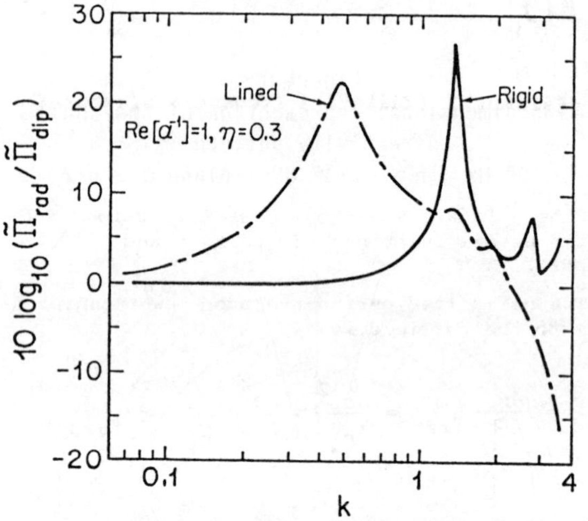

Fig. 4 Calculations of radiated power for open-ended rigid and lined cylindrical shrouds; $\hat{r}=0$, $\hat{z}=1/2$, a=1/4, $z_1=-.8$, $z_2=+.8$; Re$[\alpha^{-1}]=1$.

A simple way to establish this connection is to recall the wavenumber transform of a basic source, or freefield Green's function solution of

$$\nabla^2 G + k^2 G = \delta(\underset{\to}{x}-\underset{\to}{\bar{x}}),$$

and to note its kinship to the kernels derived in the foregoing analysis:

$$G = -\frac{e^{ik\sqrt{r^2+\bar{r}^2-2r\bar{r}\cos(\phi-\bar{\phi})+(z-\bar{z})^2}}}{4\pi\sqrt{r^2+\bar{r}^2-2r\bar{r}\cos(\phi-\bar{\phi})+(z-\bar{z})^2}}$$

$$= -\frac{i}{8\pi}\int_{-\infty}^{\infty} d\bar{k}\; e^{-i\bar{k}(z-\bar{z})} H_0^{(1)}\left(\gamma\sqrt{r^2+\bar{r}^2-2r\bar{r}\cos(\phi-\bar{\phi})}\right)$$

$$= -\frac{i}{8\pi}\sum_{\nu=-\infty}^{\infty} e^{i\nu(\phi-\bar{\phi})}\int_{-\infty}^{\infty} d\bar{k}\; e^{-i\bar{k}(z-\bar{z})} H_\nu^{(1)}(\gamma r)J_\nu(\gamma\bar{r}) \qquad (61)$$

Table 1: Balance Between Radiated Plus Dissipated Powers, and Input Power

$\hat{z}=1/2$, $a=1/4$, $\alpha^{-1}=3(1-.3i)$, $z_1=-.8$, $z_2=.8$

k	$\tilde{\Pi}_{inp}$	$\tilde{\Pi}_{rad}$	$\tilde{\Pi}_{dis}$	$\tilde{\Pi}_{rad}+\tilde{\Pi}_{dis}$	$\frac{\lvert\tilde{\Pi}_{inp}-(\tilde{\Pi}_{rad}+\tilde{\Pi}_{dis})\rvert}{\tilde{\Pi}_{inp}} \times 100\%$
0.075	0.132708E+00	0.317857E-03	0.132425E+00	0.132743E+00	0.264431E-01
0.151	0.280764E+00	0.148435E-02	0.279334E+00	0.280818E+00	0.192976E-01
0.226	0.464253E+00	0.427914E-02	0.460024E+00	0.464303E+00	0.107076E-01
0.302	0.717222E+00	0.106159E-01	0.706615E+00	0.717231E+00	0.118009E-02
0.377	0.110753E+01	0.251969E-01	0.108223E+01	0.110742E+01	0.988091E-02
0.452	0.179127E+01	0.606621E-01	0.173024E+01	0.179090E+01	0.205441E-01
0.528	0.318896E+01	0.156164E+00	0.303185E+01	0.318802E+01	0.295840E-01
0.603	0.662775E+01	0.455638E+00	0.616973E+01	0.662537E+01	0.358793E-01
0.679	0.155981E+02	0.145756E+01	0.141349E+02	0.155925E+02	0.360056E-01
0.754	0.200501E+02	0.245565E+01	0.175891E+02	0.200447E+02	0.266838E-01
0.829	0.995154E+01	0.153350E+01	0.841746E+01	0.995097E+01	0.573075E-02
0.905	0.501339E+01	0.929707E+00	0.408504E+01	0.501475E+01	0.269740E-01
0.980	0.310861E+01	0.662755E+00	0.244807E+01	0.311082E+01	0.710052E-01
1.056	0.227822E+01	0.534409E+00	0.174659E+01	0.228100E+01	0.121961E+00
1.131	0.189502E+01	0.470127E+00	0.142822E+01	0.189835E+01	0.175522E+00
1.206	0.174120E+01	0.441407E+00	0.130375E+01	0.174516E+01	0.227698E+00
1.282	0.173621E+01	0.435692E+00	0.130530E+01	0.174099E+01	0.275219E+00
1.357	0.184902E+01	0.442851E+00	0.141198E+01	0.185483E+01	0.313944E+00
1.433	0.202980E+01	0.441208E+00	0.159542E+01	0.203662E+01	0.336050E+00
1.508	0.214809E+01	0.404144E+00	0.175117E+01	0.215531E+01	0.336314E+00
1.583	0.218199E+01	0.359538E+00	0.182958E+01	0.218912E+01	0.326783E+00
1.659	0.228895E+01	0.348690E+00	0.194769E+01	0.229638E+01	0.324680E+00
1.734	0.254684E+01	0.370433E+00	0.218483E+01	0.255526E+01	0.330700E+00
1.810	0.297771E+01	0.416415E+00	0.257144E+01	0.298785E+01	0.340721E+00
1.885	0.361962E+01	0.485332E+00	0.314699E+01	0.363233E+01	0.351105E+00
1.960	0.454082E+01	0.580721E+00	0.397640E+01	0.455712E+01	0.358907E+00
2.036	0.582478E+01	0.706814E+00	0.513931E+01	0.584612E+01	0.366397E+00
2.111	0.749411E+01	0.858642E+00	0.666332E+01	0.752196E+01	0.371614E+00
2.187	0.929930E+01	0.100204E+01	0.833210E+01	0.933414E+01	0.374699E+00
2.262	0.105157E+02	0.106335E+01	0.949160E+01	0.105549E+02	0.373692E+00
2.337	0.104417E+02	0.992124E+00	0.948810E+01	0.104802E+02	0.368693E+00
2.413	0.934871E+01	0.840034E+00	0.854232E+01	0.938235E+01	0.359855E+00
2.488	0.809074E+01	0.696074E+00	0.742288E+01	0.811896E+01	0.348808E+00
2.564	0.718858E+01	0.601938E+00	0.661087E+01	0.721281E+01	0.337043E+00
2.639	0.676417E+01	0.560183E+00	0.622621E+01	0.678639E+01	0.328505E+00
2.714	0.678914E+01	0.562304E+00	0.624903E+01	0.681134E+01	0.326994E+00
2.790	0.721137E+01	0.599883E+00	0.663546E+01	0.723535E+01	0.332493E+00
2.865	0.796641E+01	0.664568E+00	0.732933E+01	0.799389E+01	0.345052E+00
2.941	0.893160E+01	0.743375E+00	0.822070E+01	0.896407E+01	0.363538E+00
3.016	0.986759E+01	0.814179E+00	0.909146E+01	0.990564E+01	0.385603E+00
3.091	0.104347E+02	0.848402E+00	0.962895E+01	0.104774E+02	0.409028E+00
3.167	0.103763E+02	0.827760E+00	0.959343E+01	0.104212E+02	0.432375E+00
3.242	0.973138E+01	0.760636E+00	0.901486E+01	0.977550E+01	0.453328E+00
3.318	0.878815E+01	0.675036E+00	0.815452E+01	0.882956E+01	0.471219E+00
3.393	0.784665E+01	0.597206E+00	0.728739E+01	0.788460E+01	0.483604E+00
3.468	0.707737E+01	0.541025E+00	0.657130E+01	0.711232E+01	0.493919E+00
3.544	0.652609E+01	0.509486E+00	0.604947E+01	0.655895E+01	0.503610E+00
3.619	0.616339E+01	0.499252E+00	0.569616E+01	0.619541E+01	0.519474E+00

where $r>\bar{r}$, as would pertain to the p_ν^o solution constructed for the spatial exterior of the diffracting geometry (Eq. (20)). Here the generalized force from the applied drive, i.e., from the insonifying dipole, acts along the cylinder's axis and so brings out only the $\nu=0$ component of the above. The zeroth, first, and second derivatives of Eq.(61) with respect to the source radial coordinate $\bar{r}(=a)$ reproduce the kernels of system equations I and II in Eqs. (29), (31).

The main point here is that Eq. (61), with $\bar{r}$ now freed to be a function of axial coordinate $\bar{z}$, and upon replacement of radial derivatives with arbitrary directional ones obtained from $\vec{n}.\nabla$, where $\vec{n}$ is the normal-to-surface vector, effectively generalizes the present formulation to thin-walled lined ducts of circular cross sections with axially varying diameter. The essential feature of subtracting out from the most problematic kernel its steady "singular" part, as was done in Eqs. (21)-(29), could follow the same route as in the current analysis but with running variable $\bar{z}$ in the integral equations replaced by arclength $\bar{s}$, measured along the wall's profile curve, or generator. Thus, for example, Eq. (28) would then be in terms of s and $\bar{s}$, rather than z and $\bar{z}$.

IV. CONCLUSIONS

This paper's immediate objective of providing solutions for the virtual sources making up the surface of an open-ended lined duct was really secondary to its settlement or re-affirmation of more general issues. Thus it demonstrated that

(1) The differentiated version of the Helmholtz integral equation is the only one one needs to solve scattering problems when the object's wall is very thin

in terms of the wavelength of the incoming signal;

(2) The approach of the object's surface by the field control point produces a kernel that may seem to be highly singular, but which is in fact perfectly regular by virtue of the self cancellation of its infinities;

(3) It is possible to check the self consistency of the solution by verifying numerically the balance between the power input at the freefield point of the insonifying source, and the sum of the powers radiated by the virtual/true source configuration and dissipated by the damped sections of the scattering surface. This diagnostic application of the principle of energy conservation appears to be novel; it could be used in general situations by exchanging the actual insonifying field for that of a test nearfield source or dipole. The corroboration of the power balance would validate the numerical generation of the integral equations' kernels and thus of the fluid's matrix system. The incident field of interest would then replace the driving vector of the test source for the final calculations.

REFERENCES

1. Reut, Z. "On the Boundary Integral Methods for the Exterior Acoustics Problem," J. Sound & Vib. 103 (2), 297-298 (1985).

2. Schenck, H.A., "Improved Integral Formulation for Acoustic Radiation Problems," J. Acoust. Soc. Am. 44, 41-58 (1968).

3. Seybert, A.F. and Rengarajan, T.K., "The Use of CHIEF to Obtain Unique Solutions for Acoustic Radiation Using Boundary Integral Equation," J. Acoust. Soc. Am. 81, 1299-1306 (1987).

4. Tobocman, W., "Calculation of Acoustic Wave Scattering by Means of the Helmholtz Integral Equation," J. Acoust. Soc. Am. 76, 599-607 (1984).

5. Burton, A.J. and Miller, G.F., "The Application of Integral Equation Methods to the Numerical Solution of Some Exterior Boundary-value Problems," Proc. Royal Soc. of London A323, 201-210 (1971).

6. Mathews, I.C., "Numerical Techniques for Three-Dimensional Steady-State Fluid-Structure Interaction," J. Acoust. Soc. Am. 79, 1317-1325 (1986).

7. Cunefare, K.A., Koopman, G. and Brod, K., "A Boundary Element Method for Acoustic Radiation Valid for all Wavenumbers," J. Acoust. Soc. Am. 85, 39-48 (1989).

8. Martinez, R., "Diffracting Open-ended Pipe Treated as a Lifting Surface," AIAA Journal 26 (4), 396-404 (1988).

9. Pierce, A.D., Acoustics, an Introduction to its Physical Principles and Applications (McGraw-Hill, NY), p. 167, Eq. 4-4.5.

10. Martinez, R., "Formulation for the Diffracted Field of an Open-ended Pipe," CAA Rept. U-1357-351, April 1986.

11. Noble, B., Methods Based on the Wiener-Hopf Technique (Pergamon, NY, 1958), p. 74, where the author points out that the jump in potential ϕ vanishes as the square root of the distance to the diffracting edge, i.e., $\phi \sim \sqrt{x}$.

12. Abramowitz, M. and Stegun, I., Handbook of Mathematical Functions (Dover, NY, 1968).

13. Kellogg, O.D., Foundations of Potential Theory (Dover, N.Y., 1953), p. 221.

CALCULATION OF SOUND FIELDS OVER AUDIENCE SEATS BY USING INTEGRAL EQUATION METHOD

Y. Kawai
Department of Architectural Engineering
Osaka Institute of Technology
Osaka, Japan

T. Terai
Department of Architecture,
Kyoto University
Kyoto, Japan

ABSTRACT

It is known that seat rows have a considerable effect on the sound transmission in auditoriums. When the direct wave propagates over the audience seats at near-grazing angle, excess attenuation appears in the low frequency range of 100-300Hz because of the interference between the direct wave and the scattered wave from the seat rows arranged periodically. The integral equation derived from normally derived Helmholtz-Kirchhoff's formula (NDF) is used to solve this problem. In numerical calculation, it is assumed that the seats are composed of rigid plates of negligible thickness and the potential is uniform in each divided surface element. Sound fields calculated were compared with 1:10 scale-model tests and showed good agreement.

1. INTRODUCTION

The problem of sound propagation over the audience seats in auditoriums so far has been reported by several authors. In 1964, Schultz and Watters[1] reported about the excess attenuation of the grazing-incident sound over seat rows in low frequencies. From the measurements of real concert halls and of a 1:10 scale model, they reported about excess attenuation over the inverse-square law, and this is not affected by the seating material or persons sitting on the seats. Sessler and West[2] also obtained the result from the measurements for a real concert hall.

Ando et al.[3] analyzed theoretically the case of plane wave incidence over seat rows which is simulated by plane periodically rugged surface extends infinitely. They showed that the results of 1:10 scale model experiments are in good agreement with that calculated.

The seat rows of concert halls, however, are finite and usually have underpass, aisles, and floors are not flat in many cases. Therefore it is necessary to use more flexible method to treat practical situations.

In this paper we use the integral equation method in order to predict the sound fields over the audience seats which are composed of rigid plates with negligible thickness. The basic integral equations derived from Helmholtz-kirchhoff's formula (BF) and from layer potential are commonly used in the acoustic scattering problem, but these method could not be used to solve for the case of thin plates, only their normally derived forms are applicable. The kernel, however, is strongly singular, and the regularization methods for this singular kernel have been proposed[4-9]. Filippi and Dumery[10] and Cassot[11] introduced the concept of the "finite part" of the divergent integral. Terai[12] presented a direct numerical method for three-dimensional case, which approximates area integral to line integral to remove the difficulty due to singularities by assuming uniform distribution. Kawai[13] also evaluated diffracted fields by finite sound barriers by using this method and verified the availability.

The problem of the sound propagation over audience seats corresponds to calculate the diffraction by multiple barriers, hence only the integral equation method is applicable but often used methods such as asymptotic evaluations are ineffective. In this paper, we apply the method[12] and also explain about the singular kernel in two-dimensional case.

2. THEORETICAL CONSIDERATION

Let P_s be a source (line source in two dimensions and point source in three dimensions) and P be a receiving point. A diffracting object is denoted by Ω with surface F and the exterior region by Ω_0 as shown in Fig. 1 . We denote here points on F by lower case letters and those in Ω_0 by upper case letters. By the application of Helmholtz-Kirchhoff's formula in the exterior region, we obtain

$$\Phi_d(P) + \int_F \left\{ \phi(q)\frac{\partial G(P,q)}{\partial n_q} - \frac{\partial \phi(q)}{\partial n_q} G(P,q) \right\} dS_q = \phi(P) \qquad (P \in \Omega_0), \quad (1)$$

where Φ denotes the velocity potential, k wave number, r=Pq, d=PPs, Φ_d=exp(-ikd)/d the direct wave, and n_q the outward normal to the surface F at q . In Eq. (1), it is assumed that Φ satisfies the radiation condition and the time factor $\exp(i\omega t)$ is neglected. The function G(P,q) is the fundamental solution to Helmholtz' equation: i.e.,

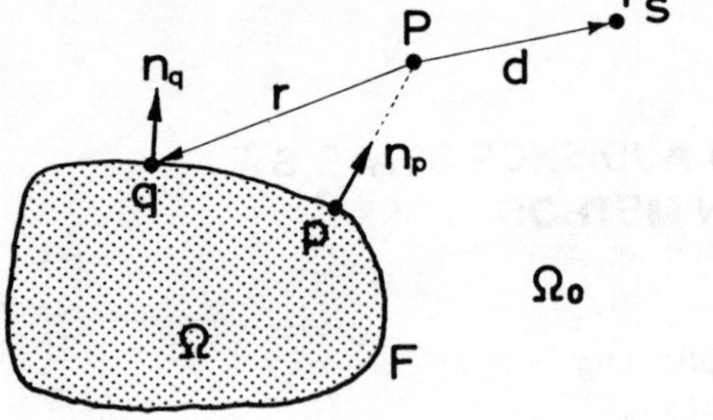

Fig. 1 Geometry of Eq.(1): a point source is at P_s and a receiving point at P. An object is denoted by Ω, with surface F, and Ω_0 denotes the exterior of Ω; n_q and n_p are the outward normals to F at the point q and p on the surface, respectively.

$$G(P,q)=\begin{cases}\dfrac{i}{4}H_0^{(2)}(kr_{Pq}) & \text{in two dimensions,}\\[2mm] \dfrac{\exp(-ikr_{Pq})}{4\pi r_{Pq}} & \text{in three dimensions,}\end{cases}$$

where $H_0^{(2)}(x)$ is the Hankel function of the second kind of order zero.

If the object Ω has a rigid surface and its thickness tends to zero (Fig.2), the velocity potential is then given by

$$\phi_d(P)+\int_F \tilde{\Phi}(q)\frac{\partial G(P,q)}{\partial n_q}dS_q=\begin{cases}\phi(P) & (P\in\Omega_0),\\ \frac{1}{2}\phi(p) & (p\in F),\end{cases} \quad (2)$$

where $\tilde{\Phi}=\Phi_1-\Phi_2$, Φ_1 and Φ_2 are the potentials on both the surfaces of F. Note only the difference $\Phi_1-\Phi_2$ is required for calculating the potential at P, usually it is not necessary to solve the potential separately on each surface.

Eq.(2) could be differentiated with respect to the direction n_p, normal to p passing P. By taking the limit $P\to p$, we obtain the following equation

$$\frac{\partial\phi_d(p)}{\partial n_p}+\int_F \tilde{\Phi}(q)\frac{\partial^2 G(p,q)}{\partial n_p\partial n_q}dS_q=0 \quad (\lim P\to p, P\in\Omega_0). \quad (3)$$

Eq.(3) is expressed simply with lower case letter p, but note the normal derivative of double layer potential could not defined on the boundary F, therefore Eq.(3) means the limit when P tends to the point p [14]. The kernels, the second-order derivatives of G(p,q), are

$$\frac{\partial^2}{\partial n_p\partial n_q}H_0^{(2)}(kr)=k\frac{H_1^{(2)}(kr)}{r}\cos(n_q,n_p)$$
$$+\{k^2H_0^{(2)}(kr)-2k\frac{H_1^{(2)}(kr)}{r}\}\cos(r,n_p)\cos(r,n_q), \quad (4)$$

$$\frac{\partial^2}{\partial n_p\partial n_q}\left\{\frac{\exp(-ikr)}{r}\right\}=-\frac{\exp(-ikr)}{r^3}[-(1+ikr)\cos(n_q,n_p)$$
$$+\{3(ikr+1)+(ikr)^2\}\cos(r,n_p)\cos(r,n_q)]. \quad (5)$$

By solving Eq.(3) for $\tilde{\Phi}$, the potential in Ω_0 could be obtained from Eq.(2).

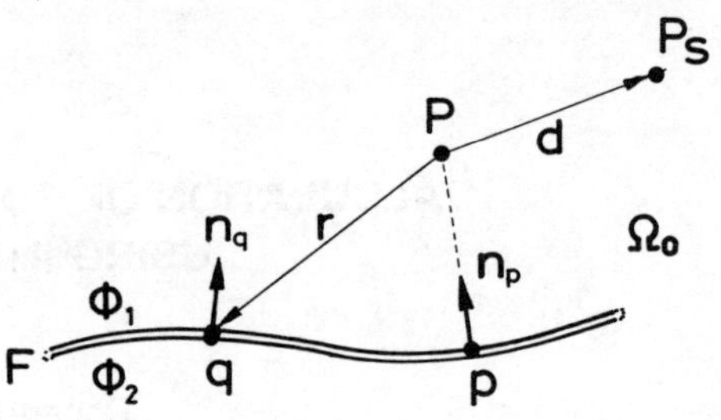

Fig. 2 Illustration where the object thickness tends to zero.

3. NUMERICAL METHOD OF CALCULATION

In order to solve numerically the above integral equation, the surface is divided into elements and approximated that $\tilde{\Phi}$ has uniform value, which is represented by the value at the center, in each element. Then the integral equation reduces to a system of equations.

For three-dimensional case, the contribution from each element can be calculated by the line-integral similarly as in [12].

Let δ be the angle between two planes F_M and F_N, on which situate the element M and N, respectively. Consider the integral with respect to q over the element M. We temporarily choose a coordinate system so that the origin is at p and the X-Y plane on F_N, the X-axis being parallel to the intersection of the two planes (Fig.3). If M is not equal to N, the contribution can be written as

$$\tilde{\Phi}_M\int_M\frac{\partial^2}{\partial n_p\partial n_q}\left\{\frac{\exp(-ikr)}{4\pi r}\right\}dS_q$$
$$=-\frac{\tilde{\Phi}_M}{4\pi}\Big[Z_p\sin\delta\int\left[\frac{\exp(-ikr)}{r^3}(ikr+1)\right]_{r_L}^{r_U}dx$$
$$-\cos\delta\oint_{\Gamma_M}\frac{\exp(-ikr)}{r}\{(ikr+1)(\frac{Z_p^2}{r^2})-1\}d\theta$$
$$+ik\cdot\exp(-ik|Z_p|)\cdot\alpha\Big], \quad (6)$$

where $\oint$ denotes the integration along the boundary Γ_M of the element M, and α takes the value 2π or zero if point P', the foot of the perpendicular from

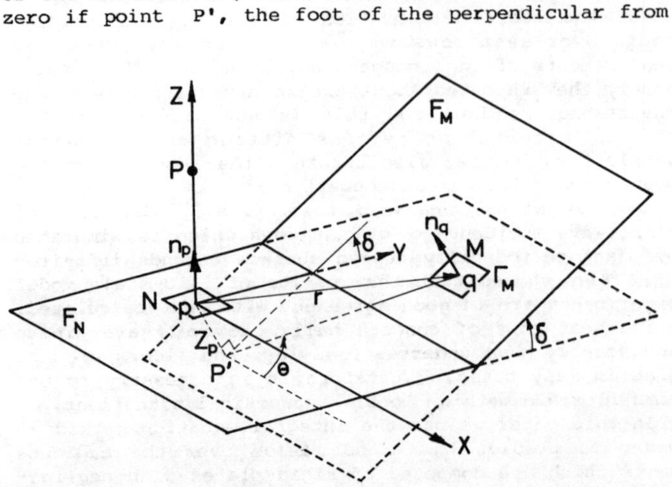

Fig. 3 Geometry of the element M to the element N containing point p.

p to F_M, lies inside or outside the element M, respectively; Z_p is the length of the perpendicular. If P' is just on the boundary Γ_M, α is the angle subtended by M at P'.

The integral for the element N containing p is

$$\tilde{\Phi}_N \int_N \frac{\partial^2}{\partial n_p \partial n_q} \left\{ \frac{\exp(-ikr)}{4\pi r} \right\} dS_q$$
$$= -\frac{\tilde{\Phi}_N}{4\pi} \left[2\pi i k + \oint_{\Gamma_N} \frac{\exp(-ikr)}{r} \, d\theta \right]. \qquad (7)$$

Thus we can evaluate the singular integrals.

For two-dimensional case, a similar procedure could be applied to evaluate $\partial^2 G / \partial n_p \partial n_q$ (Fig.4).

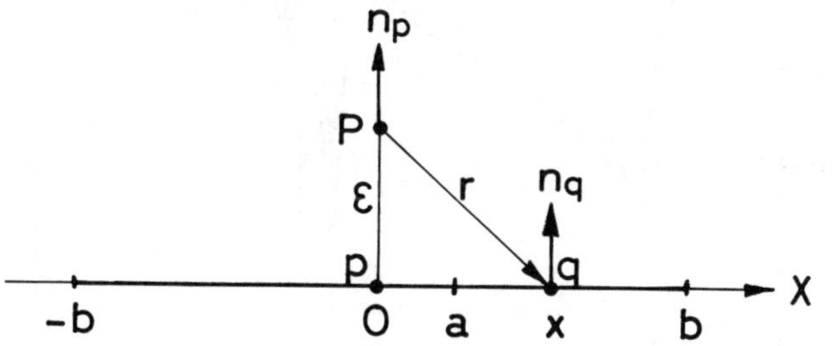

Fig. 4 Geometry in two dimensions.

In the region of the integration, a<x<b , the direct numerical integration is feasible because there exists no singularity. With respect to the region 0<x<a , Eq.(4) reduces to

$$\frac{\partial^2}{\partial n_p \partial n_q} H_0^{(2)}(kr)$$
$$= k \frac{H_1^{(2)}(kr)}{r} + \{k^2 H_0^{(2)}(kr) - 2k \frac{H_1^{(2)}(kr)}{r}\} (\frac{\varepsilon}{r})^2$$
$$\approx \frac{2i}{\pi r^2} - \frac{2ik^2}{\pi} \log(kr) (\frac{\varepsilon}{r})^2 - \frac{4i}{\pi r^2} (\frac{\varepsilon}{r})^2 \qquad (8)$$

The integral of each term in Eq.(8) reduces to

$$\frac{2i}{\pi \varepsilon} \tan^{-1} \frac{a}{\varepsilon} - \frac{2ik^2}{\pi} \varepsilon^2 \left[\frac{1}{\varepsilon} \tan^{-1} \frac{a}{\varepsilon} \cdot \log(k\sqrt{\varepsilon^2 + a^2}) \right.$$
$$\left. - \int_0^a \frac{1}{\varepsilon} \cdot \tan^{-1} \frac{x}{\varepsilon} \cdot \frac{x}{\varepsilon^2 + x^2} dx \right] - \frac{2ai}{\pi(\varepsilon^2 + a^2)} - \frac{2i}{\pi \varepsilon} \tan^{-1} \frac{a}{\varepsilon}.$$

By taking the limit we get

$$\lim_{\varepsilon \to 0} \int_0^a \frac{\partial^2}{\partial n_p \partial n_q} H_0^{(2)}(kr) dx = -\frac{2i}{\pi a} \qquad (9)$$

which corresponds to Eq.(7).

4. NUMERICAL RESULTS AND DISCUSSION

A. Results in two dimensions

(i) Seat rows in free space

Frequency responses for 20 rows of seats in free space were calculated. The frequency responses between 50 to 300 Hz at seats No. 5, No. 10, and No. 15 are shown. Zero dB denotes direct sound pressure level.

Fig. 5 shows the results for I-shaped seats, and Fig.

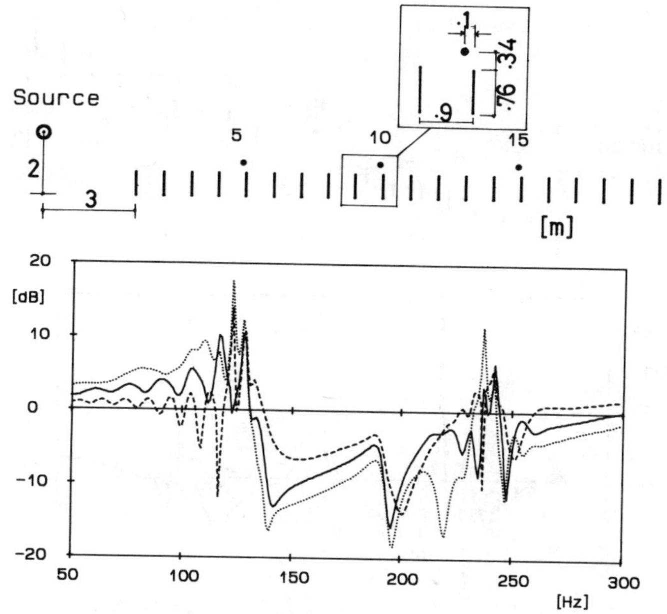

Fig. 5 Frequency responses at audience ear level: Calculation for horizontally arrangement of I-shaped seats in free space (2 dimensions); ------ , seat No.5; ———, No.10; ·········· , No.15; 3*20 elements.

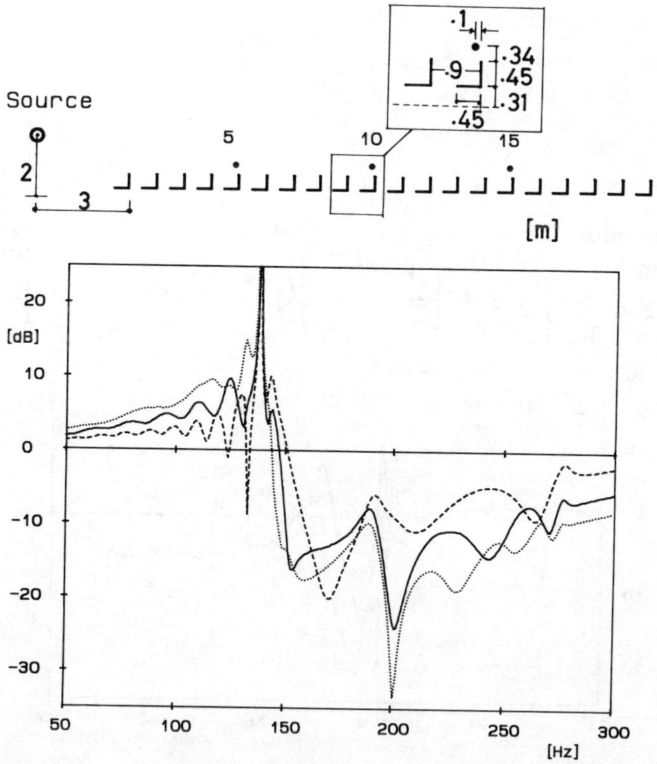

Fig. 6 Frequency responses at audience ear level: Calculation for horizontally arrangement of L-shaped seats in free space (2 dimensions); ------- , seat No.5; ———, No.10; ·········· , No.15; (2+2)*20 elements.

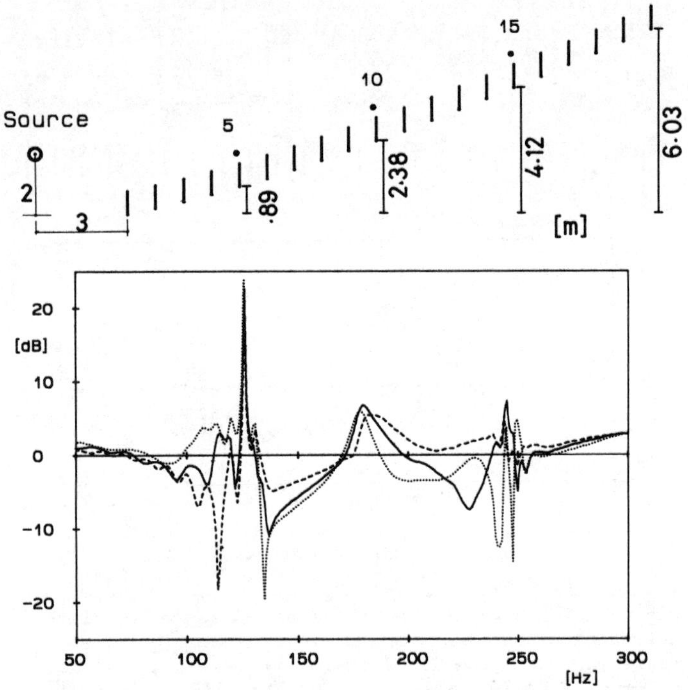

Fig. 7 Frequency responses at audience ear level: Calculation for sloped arrangement of I-shaped seats in free space (2 dimensions); ------ , seat No.5; ———— , No.10; ·············· , No.15; 3*20 elements.

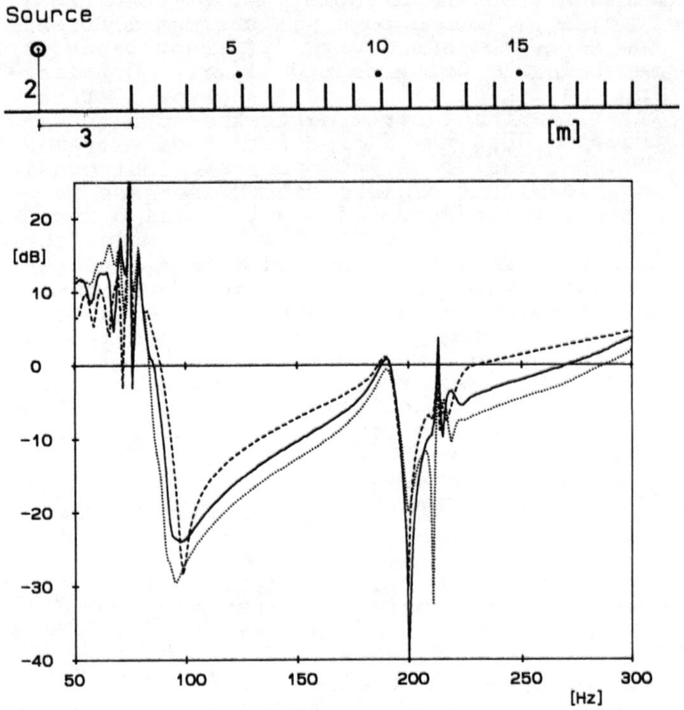

Fig. 9 Frequency responses at audience ear level: Calculation for I-shaped seats on flat rigid floor (2 dimensions); ------ , seat No.5; ———— , No.10; ·········· , No.15; 3*20 elements.

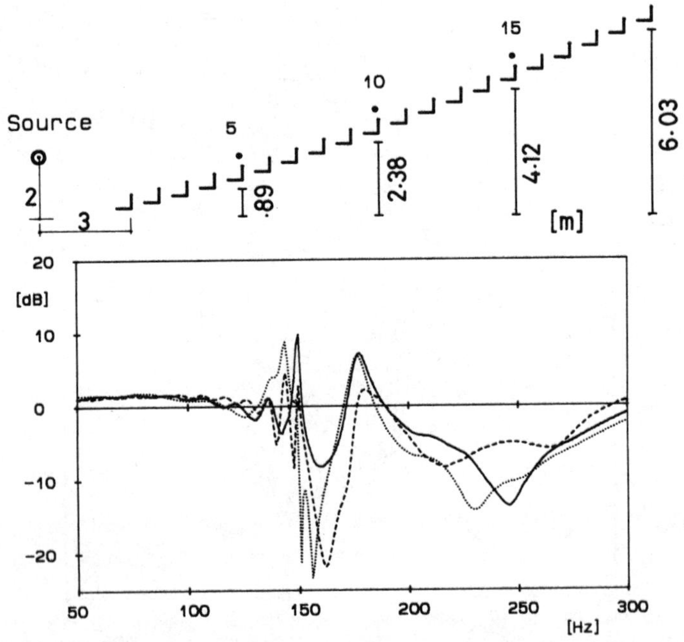

Fig. 8 Frequency responses at audience ear level: Calculation for sloped arrangement of L-shaped seats in free space (2 dimensions); ------ , seat No.5; ————, No.10; ·············· ,No.15; (2+2)*20 elements.

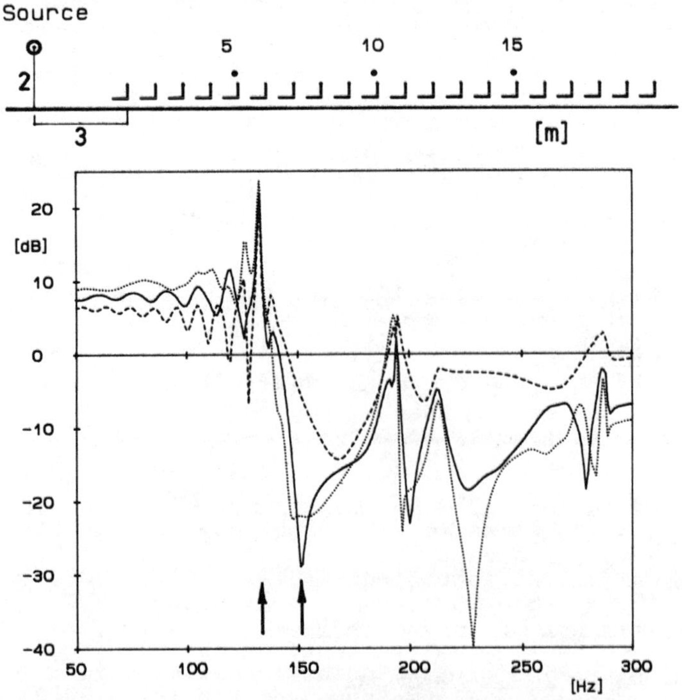

Fig. 10 Frequency responses at audience ear level: Calculation for L-shaped seats on flat rigid floor (2 dimensions); ------- , seat No.5; ———— , No.10; ············ , No.15; (2+2)*20 elements.

6 for L-shaped seats arranged horizontally. Fig. 7 and Fig. 8 show the results for sloped arrangements of seats, which simulate an auditorium floor, for I- and L-shaped seats, respectively.

It can be seen the L-shaped cases, Figs. 6 and 8, show large variation in frequency responses than I-shaped cases, Figs. 5 and 7, due to the resonance of cavities between the rows. The sloped seat arrangements, Figs. 7 and 8, show better results than horizontal arrangements both for seat positions and in frequency responses. The effect is evident for L-shaped case, Fig. 8.

(ii) Seat rows with rigid floor

Fig. 9 and Fig. 10 show the results with floor, corresponding to Figs. 5 and 6, respectively. In order to calculate these cases, images of the source and of the seat rows with respect to the floor are used. By comparing Figs. 9 and 10 to Figs. 5 and 6, we could see the variation in the frequency responses are considerably increased by the existence of the floor.

Fig. 11 shows the sound pressure distribution in the upper space of the seating area. Fig. 11(a) is calculated for the peak frequency (133 Hz) and Fig. 11(b) is for the dip frequency (151 Hz) of the solid line (seat No.10) in Fig. 10. Interference fringes caused by the reflection from the seat rows are observed.

(iii) Comparison with published result

The experimental results obtained by Sessler and West[2] on 1:10 scale model are compared with calculated values in Fig. 12. The calculations were made in two dimensions and the seats were substituted with rigid plates but the stage and the floor were taken into account. Calculated ones show similar tendency to the scale model experiments, but it seems difficult to deduce a definite conclusion about the accuracy of this calculation even taking into account the difference in three and two dimensions . In the following section therefore we compare the calculated responses in three dimensions with the scale model tests done by ourselves.

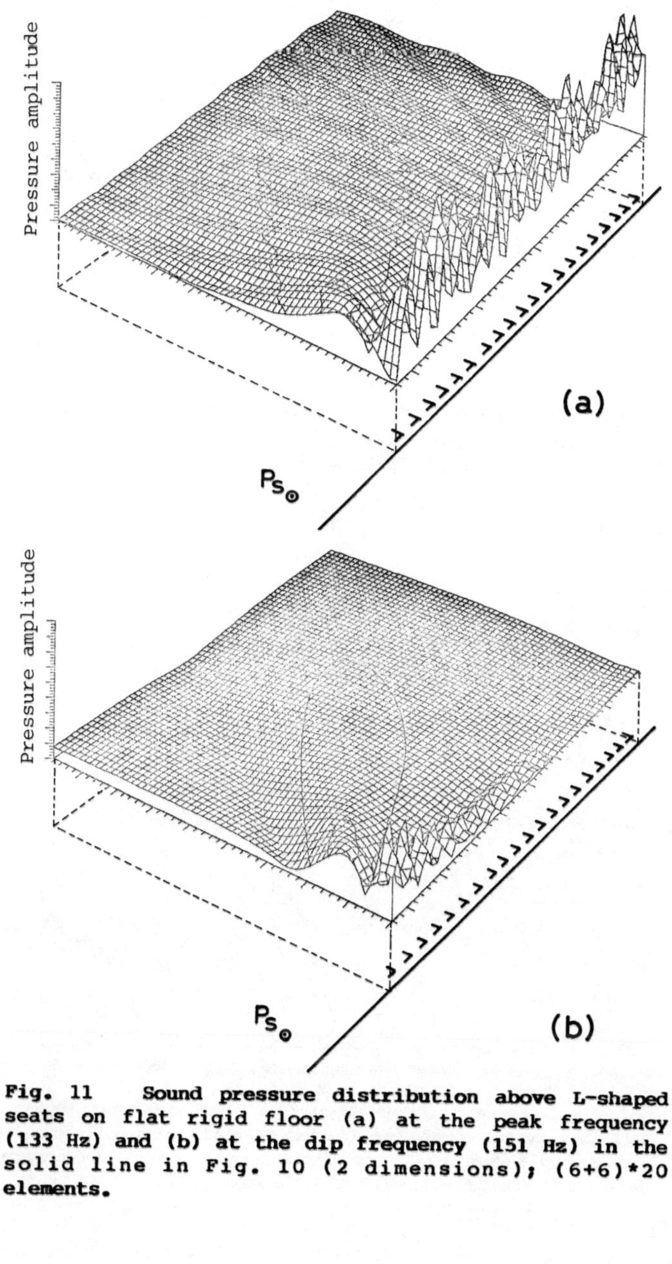

Fig. 11 Sound pressure distribution above L-shaped seats on flat rigid floor (a) at the peak frequency (133 Hz) and (b) at the dip frequency (151 Hz) in the solid line in Fig. 10 (2 dimensions); (6+6)*20 elements.

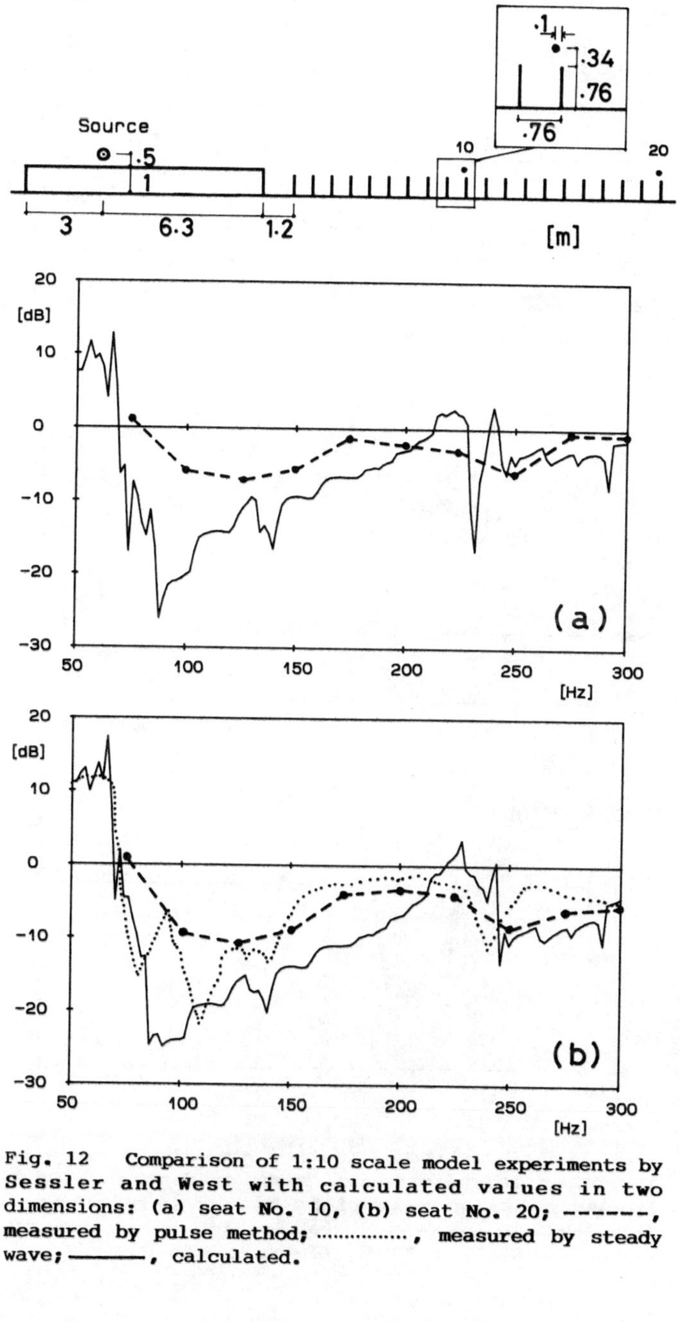

Fig. 12 Comparison of 1:10 scale model experiments by Sessler and West with calculated values in two dimensions: (a) seat No. 10, (b) seat No. 20; ------, measured by pulse method; ...., measured by steady wave; ———, calculated.

B. Results in three dimensions

(i) Effect of seat width

Fig. 13 shows the calculated frequency responses for I-shaped seats of width 3m(a) and 6m(b) on flat rigid floor. The differences between (a) and (b) are not so evident, but they differ appreciably from the two-dimensional calculations shown in Fig. 9. It can be seen smoother frequency responses are observed in three-dimensional cases than those in two dimensions. In view of this fact, it seems that the calculated values in three dimensions agree better with experiments shown in Fig. 12 than those in two dimensions.

(ii) Comparison between calculated and scale model results in three dimensions

To verify the accuracy of the method, we carried out 1:10 scale model experiments. The seats were made of iron plate of 2.5 mm thick, which is negligible compared to wavelength. An electric discharger was used to generate spherical waves on the central line. The transient responses on the central line were recorded at discrete time interval of 0.01 msec by 1/4 inch condenser microphone and were transformed into frequency responses. Results are expressed in real scale.

Figs. 14 and 15 show measured and calculated frequency responses for I-shaped and L-shaped seats with rigid floor respectively. Since the number of elements increases exceedingly in three dimensions, much computing time and storage are required. Therefore computations were not done so many frequencies as in two-dimensional cases, but the main features are expressed well. It can be seen the good agreement is obtained between the calculated and the measured results.

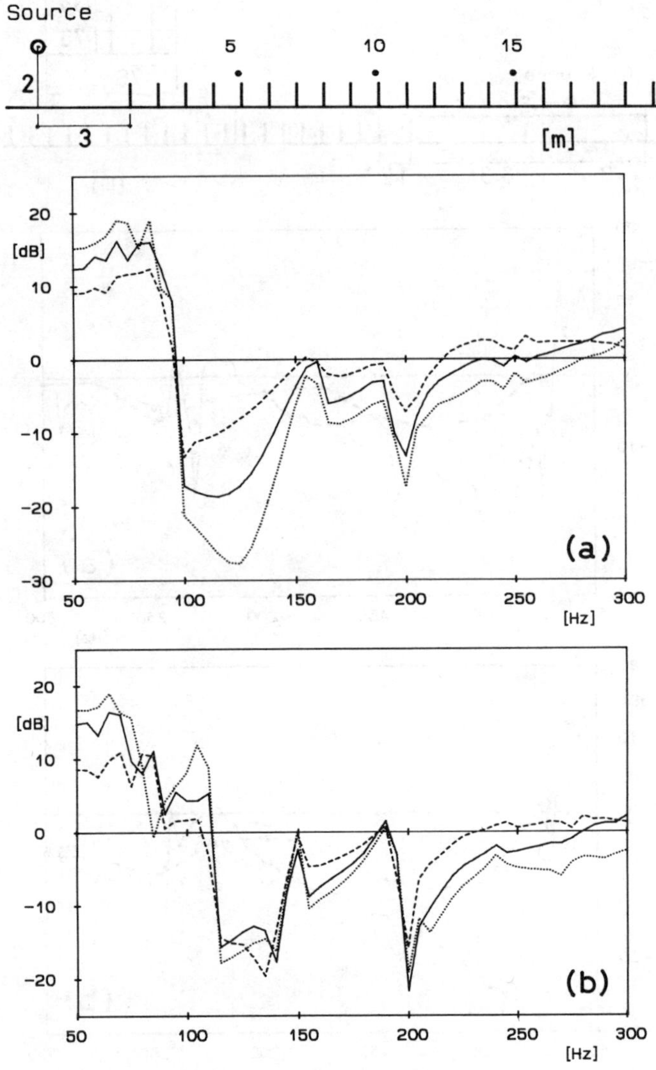

Fig. 13 Calculation of three dimensional frequency responses at audience ear level on central line I-shaped seats on flat rigid floor; (a)width 3 m, (b)width 6 m; -------- , seat No.5; ———— , No.10; ...., No.15; 3*20*12 elements.

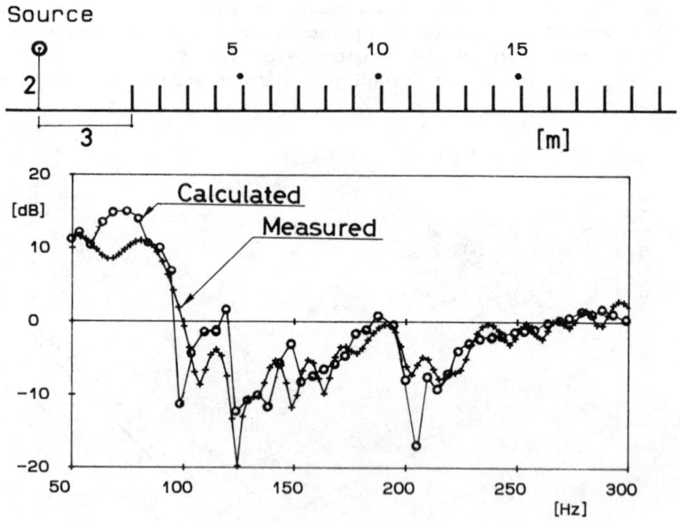

Fig. 14 Experiment for I-shaped seats of width 9 m on flat rigid floor in three dimensions: +—+, measured; o—o, calculated (3*20*18 elements).

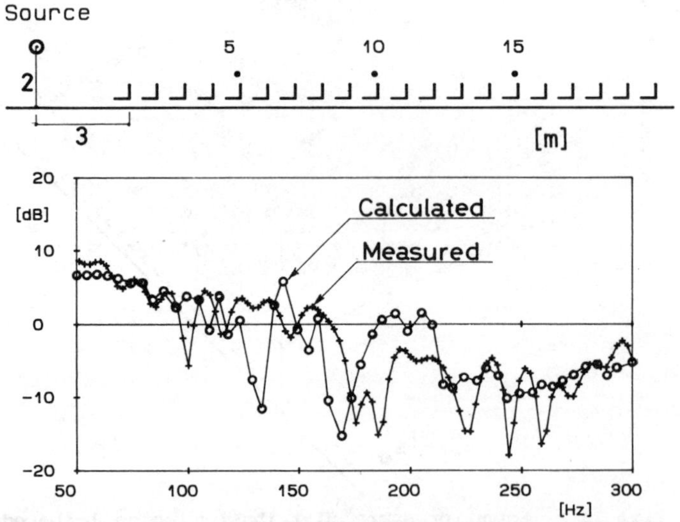

Fig. 15 Experiment for L-shaped seats of width 9 m on flat rigid floor in three dimensions: +—+, measured; o—o, calculated ((2+2)*20*18 elements).

By comparing Figs. 14 and 15 with Figs. 9 and 10, it can be seen the peaks and dips are strongly reduced in three-dimensional cases, especially in L-shaped seats.

The experiment of Sessler and West shown in Fig. 12 is somewhat different from that in Fig. 14 between 100 to 150 Hz. This could be affected by the presence of the stage on their experiment.

Further research would be necessary for the effect of stage and of sloping floor.

5. CONCLUSIONS

The integral equation of the first kind derived from normally derived Kirchhoff's formula is used by assuming the uniform distribution of variable in each element to calculate sound fields over the audience seats in auditoriums. It is shown this method could give good prediction even audience seats have the underpass and aisles, and can also be applied to the case of sloped floor.

ACKNOWLEDGMENT

We would like to thank Mr. A. Fujiwara, graduate student at Osaka Institute of Technology, for help in carrying out the measurements.

REFERENCES

1 Schultz, T. J., and Watters, B. G., 1964, "Propagation of sound across audience seating," Journal of the Acoustical Society of America, Vol. 36, pp. 885-896.

2 Sessler, G. M., and West, J. E., 1964, "Sound transmission over theatre seats," Journal of the Acoustical Society of America, Vol. 36, pp.1725-1732.

3 Ando, Y., Takaishi, M., and Tada, K., 1982, "Calculation of the sound transmission over theater seats and methods for its improvement in the low-frequency range," Journal of the Acoustical Society of America, Vol. 72, pp. 443-448.

4 Maue, A. W., 1949, "Zur Formulierung eines allgemeinen Beugungsproblems durch eine Integralgleichung," Zeitschrift fur Physik, Vol. 126, pp. 601-618.

5 Mitzner, K. M., 1966, "Acoustic scattering from an interface between media of greatly different density," Journal of Mathematical Physics, Vol. 7, pp. 2053-2060.

6 Stallybrass, M. P., 1967, "On a pointwise variational principle for the approximate solution of linear boundary value problems," Journal of Mathematics and Mechanics, Vol. 16, pp. 1247-1286.

7 Meyer, W. L., Bell, W. A., Zinn, B. T., and Stallybrass, M. P., 1978, "Boundary integral solutions of three dimensional acoustic radiation problems," Journal of Sound and Vibration, Vol. 59, pp. 245-262.

8 Burton, A. J., and Miller, G. F., 1971, "The application of integral equation methods to the numerical solution of some exterior boundary-value problems," Proceedings of the Royal Society, London A323, pp. 201-210.

9 Burton, A. J., 1973, "The solution of Helmholtz' equation in exterior domains using integral equations," NPL Report NAC 30. London: National Physical Laboratory.

10 Filippi, P., and Dumery, G., 1969, "Etude theorique et numerique de la diffraction par un ecran mince," ACUSTICA, Vol. 21, pp. 343-350.

11 Cassot, F., 1975, "Contribution a l'etude de la diffraction par un ecran mince," ACUSTICA, Vol. 34, pp. 64-71.

12 Terai, T., 1980, "On calculation of sound fields around three dimensional objects by integral equation methods," Journal of Sound and Vibration, Vol. 69, pp. 71-100.

13 Kawai, Y., 1983, "Sound attenuation by finite barriers," Proceedings of the 11th I.C.A., Paris, Vol.1, pp. 129-132.

14 Kellogg, O. D., 1929, Foundations of Potential Theory. Berlin: Springer-Verlag, 1967 re-issue.

SOUND RADIATION BY A BAFFLED CYLINDRICAL SHELL (BOUNDARY INTEGRAL EQUATIONS TECHNIQUE)

P. J. T. Filippi and D. Habault
Laboratoire de Mecanique et d'Acoustique
Marseille, France

SUMMARY :

A cylindrical thin shell of finite length, extended by two semi-infinite perfectly rigid cylinders, is immersed in a fluid of infinite extent ; it is excited by a harmonically time dependent source.

Because of the cylindrical symmetry of the geometric data, the solution of the problem (shell displacement, and sound pressure) can be expanded into a Fourier series with respect to the angular variable. To compute each angular component of the shell displacement, use is made of the infinite fluid-loaded shell Green's tensor, and a boundary sources method is adopted : this leads to a system of 8 algebraïc equations, the coefficients and second members (i.e. the terms accounting for the external excitation of the shell) of which are given by Fourier integrals. The Fourier integrals are evaluated numerically, with an accuracy given a priori. Thus, the accuracy of the solution of the problem can be estimated.

The various advantages of these developments are discussed. Examples of far-field radiation patterns are shown.

1 - INTRODUCTION :

This problem has already been studied by P.R. Stepanishen [1], who used a modal theory : the shell displacement is sought as a series of the in vacuo shell modes, the coefficients of which are solutions of an infinite system of algebraïc linear equations ; from a practical point of view, this method seems to be useful for simply supported shells only. M.A. Hamdi and P. Jean [2] propose a general method : the sound pressure is expressed by its Green's representation in terms of the normal shell displacement and its value on the shell surface ; a mixed system of partial differential and boundary integral equations is obtained ; the numerical method combines finite elements and boundary elements.

The method developed here is less general than that presented in [2], but requires much simpler programming ; it is similar to that used in a previous paper [3] devoted to fluid-loaded baffled plates. The Green's representation of the radiated pressure is used to obtain an integro-differential equation for the shell displacement. By using the Green's tensor of the fluid-loaded infinite shell equations, the shell displacement is represented by a Green's formula in terms of unknown sources, the supports of which are the shell boundaries : the sources densities are solutions of a system of 8 boundary integral equations. The cylindrical geometry allows a Fourier series expansion of all quantities with respect to the angular co-ordinate of the natural cylindrical co-ordinates system. Thus, the problem is reduced to a sequence of 8-th order systems of algebraïc equations : the coefficients are known as Fourier integrals which are numerically computed, with an accuracy given a priori ; the second members, which account for the external excitation of the shell are computed by Fourier integrals of the same type. The accuracy on the numerical solution of the problem (shell displacement, radiated sound pressure) is of the same order as the numerically computed Fourier transforms integrals. More details on this approach can be found in [4].

2 - STATEMENT OF THE PROBLEM :

Let (ρ,Φ,z) be a cylindrical co-ordinates system, and σ be the cylinder $\rho = a$. A thin elastic shell occupies the portion σ_0 defined by $[- L < z < + L]$; the complementary of σ_0 is perfectly rigid. The exterior Ω of σ is occupied by a perfect fluid, while its interior is a vacuum. The shell is mechanically defined by : h = thickness ; E = Young modulus ; ν = Poisson ratio ; μ = density ; use will be made of $\chi = h^2/12\, a^2$. The fluid is characterized by : c = sound speed ; μ_0= density. The system is excited by a harmonic $(e^{-i\omega t})$ source $F = [F_z, F_\Phi, F_\rho]$ applied to the shell.

Let $U = [u,v,w]$ denote the shell displacement, and p the radiated sound pressure. These functions are solutions of :

$$(1)\quad \begin{cases} \mathbb{C}\, U = F - \begin{pmatrix} 0 \\ 0 \\ p \end{pmatrix} & \text{on } \sigma_0 \\ (\Delta + k^2)\, p = 0 & \text{in } \Omega \\ \partial_\rho p = 0 & \text{on } \sigma - \sigma_0 \\ \partial_\rho p = \omega^2\, \mu_0\, w & \text{sur } \sigma_0 \end{cases}$$

where $\mathbb{C}$ is the Donnell-Mushtari operator [5,6]. The uniqueness of the solution is ensured by a Sommerfeld condition on p, and boundary conditions for U on $z = - L$ and $z = + L$.

Let G_p be the Green's function for the Neumann problem in Ω, satisfying the Sommerfeld condition. The sound pressure takes the form :

$$(2)\qquad p(M) = \omega^2\, \mu_0 \int_{\sigma_0} w(P)\, G_p(M)\, ds(P)$$

Thus the shell displacement satisfies the integro-differention equation :

$$(3)\quad \begin{cases} (\, \mathbb{C} + \mathbb{R}\,)\, U = F & \text{on } \sigma_0 \\ \mathbb{R}^j_i = 0 & \text{if } (i,j) \neq (3,3) \\ \mathbb{R}^3_3 w(M) = \omega^2 \mu_0 \int_{\sigma_0} G_p(M)\, w(P)\, ds(P) \end{cases}$$

3 - GREEN'S REPRESENTATION OF THE SHELL DISPLACEMENT :

The classical change of variable $s = z/a$ is introduced, and (3) becomes :

$$(4)\qquad (\, C + \mathfrak{R}\,)\, U = \frac{(1 - \nu^2)\, a^2}{E\, h}\, F \;,\; F = [-F_s, -F_\Phi, F_\rho]\,.$$

The operators C and $\mathfrak{R}$ are given by

$$(5)\quad \begin{cases} C^1_1 = \dfrac{\partial^2}{\partial s^2} + \dfrac{1-\nu}{2}\dfrac{\partial^2}{\partial \Phi^2} + \omega^2\, \mu_0\, \dfrac{(1-\nu^2)\, a^2}{E}\;; \\ C^2_1 = C^1_2 = \dfrac{1+\nu}{2}\dfrac{\partial^2}{\partial s\, \partial \Phi}\;;\quad C^3_1 = C^1_3 = \nu\, \dfrac{\partial}{\partial s}\;; \\ C^2_2 = \dfrac{1-\nu}{2}\dfrac{\partial^2}{\partial s^2} + \dfrac{\partial^2}{\partial \Phi^2} + \omega^2\, \mu_0\, \dfrac{(1-\nu^2)\, a^2}{E}\;; \\ C^3_2 = C^2_3 = \dfrac{\partial}{\partial \Phi}\;; \\ C^3_3 = 1 + \chi \left(\dfrac{\partial^2}{\partial s^2} + \dfrac{\partial^2}{\partial \Phi^2} \right)^2 - \omega^2\, \mu_0\, \dfrac{(1-\nu^2)\, a^2}{E}\;; \\ \mathfrak{R}^i_j = 0\,, \quad \forall\, (i,j) \neq (3,3)\;; \\ \mathfrak{R}^3_3 w = \omega^2 \mu_0 \dfrac{(1-\nu^2)\, a^4}{E\, h}\,. \\ \qquad \displaystyle\int_0^{2\pi} d\Phi' \int_{-L/a}^{+L/a} G(a,\Phi,s;a,\Phi',s')\, w(s',\Phi')\, ds'\;; \\ G(s,\Phi,a;s',\Phi',a) = G_p(M), \\ M = (a,\Phi,s),\; P = (a',\Phi',s')\,. \end{cases}$$

Let Γ be the Green's tensor of the operator (5) which satisfies a Sommerfeld-like condition at infinity. It is straightforward to show that U admits a Green's representation of the form :

$$(6)\qquad U = \frac{(1-\nu^2)\, a^2}{E\, h}\, \Gamma * [\, F - S^+ + S^-\,]\;,$$

where the symbol $*$ represents a convolution product. The distributions S^+ and S^- are sources, supported by the shell boundaries, with densities related to U and its successive derivatives up to order 3. Due to the boundary conditions, each source depends on 4 unknown functions only. For example, if the shell is simply supported, the components of S^+ have the following expressions :

$$\begin{cases} S^+_1 = \dfrac{E\, h}{(1-\nu^2)\, a^2}\, u(L/a,\, \Phi)\, \dfrac{d\delta}{ds}(L/a) \\ S^+_2 = \dfrac{E\, h}{(1-\nu^2)\, a^2} \left\{ \dfrac{1+\nu}{2}\dfrac{\partial u}{\partial \Phi}(L/a,\Phi) + \dfrac{1-\nu}{2}\dfrac{\partial v}{\partial s}(L/a,\Phi) \right\} \delta(L/a) \\ S^+_3 = \dfrac{E\, h}{(1-\nu^2)\, a^2} \left\{ \nu u(L/a,\Phi) + \chi\, \dfrac{\partial^3 w}{\partial s^3}(L/a,\Phi) \right. \\ \left. \quad + 2\chi \dfrac{\partial^3 w}{\partial s \partial \Phi^2}(L/a) \right\} \delta(L/a) + \dfrac{E\, h\, \chi}{(1-\nu^2)\, a^2}\, \dfrac{\partial w}{\partial s}(L/a,\Phi)\, \dfrac{d^2\delta}{ds^2}(L/a) \end{cases}$$

In these expressions, $\delta(L/a)$ is the Dirac measure at the point $s = L/a$. These expressions depend on the following independent functions :

$$
\begin{cases}
u(L/a) \\
\dfrac{1+\nu}{2}\dfrac{\partial u}{\partial \Phi}(L/a,\Phi) + \dfrac{1-\nu}{2}\dfrac{\partial v}{\partial s}(L/a,\Phi) \\
\nu u(L/a,\Phi) + \chi \dfrac{\partial^3 w}{\partial s^3}(L/a,\Phi) + 2\chi \dfrac{\partial^3 w}{\partial s \partial \Phi^2}(L/a) \\
\chi \dfrac{\partial w}{\partial s}(L/a)
\end{cases}
$$

The eight unknown functions appearing in S^+ and S^-, are solutions of a system of 8 boundary integral equations ; as in electrostatics or in acoustics, different systems of equations can be obtaineed. As an example, by taking the values at $s = L/a$ and at $s = -L/a$ of u, v, w, and $\partial u/\partial s$ such equations are obtained.

4 - FOURIER SERIES EXPANSION OF THE SOLUTION :

The shell displacement U is expanded into a Fourier series with respect to Φ

$$
(7) \qquad U = \sum_{n=-\infty}^{+\infty} U_n \, e^{in\Phi} .
$$

Then U_n takes the form :

$$
(8) \qquad U_n = \frac{(1-\nu_2)\, a^2}{E\, h} \Gamma_n * [\, F_n - S_n^+ + S_n^- \,] ,
$$

where Γ_n, F_n, S_n^+, and S_n^- are the Fourier components of Γ, F, S^+, and S^-. Γ_n is the inverse of the n-th component $[\, \mathcal{C}_n + \mathfrak{R}_n]$ of the operator $(\, \mathcal{C} + \mathfrak{R}\,)$. The operators $\mathfrak{R}_n$ are known through their Fourier transforms. Finally, it is easily seen that the boundary integral equations reduce to a sequence of algebraïc equations systems of order 8 ; moreover, each system can be split into two 4-th order systems.

Using the series expansion of w derived from (7) and (8), the expression (2) of the radiated pressure becomes a series in terms of the components w_n of the shell radial displacement. Let (R,θ,Φ) be the spherical co-ordinates of a point M ; the far-field sound pressure takes the form :

$$
(9) \quad p(M) = - \frac{e^{ikR}}{4\pi R} \frac{4 i \omega^2 \mu_0}{k \sin\theta} \sum_{-\infty}^{+\infty} \hat{w}_n\left(\frac{k \cos\theta}{2\pi}\right) \frac{e^{in(\Phi-\pi/2)}}{H_n'(ka \sin\theta)} + \mathcal{O}(R^{-2})
$$

In this expression, $H_n'(\zeta)$ is the first derivative of the Hankel function of the first kind and order n ; $\hat{w}_n(\zeta)$ is the Fourier transform of the function which is equal to the shell radial displacement extended by zero on the baffle : it is expressed through the Fourier transforms of the corresponding components of F, S^+, S^-, and Γ.

5 - NUMERICAL TECHNIQUE, AND EXAMPLE :

The only difficulty is to compute numerically the coefficients and second members of the linear systems giving the boundary sources components. They are given as inverse Fourier transforms of functions which are not always integrable, and can present singularities on or close to the real axis.

<u>5-1 Integration of a regular function</u> : it is reduced to the integration over a finite sequence of intervals

$$
\int_0^{\infty} f(\zeta)\, e^{2i\pi\zeta}\, d\zeta \simeq \sum_{q=1}^{N} \int_{a_q}^{a_{q+1}} f(\zeta)\, e^{2i\pi\zeta}\, d\zeta \; , \; a_1 = 0 \, .
$$

The length of each interval is chosen to obtain the value of the integral with a maximum of 32 Gauss points, and an accuracy ε given a priori ; the number N of intervals is chosen to get a relative accuracy ε on the integral value.

<u>5-2 Behaviour at infinity</u> : some functions to be Fourier inverted do not decrease fast enough at infinity. In such a case, their first order asymptotic term is removed and analytically inverted, while the remainder is numerically integrated.

<u>5-3 Removing poles</u> : when a pole exists close to the real axis, the numerical integration gets unstable. On a small interval around the pole, the function is approximated by the first term of its Laurent expansion which is integrated analytically ; out of this interval, the numerical method is used.

<u>5-4 Example</u> : the far-field directivity pattern of a finite-length shell is represented by the following surface :

$A(\theta,\Phi) = 20 \log[\,|p(R,\theta,\Phi)|4\pi R]$, $p(R,\theta,\Phi)$ given by (9).

The geometrical and mechanical data are : shell length $2L = 1$ m ; external diameter $2a = 12$ m ; thickness $h = 0.06$ m ; shell material density $\mu = 7850\ kg/m^3$; Young modulus $E = 2.26\ 10^{11}$ Pa ; Poisson ratio $\nu = 0.28$; fluid density $\mu_0 = 10^3\, kg/m^3$; sound speed $c = 1500$ m/s. The excitation is a point radial force $F = [0,0,\delta_{s=0} \otimes \delta_{\Phi=0}]$.

The results correspond to two different frequencies : 100 Hz (fig. 1 to 3), and 1 000 Hz (fig. 4 to 6). For each frequency, the first diagram represents the directivity pattern (logarithmic scale) in the plane $\theta = \pi/2$ as a function of Φ ; the second one is the directivity pattern in the plane $[\Phi = 0,\ \Phi = \pi]$ as a function of θ ; the third drawing is a three-dimensional representation of the directivity pattern (surface defined by expression (9)). The Fourier series expansion is reduced to 13 terms $(-6 \leqslant n \leqslant +6)$ for the first case, and to 41 terms ($-20 \leqslant n \leqslant +20$) for the second one.

6 - CONCLUDING REMARKS :

First of all, it must be mentionned that the method presented here is much more general than the modal technique. The *in vacuo* modes of the shell can be computed for particular boundary conditions only : this is one of the main limits of modal approaches. The boundary integral equations method allows for any kind of boundary conditions, including energy loss or non local conditions.

Modal methods lead to an infinite system of linear equations : assuming that this system is convergent, it is not easy to choose the set of equations and unknowns which approach accurately enough the exact system ; furthermore, it seems difficult to estimate the error. In the present method, the accuracy of the result can be decided *a priori* : indeed, the accuracy in the final result is of the same order as that of the numerically computed Fourier transforms which can be completely controlled.

The third advantage of the boundary elements method is that it can be implemented on small computers, and does not require a very long computation time. For example, on a 0.5 mips computer, the calculation and drawing time at 1000 Hz is about 60 mn.

This method is now being extended to the case of a stiffened fluid loaded shell. For a sufficiently narrow stiffener, its influence on the shell can be modelized as a line of forces (or forces and moments) ; the stiffener excitation is the force that the shell exerts on it. To determine the force density, a system of boundary and line integral equations is established. To illustrate this, let us first consider the case of an annular stiffener, located at $z = z_0$. If U' denotes its displacement, the complete set of equations is :

$$(10) \quad \begin{cases} (\mathbb{C} + \mathbb{R})\, U = F + f \otimes \delta_{z_0} \\ \mathbb{C}'\, U' = f \\ U'(\Phi) = U(z_0, \Phi) \end{cases}$$

where $\mathbb{C}'$ denotes the operator governing the stiffener displacement. Let Γ' be the Green's tensor of the stiffener equation. The solution of the system writes :

$$(11) \quad \begin{cases} U = \Gamma * \left[F + f \otimes \delta_{z_0} - S^+ + S^- \right] \\ U' = \Gamma' * f \end{cases}$$

The boundary conditions on U, and the continuity relationship between U and U′ lead to a system of 11 integral equations to determine the unknown sources densities. If a Fourier series expansion with respect to Φ is used, the system of integral equations reduces to a sequence of linear systems of order 11.

For a longitudinal stiffener, located along the line $\Phi = \Phi_0$, the governing equations write as follows :

$$(12) \quad \begin{cases} (a) & (\mathbb{C} + \mathbb{R})\, U = F + f \otimes \delta_{\Phi_0} \\ (b) & \mathbb{C}'\, U' = f \\ (c) & U'(z) = U(z, \Phi_0) \end{cases}$$

where $\mathbb{C}'$ denotes the operator governing the stiffener displacement. Let Γ' be the Green's tensor of equation (12,b) for a stiffener of infinite extent. The Green's representation of the displacement of the finite length stiffener is

$$(13) \quad U' = \Gamma' * [\, f - s^+ + s^- \,] ,$$

where s^+ and s^- are boundary sources accounting for the boundary conditions imposed to the stiffener ; each of them is defined by 3 constants. The shell displacement Green's representation is

$$(14) \quad U = \Gamma * \left[F + f \otimes \delta_{\Phi_0} - S^+ + S^- \right]$$

The unknown functions S^+, S^- and f, and the constants defining s^+ and s^- are determined by the boundary conditions satisfied by the shell and the stiffener displacements, and by the continuity conditions between U and U′. In that case, the system of integral equations does not reduce to a sequence of algebraïc systems.

As a final remark, it is to be noticed that the technique of boundary integral equations can be used whatever the shape of the shell boundary is.

REFERENCES

[1] P.R. STEPANISHEN, 1982, Journal of the Acoustical Society of America, 71, 813-823. Modal coupling in the vibration of a fluid loaded cylindrical shell.

[2] M.A. HAMDI, and P. JEAN,1985, IUTAM Symposium on Aero- and Hydro- Acoustics, Lyon, France. A mixed functional for the numerical resolution of fluid structure interaction problems.

[3] P.J.T. FILIPPI, 1985, Journal of Sound and Vibration, 100, 69-81. Sound radiation by baffled plates and related boundary integral equations.

[4] P.J.T. FILIPPI, and D. HABAULT, 1989, Journal of Sound and Vibration, 130(1), 13-36. Sound radiation by

a baffled cylindrical shell : a numerical technique based on boundary integral equations (parts one and two).

[5] A.W. LEISSA, 1973,Vibrations of shells. N.A.S.A. SP-288, Washington D.C., U.S.A.

[6] L.D. LANDAU, and E.M. LIFSHITZ, 1970, Theory of elasticity. Pergamon Press, Oxford.

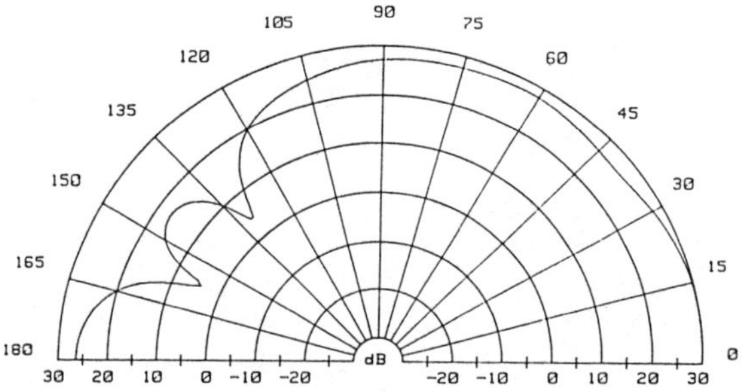

Figure 1
Two-dimensional represention of the shell directivity pattern :
100 Hz ; n_{max} = 37 dB ; plane $\theta = \pi/2$

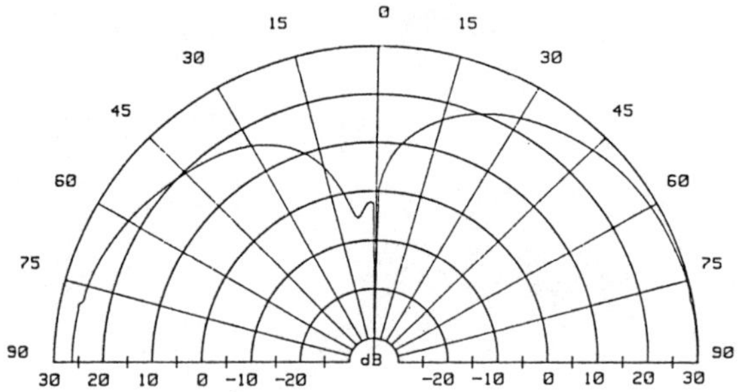

Figure 2
Two-dimensional represention of the shell directivity pattern :
100 Hz ; n_{max} = 37 dB ; plane [$\Phi = 0$, $\Phi = \pi$]

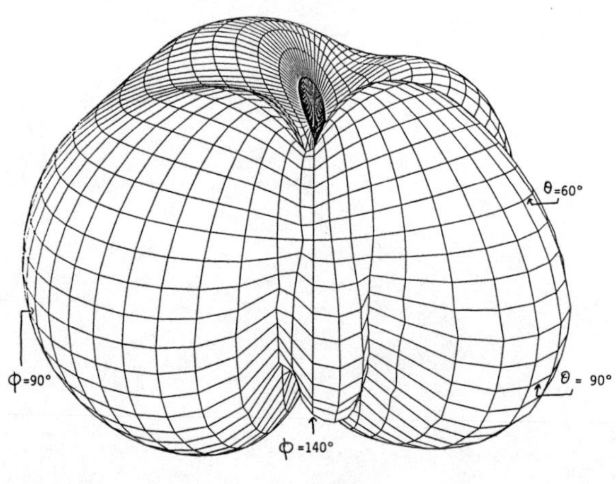

Figure 3
Three-dimensional representation do the shell directivity pattern : 100 Hz

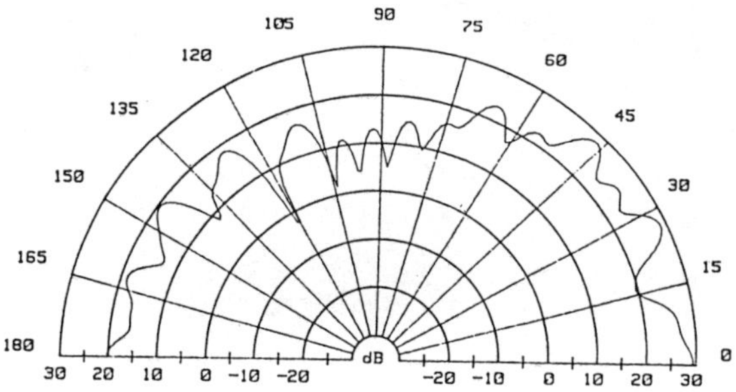

Figure 4
Two-dimensional represention of the shell directivity pattern :
1 000 Hz ; n_{max} = 58 dB ; plane $\theta = \pi/2$

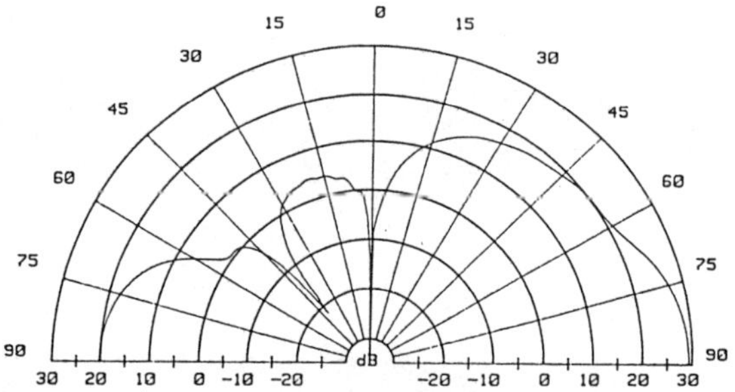

Figure 5
Two-dimensional represention of the shell directivity pattern :
1 000 Hz ; n_{max} = 58 dB ; plane [$\Phi = 0$, $\Phi = \pi$]

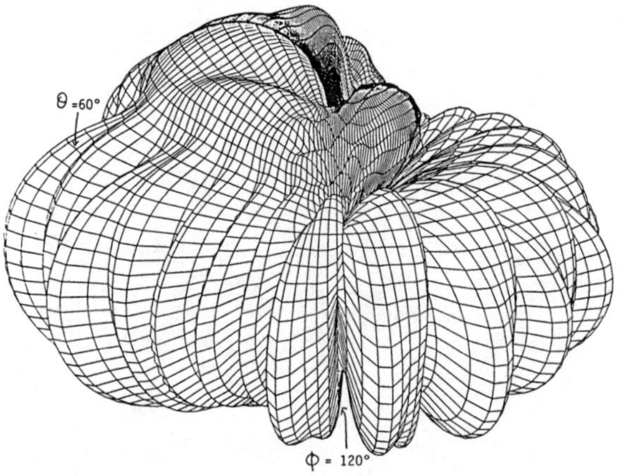

Figure 6
Three-dimensional representation do the shell directivity pattern : 1 000 Hz

INFLUENCE OF TAPER ON THE ACOUSTIC RADIATION

A. Ramachandraiah, N. Ganesan and B.V.A. Rao
Indian Institute of Technology
Madras, India

ABSTRACT

Study of noise radiation of structures and an insight into the problem especially at resonant modes is of considerable interest. In the present work the authors have studied the acoustic radiation of a cantilever beam of tapered cross sections of different types of structures. The surface velocity of the beam is obtained by the finite element approach and the acoustical parameters such as sound pressure, sound power are evaluated. This particular study reveals the qualitative and quantitative aspects of acoustical parameters of different types of structures involved in the study.

NOMENCLATURE

p	acoustical pressure, N/m^2
P	acoustical power
$V_n(rs)$	surface velocity
L	length of the beam
k	wave number, m^{-1}
x	distance from the clamped end
T_o	thickness of the beam at n = 0
$T_{uniform}$	thickness of the uniform beam
U_o	amplitude of velocity of the beam m/sec.
r	distance from the vibrating surface to field point
λ	wavelength of sound in air, m
ρ	mean density of air, kg/m^3
α	taper parameter
ω	vibrating frequency, rad/sec.

INTRODUCTION

The study of acoustic radiation from vibrating structures has been of some interest in view of its importance as a guiding tool at the manufacturing stage of any noise producing source. Structures that radiate sound through vibrations are extremely diverse in their geometric forms material properties, shapes etc. Some important studies that have appeared in this direction are due to Junger [1] , Koopmann and Benner [2] , Keltie [3] , Prasad and Jiang [4] . Koopmann and Benner have presented a method to calculate the sound power of machine based on surface Helm Holtz integral equation and that they have considered piston and beam type structures.

From the literature it is observed that there is no definite work to indicate whether it is possible to change the acoustical pattern of beam type structures by considering its taper variation. Hence in this work we have considered this particular aspect. The main purpose of the study being to evaluate the sound radiation at and off the resonance frequencies. For this purpose the mass of the beam kept constant in order to find out the feasibility of better design of such a beam with respect to noise emission.

THEORITICAL ASSUMPTION

When the vibrating surface is planar and infinitely extended the Kirchoff-HelmHoltz integral reduces to Rayleigh integral. In this study the structure is assumed to have an infinite baffle backing and the acoustical field is computed through Rayleigh integral.

$$p = -\frac{i\omega\rho}{2\pi}\int_S \frac{U_o e^{ikr}}{r}\,ds \qquad (1)$$

The acoustical power radiated 5 by the structure is

$$P = \int_S \frac{1}{2}\,\mathrm{Re}\;\; p(r_s)\; V_n^*(r_s)\;\; dS \qquad (2)$$

where $p(r_s)$ is the surface pressure

$V_n^*(r_s)$ is the complex conjugate of surface velocity.

The singularity problem of calculating the surface pressure is circumvented by using a different Gaussian integration scheme for evaluating p and P. Incidentally, this evaluation also avoids the dipole coefficient since it is on the surface

NUMERICAL COMPUTATION

The numerical computation is made by discretizing the structure into ten beam elements. The element is one of two noded beam element with two degrees of freedom per each node namely displacement and slope. The beam chosen has dimensions of .1 m length and width and thickness of .005 m. The lateral surface is assumed to have no velocity. The length of the element discretized within 0.2 where is the wave length of the sound being radiated from the beam, so that better accuracy is achieved keeping in with the assumption that the surface velocity is constant within the area of the element. The surface velocity is obtained in this case by modal superposition approach. All the computations have been carried out in system Siemens 9748. The various structures studied in this and which are shown in Fig.(1) are governed by the following equations.

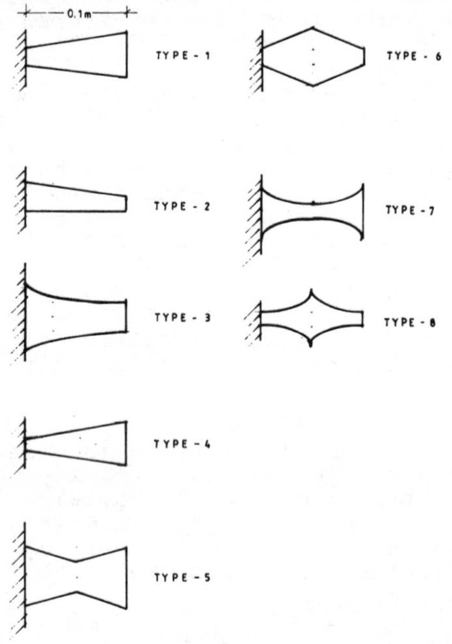

FIG.1. DIFFERENT PROFILES OF THE STRUCTURES.

Type 1 $\quad T_o = \dfrac{T_{uniform}}{(1 + .5\alpha)}$

$$T_{uniform} = T_o\,(1 + \alpha\,\tfrac{x}{L})$$

Type 2 $\quad T_o = \dfrac{T_{uniform}}{(1 - .5\alpha)}$

$$T_{uniform} = T_o\,(1 - \alpha\,\tfrac{x}{L})$$

Type 3 $\quad T_o = \dfrac{T_{uniform}}{(1 - .33\alpha)}$

$$T_{uniform} = T_o\left[1 - \alpha\,(\tfrac{x}{L})^2\right]$$

Type 4 $\quad T_o = \dfrac{T_{uniform}}{(1 + .5\alpha)}$

For $\frac{x}{L} \leq 5 \quad T_{uniform} =$

$$T_o\,(1 - \alpha\,(\tfrac{x}{L} - .5) \;/\; .5$$

$\frac{x}{L} \geq 5 \quad T_{uniform} =$

$$T_o\,(1 + \alpha\,(\tfrac{x}{L} - .5) \;/\; .5$$

Type 5 $\quad T_o = \dfrac{T_{uniform}}{(1 - .5\alpha)}$

$\frac{x}{L} \leq 5 \quad T_{uniform} =$

$$T_o(1 + \tfrac{x}{L} - 0.5)^2\,\alpha \;/\; .5^2$$

$\frac{x}{L} \geq 5 \quad T_{uniform} =$

$$T_o\,(1 - \tfrac{x}{L} - .5)^2\alpha \;/\; .5^2$$

Type 6 $\quad T_o = \dfrac{T_{uniform}}{(1 - .33\alpha)}$

For $\frac{x}{L} \leq 5 \quad T_{uniform} =$

$$T_o\,(1 + \tfrac{x}{L} - .5)^2\alpha \,/ .5^2$$

For $\frac{x}{L} \geq 5 \quad T_{uniform} =$

$$T_o\,(1 - \tfrac{x}{L} - .5)^2\alpha \,/ .5^2$$

Type 7 $\quad T_o = \dfrac{T_{uniform}}{(1 + .33\alpha)}$

$$T_{uniform} = \frac{T_o}{1 + (\tfrac{x}{L})^2\alpha}$$

Type 8 $\quad T_o = \dfrac{T_{uniform}}{1 + .33}$

For $\frac{x}{L} \leq 5 \quad T_{uniform} = \dfrac{T_o\,(1 - \tfrac{x}{L} - .5)^2_{\alpha}}{.5^2}$

$$T_{uniform} = \frac{T_o\,(1 + \tfrac{x}{L} - .5)^2_{\alpha}}{.5^2}$$

In the above α varying from 0 to .95 in steps 0.2 All the other descriptions of the parameters are given in nomenclature.

The results obtained from these computations have been processed as under.

i) modal frequencies

ii) sound pressure level radiation at different distances from the horizontal axes of the beam

iii) sound powers of structures at the three modal frequencies.

RESULTS AND DISCUSSION

The sound radiation values obtained as per the computation outlined above are compared with respect to ref [4] . Reasonably good comparison is obtained from experimental values and predicted values Fig. (2) . The coincidence of dip is reasonably good.

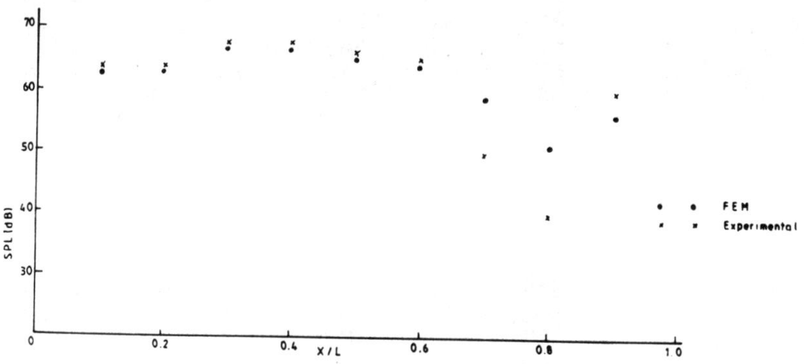

FIG 2 HORIZONTAL FIELD VARIATION ALONG THE SQUARE CROSS SECTION BEAM

The three different modal patterns and the corresponding SPL variation along the horizontal axes of the beam are shown in Fig.(3). It may be observed from these figures that the taper influences the location of occurence of dip. The pattern of radiation along the axis is considerably different especially for the higher order modes.

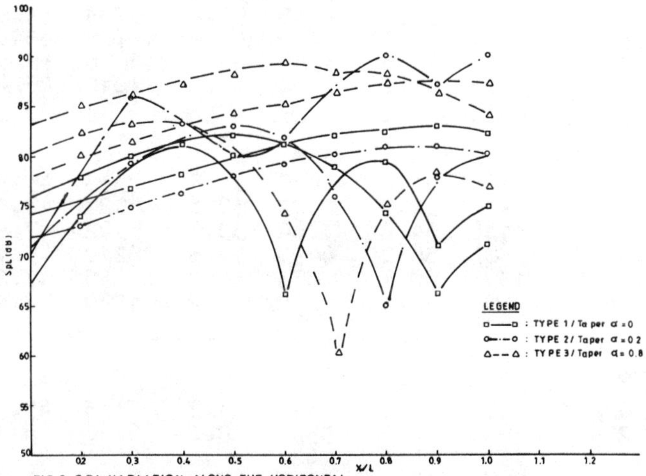

FIG 3 SPL VARIATION ALONG THE HORIZONTAL

Fig.(4) shows the comparitive account of I mode for Type 1,2,3, and 8 structure. The taper parameter chosen is   = .8 which happens to be the extreme. It may be observed from these that the Type 8 tends towards a high radiating low frequency component. Also it is observed that a high frequency modal pattern radiates comparitively lesser noise.

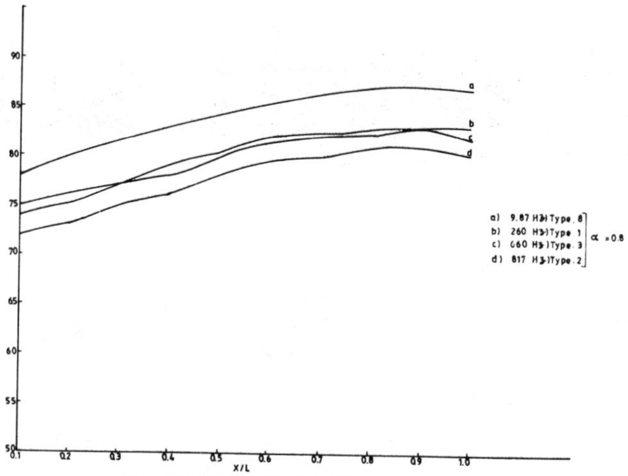

FIG.4. COMPARISON OF I MODES

Fig.(5) shows the sound field variation along the vertical distances from the beam. At .02 m distance from the beam the pattern of radiation exhibits vibration pattern of radiation and at .5 m distance the SPL value remains constant along the horizontal axes and is not recognizable as vibration signature. In other words the near field effects tend to disappear.

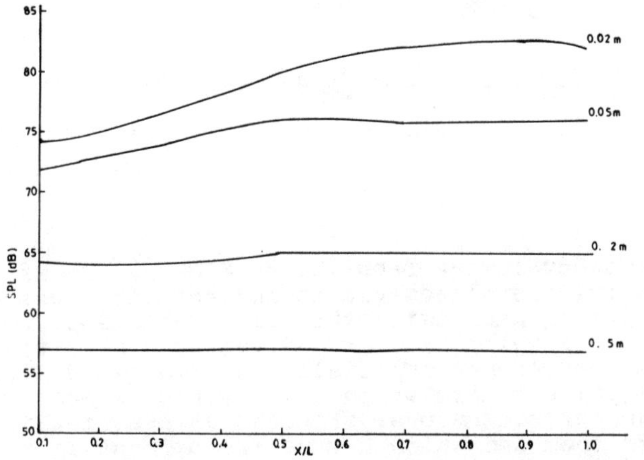

FIG 5 SOUND PRESSURE LEVEL VARIATION FROM NEAR TO FAR FIELD

Fig.(6) and (7) show the variation of SPL with respect to resonant and non resonant frequencies of Type (3) and Type (4) structures respectively. These figures indicate the way in which frequency gets altered to a higher or lower value as the taper varies. For examples in case of Fig.(6) the frequency changes from 400 Hz to about 670 Hz. The shift takes place from lower to higher frequency as shown by the arrow on the top of the curves. Also it is observed that at off resonance frequencies the sound radiated is usually less in a comparitive sense. In the case of results depicted in Fig.(7) the

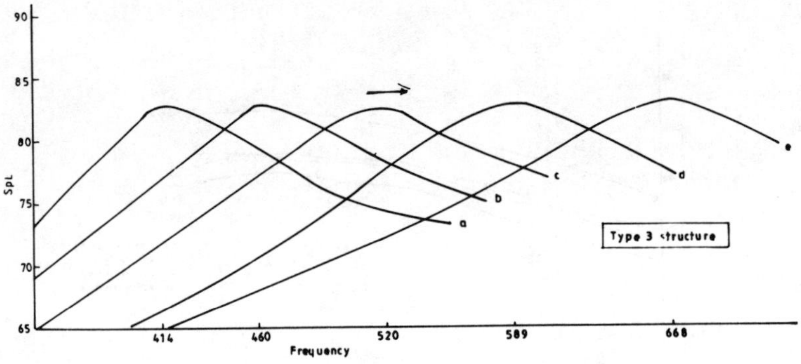

FIG 6. SPL FOR RESONANT AND NON RESONANT FREQUENCIES

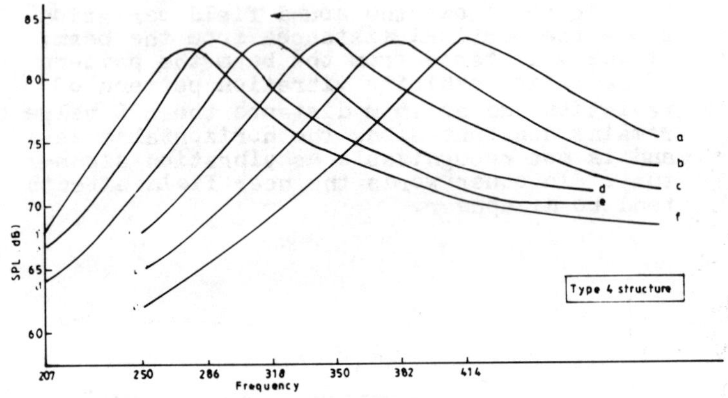

FIG 7 SPL FOR RESONANT AND NON RESONANT FREQUENCIES

frequency lowers from 400 to 250 Hz. These results visualises one to indicate the possibility of minimising the noise radiation from a source which excites indirectly some attached components especially in situations like the teeth of stator in a motar etc. which have certain natural frequencies associated with them and which become predominant in certain situations of over all noise emission of the machine.

Table 1 Sound Powers and Frequencies Radiated by different Types of Structures

	I mode		II mode		III mode	
	f(Hz)	SWL	f(Hz)	SWL	f(Hz)	SWL
Type 1						
0	417	51.6	2619	51.7	7335	45.3
.2	366	51.8	2329	52.1	12979	34.1
.4	325	52.0	2096	52.3	11819	32.5
.6	291	52.1	1903	52.6	5449	45.6
.8	260	52.3	1743	52.8	5020	45.7
.95	240	52.4	1639	52.9	4739	45.7
Type 2						
0	417	51.6	2619	51.7	7335	45.3
.2	479	51.3	2986	51.3	8294	45.2
.4	561	51.0	3467	50.7	9536	45.1
.6	670	50.5	4117	49.9	11171	45.4
.8	817	50.0	5027	49.6	12430	50.8
.95	969	48.8	6042	45.0	4030	60.6
Type 3						
0	417	51.6	2619	51.7	7335	45.3
.2	462	51.6	2877	51.6	7992	45.6
.4	517	51.7	3195	51.4	8789	46.0
.6	583	51.7	3593	51.3	9750	47.0
.8	667	51.8	4098	51.8	13972	44.1
.95	741	50.3	3051	65.9	4730	45.3
Type 4						
0	417	51.6	2619	51.7	7335	45.3
.2	378	51.6	2404	51.9	6786	45.1
.4	344	51.5	2222	51.9	16554	34.1
.6	314	51.5	2066	52.1	5913	44.9
.8	289	51.4	1931	52.1	5560	45.0
.95	272	51.4	1841	52.2	5322	44.8
Type 5						
0	417	51.6	2619	51.7	7335	45.3
.2	453	50.9	2857	51.0	8002	44.4
.4	486	50.3	3055	50.4	16783	33.9
.6	512	49.9	3223	49.9	17705	35.4
.8	535	49.5	3367	49.5	18496	35.5
.95	551	49.2	3462	49.2	17210	25.4
Type 6						
0	417	51.6	2619	51.7	7335	45.3
.2	371	52.6	2328	52.6	6521	46.3
.4	313	54.0	1964	53.6	5502	47.9
.6	238	56.2	4490	54.5	8212	45.1
.8	139	60.4	873	54.2	2445	47.2
.95	39	66.1	249	79.2	1370	70.1
Type 7						
0	417	51.6	2619	51.7	7335	45.3
.2	569	50.5	3524	50.2	9726	44.1
.4	752	49.4	4621	48.5	12571	43.8
.6	967	48.2	5947	47.0	15563	45.9
.8	1226	46.9	7480	49.2	17876	51.2
.95	1452	37.3	3071	61.5	9168	39.0
Type 8						
0	417	51.6	2619	51.7	7335	45.3
.2	289	52.6	1872	53.2	6408	48.5
.4	182	53.5	1256	53.6	3664	45.4
.6	94	54.3	748	52.8	2339	47.2
.8	9.7	55.9	340	56.5	1023	59.3
.95	-	56.8	-	55.2	-	56.5

The sound powers radiated by these structures are shown in table (1). Type 1 to Type 8 . Sound power is estimated for all the three modes and with different taper parameters ranging from 0 to .95. Considering the I mode from type 1 to Type 5 structures, as the taper changes the sound power radiated does not differ much from extremes of taper variation incorporated. But considerable amount of increase or decrease is noticed in the case of type 6,7 and 8 structures. In the case of II and III modes there is a variation even in Type 1 to 5 structures. However considering the I mode it can be said that the influence of taper is more in the case of extreme geometric forms as compared to that of ordinary forms, so as to alter the sound power. Some of the higher modes radiate less power compared to I and II modes which can be due to the same point of excitation which could have been nearer to the mode compared to lower modes.

In essence it can be told that tapering the structure can conv niently be used to alter the noise resonant frequency of the structure. With respect to the absolute sound power radiation the extreme geometric forms of a given structure get affected by the influence of taper. The influence of taper is also to shift the nodal points of structure which can influence the power radiated by the structure with respect to point of excitation.

CONCLUSIONS

A numerical study accounting for the acoustic radiation characteristics such as sound pressure level and sound power level for eight different types beam with various geometric forms has been studied. This has been done on the simple source formulation using Rayleigh integral. The results indicate the possibilities of studying the acoustic radiation of tapered structures through a computation of FEM formulation. The procedure employed in this also enables one to study the SPL values both at resonance and o f resonance points. The influence of taper to alter the resonant frequency and the nodal points of sound radiating component is also demonstrated.

REFERENCES

1 Junger M.C., "Sound Radiation by Resonances of Free - Free beams" Jl. of Acoustical Society of America Vol.52(1), 1972, pp 332-334

2 Koopmann, G.H. and Benner, H., "Method for computing the Sound power of machines based on the Helm Holtz Integral", Jl. of Acoustical Soc.m. 71(1), 1982, pp 78 - 89.

3 Keltie, R.F., A note on the acoustic radiation from point forced elastic beams', Jl.of Sound and Vibration, 92(2) 1984, pp 253 - 260

4 Jiang, J.K. and Prasad M.G., "A parametric study of acoustical fields of Vibrating Structures for noise control applications' Jl. of Acoustical Society of America, 83, 1988, pp 1420 - 1430

5 Frank Fahy, Sound and Structural Vibration, Academic Press 1985.